DUCKTOWNSMOKE

DUCKTOWN SMOKE

The Fight over One of the South's Greatest Environmental Disasters

Duncan Maysilles   *The University of North Carolina Press Chapel Hill*

THIS BOOK WAS PUBLISHED WITH THE ASSISTANCE OF THE
Thornton H. Brooks Fund of the University of North Carolina Press.

© 2011 The University of North Carolina Press

All rights reserved. Designed by Kimberly Bryant. Set in Merlo with Avenir display by Tseng Information Systems, Inc. Manufactured in the United States of America. The paper in this book meets the guidelines for permanence and durability of the Committee on Production Guidelines for Book Longevity of the Council on Library Resources. The University of North Carolina Press has been a member of the Green Press Initiative since 2003.

Library of Congress Cataloging-in-Publication Data
Maysilles, Duncan.
Ducktown smoke : the fight over one of the south's greatest environmental disasters / Duncan Maysilles.
p. cm.
Includes bibliographical references and index.
ISBN 978-0-8078-3459-6 (cloth : alk. paper)
ISBN 978-1-4696-2987-2 (pbk. : alk. paper)
1. Liability for environmental damages — Appalachian Region, Southern. 2. Liability for environmental damages — Tennessee — Ducktown Region. 3. Copper mines and mining — Environmental aspects — Tennessee — Ducktown Region. 4. Georgia — Trials, litigation, etc. I. Title.
KF1298.M39 2011
333.76′513709768875 — dc22

2010047549

To Teresa

Acknowledgments ix

Contents

INTRODUCTION The View from the Mountain 3

1 The Setting, the Cherokees, and the First Era of Ducktown Mining, 1843–1878 14

2 The Revival of Ducktown Mining and the First Smoke Suits, 1890–1903 36

3 The Farmers and the Copper Companies Wage Battle in the Tennessee Courts 58

4 Georgia Enters the Fray 81

5 The Ducktown Desert and Georgia's First Smoke Suit 105

6 Will Shippen, Forestry, and Georgia's Second Smoke Suit, 1905–1907 141

7 Attorney General Hart, the National Farmers Union, and the Search for a Remedy, 1907–1910 170

8 The Smoke Injunction and the Great War, 1914–1918 195

9 Power Dams, Whitewater Rafting, and the Reclamation of the Ducktown Desert, 1916–2010 222

EPILOGUE The View from the Mountain 251

Notes 259 *Bibliography* 301 *Index* 321

A map of Ducktown and vicinity and a section of photographs follow page 128.

Acknowledgments

The present work delves deeply into four distinct, though overlapping, fields of history: legal, environmental, southern, and Appalachian. The History Department at the University of Georgia proved to be a wonderful place to pursue such a project because of its strengths in each of these fields. Peter Hoffer, James Cobb, John Inscoe, Edward Larson, and Paul Sutter freely provided their valuable encouragement, insights, and critique. Moreover, Dr. Cobb and Dr. Hoffer have both written on Ducktown but nonetheless gave me their enthusiastic support and shared their materials to further my own work. My experience with these scholars, individually and collectively, was personally and professionally rewarding.

Monograph history depends on skilled archivists, and this project was no exception. I enjoyed the kind assistance of archivists at the Georgia Department of Archives and History, the Tennessee State Library and Archives, the Historical Branch at the Cleveland (Tennessee) Public Library, the National Archives in Washington, and the University of Colorado at Boulder. I owe thanks to Jane Adams of the Georgia Archives. When working on another project, I came across her research guide, *The Calendar of Incoming Correspondence, Governor Hugh M. Dorsey, 1917–1921*, in which she wrote, "Possible topics of research interest in the Dorsey correspondence include the First World War, prohibition and reform movements and the Tennessee Copper Company case." Though I never met her, I made it my purpose to follow her suggestion about the copper case.

Special mention goes to Ken Rush, Richard Estes, and the staff of the Ducktown Basin Museum. I was the beneficiary of their successful efforts to rescue and preserve documents from the era of smoke litigation. A mass of century-old documents from the Tennessee Copper Company and the Ducktown Sulphur, Copper & Iron Company had been stored haphazardly in an abandoned store near the former DSC&I complex at Isabella. Years of neglect under a leaking roof reduced the papers to a sodden, moldy pile that was destined to be tossed as rubbish down a mineshaft except for the museum's timely intervention. The rescued materials are rich in behind-the-scenes correspondence from attorneys, corporate officers, farmers, loggers, and others on both sides of the litigation — the sort of documents that discretion and the conventions of litigation usually omit from the formal papers filed with the courts.

Ken cheerfully set me up at a folding table in one of the display rooms and brought out one box of documents after another from the museum vault. It is a heady experience for a historian to be the first to use an unexamined cache of important original correspondence, all the more so when surrounded by the well-mounted exhibits of Ducktown's mining history. Ken and his assistants, Dawna Standridge and Joyce Allen, added to the pleasure of my visits by extending their hospitality and by sharing their deep knowledge of the area's mining heritage. I also enjoyed many stimulating conversations with Richard Estes, the museum's historian.

I also appreciate the strong support of Mark Simpson-Vos and the staff of the University of North Carolina Press. Mark encouraged me to write what is now Chapter 9 regarding the reclamation of the Ducktown Desert and helped me cut what amounted to a chapter's worth of surplus text. The two academic readers also provided valuable editorial comments and encouragement. Together, Mark, his colleagues at the press, and the readers helped to make this a much better book.

My dear wife, Teresa, and our children, Andrew, Bryce, Mary, Chara, and Emily, have my heartfelt thanks for their love and encouragement. I also thank Dan Willoughby, John Tucker, Rose Jones, Alex Panos, Jason Mattox, Greg Antine, Lisa Smith, Charles Strickland, and my other colleagues at the law firm of King & Spalding for the courtesies and flexibility they extended to me toward the completion of this work. I hasten to add that this book is a personal venture for which I alone am responsible; it was written independently of the business of the firm and its clients.

DUCKTOWNSMOKE

*. . . a land in which you will eat bread without scarcity,
in which you lack nothing,
a land whose stones are iron,
and out of whose hills you can dig copper.*
—*Deuteronomy*

INTRODUCTION THE VIEW FROM THE MOUNTAIN

He could see the smoke coming, and he knew what it was. B. H. Sebolt, a septuagenarian farmer with a little spread in the Blue Ridge Mountains of Fannin County, Georgia, recognized its appearance and odor: a dark, bluish smoke with the rotten-egg stench of sulfur. He could also identify its source. From Mr. Trammel's apple orchard on the top of the mountain, he looked north and saw the smoke gushing from the copper smelters ten miles away, across the Tennessee border. He testified about the smoke in 1914: "It come right through this county and scattered every way. . . . When we have had a shower of rain and the smoke comes in, it settles down heavy and look like a fog, and when that dries off you can see the damage." He explained that the smoke caused the leaves on his peas, beans, potatoes, cabbage, and corn to draw up and turn brown as if "you take fire and hold it close" to them.[1]

The smoke damage to Sebolt's garden was but a hint of the far greater havoc inflicted in areas closer to the smelters of the Ducktown Sulphur, Copper & Iron Company (DSC&I) and the Tennessee Copper Company (TCC). Seen from Trammel's apple orchard or from any of the surrounding peaks, the heart of the Ducktown Basin was a fifty-square-mile expanse of heavily eroded badlands fringed on the margins by patches of grass and dying forests. Sebolt could see miles of reddish-orange gullies, carved by water into dendritic patterns, where farms and forests stood not long before.[2]

Such a view would not be out of place in the arid regions of the American West, but set amid the lush hardwood forests of the Southern Appalachian Mountains, it was a bizarre and shocking vista that excited the comments of visitors. In 1902 a mining engineer described it as a "desolate barren waste." Stuart Chase, author of *Rich Land, Poor Land*, a 1936 manifesto for soil conservation written during the Dust Bowl years of the Great Depression, called Ducktown "badlands without the balance and natural composure of a desert." He made it exhibit A in his case against human abuse of the land by asking, "What happens when a continent is one great Ducktown?" A scientist for the Tennessee Valley Authority wrote of it in 1938 as "the most severe denudation east of the Black Hills [of South Dakota]." In 1967 Grady Clay, a landscape architect, said it was "variously described as: a hellhole, a blister, a desecration, something out of Dante's *Inferno*, the Tennessee Badlands, the ugliest place in the South," and, in less hostile terms, "a ravaged wonder."[3]

To many, it was simply a desert, known as the Ducktown Desert or, later, as the Copper Basin Desert, even though the term was inaccurate from the perspective of the life sciences. Technically, a desert is defined in terms of aridity: an ecosystem of plants and animals adapted to a landscape that receives less than ten inches of annual rainfall. The Ducktown Desert was shaped not by a shortfall of rain but rather by its abundance. The erosive power of sixty inches of annual rainfall sculpted the landscape wherever human activity stripped the surface of vegetation. A new environment of naked red hills and gullies replaced the green hardwood forests typical of the region. If not a desert in biological terms, it was a desert in the popular sense, a meaning that turned more on the appreciation of emptiness and desolation than on diminished rainfall. The Ducktown Desert earned its name because it was largely empty of the plants and wildlife that people expected to find there. R. E. Barclay, a contributor to the Federal Writer's Project, called the Ducktown Basin a "strangely misplaced desert." It was as if a section of lush Southern Appalachian Mountains had been exchanged for a barren section of Arizona landscape.[4]

Commentators from outside the basin tended to describe the Ducktown Desert in starkly negative terms. Jesse C. Burt penned a 1956 article for *Nature Magazine* under the title, "Desert in the Appalachians," where he blamed the formation of the Copper Basin Desert on activities in the previous century "when man's attitude towards the land was largely that of a predator." Three decades later, Wilton Barnhardt adopted a similar tone in a piece for *Discover*, writing that "the Copper Basin is the only bona fide desert east of the Mississippi," a "vast, raw plain, cooking in the summer sun," that served as "a paradigm of environmental disaster."[5]

Others expressed a more benign view of the landscape. In 1936, one eager local resident, Louise McKinney, touted "our unique and attractive mountain desert" as a tourist destination. She urged that cactus and other desert plants be imported for the benefit of "the thousands of people" who "would like to include the Beautiful Copper Basin Desert" in their travel plans. Though amusing as an ecological proposition, McKinney's enthusiasm spoke to the pride that people often take in points of locally distinctive landscape, whatever cause may lay behind the point of distinction. Yes, it was a desert in the Southern Appalachians, but as M.-L. Quinn noted, it gained the patina of familiarity to generations of miners and their families. The great expanse of red gullies provided a distinctive sense of home and of place. The landscape was a visible testament to the mighty industry that shaped the lives of all who lived and worked there. And it was not without some benefits: longtime residents remember that the barren hills left them free of the mosquitoes, chiggers, and snakes so common in moist woodlands.[6]

It is hard to make a desert in a region that receives almost sixty inches of rain each year, but that is what happened in the Ducktown Basin. The local copper industry operated within an environment that was peculiarly susceptible to the sustained impact of sulfur dioxide pollution and the sustained logging done to support it. Human conduct interacted with natural environmental factors to create a landscape previously unknown in the Southern Appalachians or, for that matter, in the entire eastern half of the United States.[7]

The Ducktown Mining District is within a topographic basin surrounded on all sides by mountains. Located at the extreme southeastern corner of Tennessee where it abuts western North Carolina and northern Georgia, the basin floor was originally covered with the mixed forests of oak, chestnut, tulip poplar, white pine, and hemlock typical of the Southern Appalachians. But underneath it lay vast deposits of sulfur-rich copper ore. For the first half century of Ducktown mining, miners expelled the sulfur by roasting it in open heaps. This approach involved piling heaps of crushed ore in a mound atop stacked logs. The logs were then ignited and allowed to burn for two or three months, with the heat roasting away most of the sulfur. Additional processing in blast furnaces removed the remaining sulfur, leaving the more desirable copper for further processing. Smoke from both processes entered directly into the atmosphere, but the mountains trapped and concentrated the fumes within the basin rather than allowing them to disperse over distant lands. Airborne sulfur dioxide readily combined with moisture from abundant rain and fog to precipitate back to earth as acid rain, killing existing vegetation and suppressing new growth.

Extensive logging compounded the effect of sulfur dioxide gases. Over time, loggers felled some fifty square miles of timber for use by the copper industry as charcoal, firewood, and construction lumber. Decades of logging and sulfur fumes left great portions of the Ducktown Basin bare to drenching mountain rains that washed denuded soil and rock down Pumpkin Creek and Davis Mill Creek into the Ocoee River. All that remained was a naked expanse of eroded red gullies, void of soil, depleted of nutrients, and empty of all vegetation except for scraggly patches of sedge grass and cat briar. Most native wildlife fled the barren land. Fish died in streams poisoned by mineral-laden runoff.[8]

The denuded areas caused by Ducktown's first era of mining, from 1843 to 1878, were relatively small. They greatly expanded during the industry's second era, beginning with the arrival of the railroad in 1890. Rail access allowed Ducktown Sulphur, Copper & Iron and its younger rival, the Tennessee Copper Company, to increase the amount of ore mined and smelted, and with increased production, the volume of sulfur dioxide fumes grew in proportion. The badlands soon began to expand into farms and forests that had been largely spared during the first era. Alarmed farmers and timber owners saw the widening zones of devastation and turned to the courts for recourse.

The first Ducktown smoke cases were filed in 1896 and 1897. In the following years, scores of claimants, ranging from semisubsistence hill farmers to wealthy lumber barons, filed more than two hundred lawsuits seeking damages for crop and timber losses against the two copper companies. Other suits, of greater consequence to the industry, were for injunctions to abate the smoke by curtailing or even terminating smelting operations.[9]

Looming above the private lawsuits was *Georgia v. Tennessee Copper Co.*, filed by the state of Georgia in the United States Supreme Court. The case began in 1904 and lasted in its active phase until 1918 (it was finally dismissed in 1937). The state initially sued for the limited purpose of forcing the copper companies to reduce smoke damage by ending the practice of open heap roasting of copper ores. It would not remain that simple. The law was too uncertain, and there were too many stakeholders to allow for simplicity. The state sought to extend the Court's original jurisdiction to embrace, for the first time, a case of transborder air pollution. The case also stretched the limits of the common law of nuisance. The Supreme Court had never before adjudicated a case of air pollution; it was, as lawyers say, a case of first impression.

The Court rose to the challenge by ruling for Georgia in the landmark 1907 opinion by Justice Oliver Wendell Holmes. He first acknowledged Georgia's right to assert Supreme Court original jurisdiction over the matter. He then shifted the analysis away from traditional nuisance law, with its emphasis on offended property rights, to instead recognize Georgia's constitutional interest in the care and control of natural resources within its borders. As he phrased it, "This is a suit by a state for an injury to it in its capacity of quasi-sovereign. In that capacity the state has an interest independent of and behind the titles of its citizens, in all the earth and air within its domain. It has the last word as to whether its mountains shall be stripped of their forests and its inhabitants shall breathe pure air."[10]

It was the first time the Court had acknowledged a state's right to seek constitutional redress to preserve its air and natural resources from transborder pollution. The ruling reflected the understanding that interstate pollution required a federal perspective, anticipating what Richard Lazarus later termed "the spatial and temporal dimensions that especially challenge the fashioning" of environmental legal policy. The opinion gave rise to the federal common law of nuisance that served to resolve matters of interstate pollution for the next half century.[11]

All of this occurred in a context that pitted the need for effective smoke abatement measures against the desire to preserve TCC and DSC&I as viable businesses. The damage to the forests and landscapes was obvious; so was the risk that a draconian antismoke injunction would force closure of the industry and terminate the jobs of thousands of workers. When the Supreme Court granted Georgia the right to an injunction, the ruling came with a note of solemn caution from Justice Holmes: "Whether Georgia, by insisting upon this claim is doing more harm than good to her citizens is for her to determine," adding that "the possible disaster to those outside the State must be accepted as a consequence of her standing upon her extreme rights."[12]

Georgia attorney general John C. Hart understood the warning. He repeatedly delayed or suspended the state's case in the hope that new technology would preclude the need for an injunction. The many stages of Georgia's Supreme Court case were marked by development of the new pyritic smelting process, the use of high smokestacks for attempted dispersal of sulfur fumes, and the development of acid condensation plants to convert toxic fumes into marketable sulfuric acid. Some of the advances helped. Others failed miserably as pollution control measures, however much they improved production and profits. The failures were punished with return trips to the Supreme Court.

Overall, the Supreme Court litigation helped to bring about a thorough transformation of the copper industry in Ducktown. The acid plants provided the greatest advance by extracting a significant portion of sulfur dioxide from the atmosphere—even if the reduction of smelter fumes was offset by the growing volume of smoke from increased production. When the suit began, copper was valued, and sulfur was wasted. When it ended, both companies found the sulfur content of local ore to be more valuable and profitable than the copper. Improved technology enabled the two copper companies to survive the era of the smoke suits; they eventually merged to become a mighty mining and chemical concern that, with various corporate changes, remained in business for the rest of the twentieth century.

Law and technology ended the smoke wars without killing the industry, but the Ducktown Desert remained for most of the twentieth century. Reclamation of a man-made desert larger than New York's Manhattan Island required different technology to return trees and plants to a landscape that had been rendered toxic by decades of smelter fumes and scoured by decades of mountain rains. In some places, more than sixteen feet of soil had been washed away into the Ocoee River. The environmental damage was so great that it required a multigenerational effort by the copper companies, the Tennessee Valley Authority, Soil Conservation Service, U.S. Forest Service, and other entities to reverse it. The work required the participation of foresters, biologists, engineers, and other specialists to devise methods of helping trees to grow in the forbidding environment. The rationale for the project shifted over the decades as the nation moved beyond the goal of resource conservation to the more encompassing environmental goal of restoring ecological integrity to land and waters of the basin. Substantial progress has been made. Little of the former desert is still visible; pine trees and grasses cover most of the formerly barren hills.[13]

The reclamation work continues, but the industry that created the Ducktown Desert is now gone. Mining and smelting ended in 1987, followed in 2000 by the acid industry. The last train of tank cars bearing sulfuric acid left Copperhill on March 6, 2001. The dominant local industry is now recreation. Vacation homes dot the slopes of the mountain rim to the north, east, and south. To the west, the Cherokee National Forest and three federal wilderness areas provide locations for hiking, camping, fishing, and mountain biking in the Unicoi and Cohutta Mountains. The greatest recreational draw is the river: the Ocoee Gorge provides some of the finest whitewater in the Southeast and served as the venue for whitewater sports during the 1996 Atlanta Olympics. A constant procession of wagons once traveled through the gorge on the old Copper Road, built in 1853 to service the mining industry. Now, Highway 64 follows the path of the Copper Road, and traffic today is notable for the number of cars and converted school buses carrying rafts, kayaks, and their excited passengers.[14]

Law shaped almost every aspect of Ducktown's history, from the Cherokee Removal, the rise of the copper industry, and the quarter century of litigation to the reclamation of the badlands. Richard N. L. Andrews, a scholar of environmental law, has argued that environmental policy might be broadly defined to include "all the policies by which Americans have used the powers of government to exploit, transform, or control their natural surroundings." This was true in Ducktown. The many actors in its story were making environmental policy but did so without using the words "environmental" and "ecology" and their cognates. Even the use of the term, "pollution" to describe dirty air was rare at the time. The lexicon of environmentalism and ecology did not yet exist because the necessary science and conceptual framework had yet to be developed. In Ducktown, people spoke instead of poisonous vapors and fumes. A few spoke of sulfur dioxide. For most, it was simply smoke—Ducktown smoke.[15]

There were two great battles over Ducktown's environmental policy, the first being the struggle between the Cherokee Nation and Georgia over the sovereign control and the ownership and use of the southern mountains. The tribe won the legal battle in the U.S. Supreme Court in the case of *Worcester v. Georgia* (1832) but lost the political war against the state and against Andrew Jackson. The Cherokees were then expelled from their Appalachian homeland in 1838 and forced to live in the alien landscape of Oklahoma, while suffering the miseries of the Trail of Tears in the process. The land gained new owners and was soon put to new uses by the copper industry.[16]

The other great environmental battle in Ducktown occurred sixty-seven years after the Cherokee Removal when the state of Georgia returned to the Supreme Court in 1904 in *Georgia v. Tennessee Copper Co.* This time, the state sought to assert its rights as a sovereign to defend against what it viewed as an invasion of its territory in the form of sulfur dioxide smoke arising from operations just across the border in Tennessee. Arising as it did out of the welter of private lawsuits, Georgia's Supreme Court case involved a wide array of interest groups: mountain farmers, timber barons, forest scientists, mine workers, engineers, merchants, investors from New York and London, New South business promoters, politicians and lawyers from two states, and the National Farmers Union (NFU). All had a stake in Ducktown environmental policy. And so did President Theodore Roosevelt and members of his administration. With the president's encouragement, Attorney General Charles J. Bonaparte hired Ligon Johnson, special counsel for Georgia, to pursue similar remedies on behalf of the federal government against western smelting companies. Even war became a vector shaping environmental policy when Bernard Baruch of the War Production Board and Secretary of the Navy Josephus Daniels intervened in Georgia's case in 1918.

During the generation-long struggle over the smelter smoke, lawyers for the farmers, loggers, and mining industry shaped environmental policy as they maneuvered in the Tennessee courts and legislature over alterations to the law and procedure of nuisance cases. It was a contest that weighed the respective strengths of industrial capitalism and agrarian populism in the South. When progress became stymied in Tennessee, the action shifted to Georgia and ultimately to the U.S. Supreme Court. Whether in the local courts, the legislatures of Georgia and Tennessee, or in Washington, D.C., the battle over Ducktown smoke brought about changes to the law that were as great as the technological changes imposed on the copper industry. The ancient common law of nuisance was transformed. For centuries, it had been a judicial vehicle for resolving small-scale local disputes between neighbors concerning offensive activities. Tennessee's law of nuisance was permanently altered in subtle but important ways to reflect the regional scale of smokestack industry.

At the national level, the landmark ruling issued by Justice Holmes in the Georgia case transformed the private law of nuisance into a constitutional remedy to advance a state's interests in its natural resources against the impact of transborder pollution. The decision and the trial techniques employed to achieve it served for several generations as the template in future cases of interstate pollution, both within the smelting industry and increasingly outside of it. Later, in the 1930s, the case also shaped the resolution of an international dispute between the United States and Canada regarding the Trail Smelter in British Columbia. After a period of dormancy, the Georgia case gained new life with the rise of the environmental movement, when states once again appreciated the words of Justice Holmes as a means to protect their interests in their natural resources.[17]

The ringing words of Justice Holmes still have power. The U.S. Supreme Court recently quoted and applied them in its first case about global warming, *Massachusetts v. Environmental Protection Agency* (2007). In an opinion issued exactly one hundred years after the 1907 decision in *Georgia v. Tennessee Copper*, the Supreme Court quoted the key ruling of the copper case to uphold the right of Massachusetts and other states to sue the Environmental Protection Agency (EPA) for its failure to regulate greenhouse gases, specifically automobile tailpipe emissions. The Court declared for the first time that carbon dioxide and greenhouse gases were pollutants within the meaning of the Clean Air Act and thus the EPA "has the statutory authority to regulate the emission of such gases from new motor vehicles." The *Massachusetts* decision and related global warming cases are the remarkable progeny of the century-old *Georgia* case, a suit that arose from a wave of private smoke suits started by ten Southern Appalachian mountaineers on the eve of the Spanish-American War.[18]

For all of its national importance to the foundation and development of American environmental law, the origins of the Supreme Court's first air pollution case are nevertheless rooted in the South. The Ducktown story involves the remarkable convergence of several major themes of the region's history: the early prominence of the Ducktown copper industry, New South industrialization, the power of Georgia's states' rights ideology, and agrarian populism. It also connects with the national movement to create federal forest reserves in the Southern Appalachians.

A dominant theme of Appalachian environmental history is the impact of extraction industries, particularly mining and logging. Both activities occurred in the mountains from colonial times but were generally done on a small scale until the penetration of railroads and steam-powered machinery into the mountains after the Civil War. The experience at Ducktown was different: copper mining began soon after the 1838 Cherokee Removal and remained, with brief gaps, the dominant economic activity in the area for one and a half centuries. Though the mines and smelters lay on the Tennessee side of the border, Georgia's leaders actively promoted the industry over a period of fifty years with a series of bills to promote construction of a railroad to serve them. The idea was that a railroad from Atlanta to the mines would open up a significant portion of the North Georgia Mountains to the economic development envisioned by the state's probusiness New South leaders.[19]

Georgia policy took a sudden turn in 1903 when the legislature adopted its first resolution directing the governor and attorney general to investigate and litigate, if necessary, the smoke damage inflicted by the copper companies to the farms and timber of Georgia citizens. The damage from cross-border industry was so extensive that it stirred Georgia's historically powerful ideology of state sovereignty, a theme that had found expression in the American Revolution, the great Supreme Court case of *Chisholm v. Georgia* (1793), the Cherokee Removal, and the Civil War. Georgia's fight against the copper companies was, in essence, a border war to repel the invasive presence of smelter smoke. States' rights ideology would prove critical to the legal and political rationale of the case.[20]

The political forces behind the case intersected with another major initiative: the movement to create a national forest in the southern mountains. Toward that end, Theodore Roosevelt's administration conducted a major study to assess forest conditions throughout the Southern Appalachians, including the Ducktown Basin. The work of federal foresters, geologists, and scientists working under the direction of Secretary of Agriculture James Wilson and Gifford Pinchot, head of the Forest Service, was of pivotal importance to Georgia's Attorney General John C. Hart in the prosecution of the state's case against the copper companies.

In another surprising turn, the state's suit against the copper companies brought its New South politicians into alignment with another and often antagonistic political force: agricultural populism. Ducktown's mining and farming communities had developed a symbiotic relationship in which the former provided a ready market for the meat and produce of the latter. Farmers lamented the loss of the industry during the hiatus of the 1880s and rejoiced with its resumption in 1890—until the volume of smoke generated in the industry's second era threatened the ability of farmers to grow and sell their crops. The earliest smoke suitors reacted with complaints redolent of Jeffersonian agrarianism. Then, as the litigation grew in scope and complexity, the farmer litigants from the Georgia side of the border affiliated themselves with the National Farmers Union for its legal and political clout. The many governors and attorneys general who prosecuted the case were quick to accept the help of the NFU, being well aware that it was a favored cause of Georgia's arch populist, Tom Watson.

The organization's support was unusual because the mountain farmers most affected by the smoke were outside of the NFU's core constituency of cotton growers. Historian Robert McMath explained the growth of agrarian movements by pointing to the "interlocking rural social networks of kinship, church, and face-to-trade." This proved true in Georgia's Fannin and Gilmer counties. For its part, the NFU had its own financial interests: it operated a cooperative fertilizer plant, and sulfuric acid from the Ducktown copper industry was critical to the manufacture of the superphosphate fertilizer needed to replenish worn-out cotton fields.[21]

Farmers pursued smoke litigation in other smelting regions of America, notably in Butte, Montana; Northport, Washington; and the Great Salt Lake Valley of Utah. The Ducktown battle preceded all but the contest in Butte and was ultimately the most influential. The powerful themes of agrarian populism, New South business ideology, and forest conservation linked in a way that gave the Ducktown smoke litigation the staying power to last for a quarter century. The industry's location on a state border created an interstate constitutional crisis that led to the Court's landmark decision. The visual power of the Ducktown Desert gave the case the shock value that compelled the intervention of the Supreme Court.

It was a southern story that resulted in a landmark decision, a case that shaped and still shapes American environmental law. Yet, for all of that, *Georgia v. Tennessee Copper Co.* and the hundreds of private lawsuits that preceded it were never solely about the health of the local gardens, fields, orchards, forests, and streams. The cases were also a struggle over human livelihoods. The battle over Ducktown smoke began when ten mountaineer farmers of mostly humble means decided they could tolerate the smoke no longer and instead filed suit. They fought the copper interests for half a decade before the state of Georgia entered the case to make it a weighty constitutional issue in the Supreme Court. Their initial lawsuits were propelled by economic frustration rather than by a hatred of industry and technology. They recognized that the industry provided a nearby market for their meat, produce, and fruit, a benefit that many more-isolated Appalachian farmers could only envy. The industry also provided contract work and seasonal labor to supplement farm incomes. But the smoke killed their crops and, with it, much of their ability to rise with the copper economy.[22]

No one could ignore the extraordinary damage that copper mining caused to the basin landscape. The expanding reach of the Ducktown Desert was there for all to see. But the battle over Ducktown smoke was also a fight to protect livelihoods. The farmers fought to preserve one way of life. As their cause grew in scope and power, they in turn threatened another way of life, that of not only the thousands of mine workers, suppliers, and merchants who owed their jobs and businesses to the copper companies but of their dependents as well. The farmers nearly shut the industry down with their quest for an injunction to reduce or eliminate the smoke. With the stakes that high, it is no wonder that it took more than twenty years to achieve a workable compromise between the farming and mining communities. It would take another seventy-five years to restore something of a forest to the Ducktown Desert.[23]

1 THE SETTING, THE CHEROKEES, AND THE FIRST ERA OF DUCKTOWN MINING, 1843–1878

In 1837, on the eve of the Cherokee Removal, there was only one way to approach Ducktown from the west: it was along a footpath that began in the Tennessee Valley and then climbed up and over Little Frog Mountain before descending into the Ducktown Basin. It is easy to imagine that a weary traveler might have stopped at some open place along the crest to take a breather and to scan the vista before moving on. The traveler would have seen a great expanse of Southern Appalachian hardwood forest, predominantly oak, chestnut, hickory, and tulip poplar, stretching across the floor of the Ducktown Basin and up the slopes of the surrounding mountains. The sweep of the woods was occasionally broken by the cabins and fields of the few Cherokees who dwelt there, a limited sign of human presence that served to emphasize the dominance of forest and mountains. The only smoke to be seen would have been the wispy columns rising from domestic hearths, and perhaps from a few blacksmith forges.[1]

Two decades later, Hardin Taliaferro, a Baptist preacher and nationally popular humorist, made his own journey from the Tennessee Valley to Ducktown. His journey was not on the old mountain footpath; rather, it was on the new Copper Road that had been blasted through the rocks of the Ocoee Gorge. He set forth his impressions through the voice of a fictional narrator in an exuberant 1860 piece in the *Southern Literary Messenger*, "Ducktown, by 'Skitt' Who Has Been 'Thar.'" As Skitt told it, the road alongside the crags and waterfalls of the gorge was "romantic in the extreme," though crowded with "Wagons! Wagons!! Wagons!!!" loaded with copper ingots for the westward journey to the railroad in the valley and with supplies and equipment for the eastward journey back to the mines. Like earlier travelers, he stopped at the crest of the mountain to take in the view: "Look down about the centre of this basin and behold those huge columns of smoke ascending towards heaven, spreading out at top like vast sheaves." He noted that wherever copper ore was roasted and smelted came smoke "covering the heavens with a smoky pall." Turning his gaze from the darkened skies to the valley floor, he cried out, "See! Their sulphurous smoke has killed most of the timber near them!"[2]

Ducktown's transition from a wooded, lightly populated corner of the Cherokee Nation to a smoke-filled, deforested copper boomtown took only two decades. The transformation involved more than smoke and logging; it involved a wholesale change of people and of sovereign control over the use of the land. The change was mirrored in the name of the region, as industry largely erased the area's Indian heritage. Throughout the nineteenth century and into the years of the smoke litigation, it was called Ducktown, said to derive from a Cherokee chief or village of that name. In the twentieth, the area came to be called the Copper Basin.[3]

The momentous changes took place within a well-defined natural basin, ten miles wide from east to west, and twenty miles long, north to south. Its rumpled floor varies from 1,600 to 1,800 feet above sea level and, as a whole, lays a thousand feet above the Tennessee River Valley. Mountains rim the Ducktown Basin on all sides. To the west is the fifteen-mile width of the Unaka Mountains (now more commonly called the Unicoi) dominated by the peaks of Big Frog (4,224 feet) and Little Frog (3,322 feet). Together, they and the outlying ranges presented a formidable barrier to travelers between the Ducktown Basin and the Tennessee River Valley, a barrier pierced in only two places by the rugged gorges of the Ocoee and Hiwassee Rivers.[4]

To the east, the Pack and Angelico peaks rise to 3,459 feet to separate the basin from North Carolina. Beyond those peaks, the eastbound traveler must cross a hundred miles of mountains before descending to the lowlands of the eastern seaboard. On the north, Stansbury and Threewit reach to about 2,580 feet and run east and west to divide the Ducktown Basin from the Hiwassee Plateau. Beyond the Hiwassee Plateau are the mighty peaks of the Great Smoky Mountains and the great sweep of the Appalachian mountain system that stretches a thousand miles all the way to Maine.[5]

To the south, the slope rises gradually from the basin floor before reaching the Blue Ridge Mountains as they arc through Fannin and Gilmer counties in Georgia. There, the Ellijay Valley provides the best access to the basin. The long valley traces an ancient fault line from Atlanta and Marietta by way of Ellijay (formerly a Cherokee town and now the seat of Gilmer County), through the mountains to the town of Blue Ridge (seat of Fannin County), before descending into the basin. This was the route eventually followed by the railroad to reach the copper works in 1890. At the junction of the Blue Ridge and the Unakas stand the jumbled peaks of the Cohutta Mountains, capped by the summit of Cowpen, which rises to 4,139 feet. Though rugged, the ranges were never totally inaccessible and were heavily logged in the early decades of the twentieth century. Even so, it is difficult and sparsely populated terrain, especially where the Unaka and Cohutta Mountains form the western rampart. Congress took notice of this under the Wilderness Act of 1964 by designating three contiguous wilderness areas along the western edge of the basin: the Cohutta Wilderness Area in 1975, the Big Frog in 1984, and the Little Frog in 1986.[6]

The Toccoa-Ocoee River bisects the basin, beginning as the Toccoa River at its headwaters in Georgia's Blue Ridge Mountains and continuing northwest across the Tennessee line where it changes names to become the Ocoee at the bridge linking McCaysville, Georgia, and Copperhill, Tennessee. As the river flows gently over the plateau, its receives the waters of Hot House Creek, Wolf Creek, Pumpkin Creek, Fightingtown Creek, and other basin streams. All of them are fed by the more than sixty inches of rain that fall each year from the frequent weather fronts that collide against the mountains, and from warm, humid air masses from the Gulf of Mexico that bring frequent summer thunderstorms. Thus nourished, the Ocoee River then leaves the basin by careening down a twenty-five-mile chasm guarded by the peaks of Big Frog and Little Frog. At the bottom of the gorge, the river meanders across the valley to join the Hiwassee River, which then meets the Tennessee River on its way to the Ohio and Mississippi Rivers. The falls and rapids of the Ocoee Gorge prevented its use for waterborne commerce but later proved to be important for the hydroelectric industry and whitewater recreation.[7]

The physical geography of the basin is easy to appreciate from a number of places in the mountains that define it. Angelico Peak and other slopes are now dotted with vacation homes situated to capture the vista. A fine view can also be had from the Ducktown Basin Museum, located atop the hill of the old Burra Burra copper mine. At the edge of the parking lot is a precipice formed when half of the hill collapsed in 1925 as a result of extensive underground mining. A viewing platform at the edge allows visitors to see the basin and the surrounding mountains, all the way to the Cohuttas.

The equally important political geography is not so readily perceived. Some state boundary lines follow natural features. The Ohio River separates Kentucky on its south bank from Ohio, Indiana, and Illinois on the other side. New York and Vermont are separated by Lake Champlain. The situation is different at Ducktown. There, the common borders of Tennessee, Georgia, and North Carolina were established with little regard for the natural features. The Georgia-Tennessee line runs due east and west without reference to either the mountains or the Toccoa-Ocoee River. The local segment of the

Tennessee–North Carolina line is equally arbitrary with regard to the landscape. Most of the border follows the crest of the Great Smoky Mountains in a southwesterly direction, but a few miles above the Ducktown Basin it takes a sudden angle to run almost due south, as if in a hurry to meet the Georgia line.[8]

A geographer might have preferred to run the borders so that Ducktown Basin fell completely within just one of the three states. Had that been done, the problem of Ducktown smoke could perhaps have been resolved by the government of a single state. Instead, the tristate corner is in the northeast quadrant of the basin, so that roughly three-quarters of its floor lies within Georgia, and most of the northern quarter within Tennessee, except for a small portion in North Carolina. The failure of the borders to respect the topography eventually became a matter of great legal importance, for although the smoke from the copper smelters of Ducktown usually respected the constraints of the surrounding mountains, it paid no heed to arbitrary political lines. Clouds of sulfur dioxide generated in Tennessee flowed without physical impediment into Georgia and, to a lesser extent, up stream valleys into North Carolina. It was the failure of Ducktown smoke to obey the lines on a map that eventually created an interstate crisis of constitutional law.[9]

The problem of state borders also bedeviled the Cherokees, whose ancestral lands embraced all of the Ducktown Basin. At the outset of the 1830s, the tribe still held a great expanse of the southern mountains and adjacent lowlands covering large sections of western North Carolina and eastern Tennessee. The tribe held all of northern Georgia down to the Chattahoochee River, just above present-day Atlanta. It even held the northeast corner of Alabama. The tribe had thus maintained its territorial integrity long after most tribes in the original thirteen colonies had been uprooted or extinguished by war, disease, and relentless white settlement. Unfortunately for the Cherokees, their lands straddled the newly formed boundaries of four states in a way that would subject them to a complicated and ultimately doomed battle to maintain their sovereignty.

The tenacious grip of the tribe upon its lands was considerably strengthened by the geographic advantages of its mountainous homeland. The Southern Appalachians are modest in height compared to the great mountain ranges of the American West. Dozens of peaks in the Rocky Mountains exceed 14,000 feet, but the tallest peaks in the South, such as Mount Mitchell, Mount Guyot, and Clingman's Dome, are just over 6,600 feet. Nonetheless, the rugged terrain and great width of the southern mountains impeded the movement of people and goods. The initial path of white settlement skirted them in favor of the easier topography and gentler climate of the Georgia Piedmont and the Tennessee Valley. Major thoroughfares followed the same path. The Federal Road, built to connect Augusta, Georgia, with frontier Nashville and Knoxville in Tennessee, followed this route, as did later highways and railroads. Atlanta owes its importance as a rail and interstate highway hub to the same tendency to circumvent the one-hundred-mile width of the Southern Appalachians.[10]

Climate was another deterrent. White settlers had an insatiable hunger for land suitable to grow cotton, but the crop did not flourish in the mountains. Successful crops could be raised in the lowlands but not in the heights above. Surrounded by the arms of the Unakas and the Blue Ridge, the Ducktown Basin was too high and too cold for the cotton plants so dear to southern growers and too inaccessible for significant commerce. The copper that would make the basin one of the richest districts in the South had yet to be discovered by whites.

Even the Cherokees chose to establish many of their larger towns in the lowlands at elevations suitable for cotton. The Cherokee capital of New Echota was in the Georgia piedmont, and the towns of Chota, Great Tellico, and Hiwassee Old Town were at the edge of the Tennessee Valley. By 1830 many Cherokees had adopted cotton agriculture, and some of their elites owned and worked black slave labor on the plantation model. At work behind the changes was a combination of tribal initiatives and the civilization policy devised by George Washington's first administration and adopted by Congress in the Indian Trade and Intercourse Acts of 1790. The policy was an attempt to reconcile expansionist settlement policy with the Enlightenment attitudes of the Founders toward the Indians by helping the Cherokees, the Creeks, and other Indian nations in the Southeast to acquire livestock, tools, new agricultural methods, skills in the manual arts, and education. The administration also encouraged the participation of church missionary societies to spread the creeds and culture of white society.[11]

The hope of the federal government was that, as the Indians gained the benefits of civilization, they would be willing to relinquish much of their former hunting lands for eventual white settlement. The Cherokees had a different idea: they freely adopted aspects of white civilization to secure their hold upon their ancient lands but without any intention of agreeing to further land transfers to white governments. After learning new ways of agriculture and the manual arts, they developed literacy in the Cherokee language, using the syllabic Cherokee text invented by Sequoyah; others also gained literacy in English. Literacy led to a formalization of tribal law from an oral tradition administered by autonomous town councils into a centralized republican government framed by the 1827 Cherokee Constitution. The new tribal government prohibited the further sale or cession or tribal land to non-Cherokees.[12]

While their kinsman grew cotton in the lowlands, the Cherokees of the Ducktown Basin practiced a largely subsistence form of agriculture in a lightly populated district valued by the tribe mostly for its qualities as a hunting ground. Though it was one of the quieter areas of the tribe's domain, the Cherokees had lived, farmed, and hunted there from time immemorial. Two small villages, Ducktown and Fighting Creek, were within the basin, and Turtletown was nearby on the other side of Stansbury Mountain. Some sources trace the Ducktown name to Chief Duck and his village at the foot of Little Frog Mountain. Others point to federal records that mentioned a village named Kawa'na (spellings vary), translated into English as duck, and hence Ducktown. Either way, a Cherokee village of that name and in that vicinity was included on a federal list of tribal settlements in 1799.[13]

Despite the isolation of the basin, the impact of the civilization program was evident among the few Cherokees who lived there. The 1835 census of the Cherokee Nation found only forty-four households within the basin with a total of 312 individuals scattered over its ten-by-twenty-mile area. A typical family entry for Ducktown read, "Tan A De He: Twenty fullbloods. Five farmers. One mechanic. Two read Cherokee. Two weavers and three spinners." Another for Turtletown mentioned "Ark A Lu Ka: Ten fullbloods. Three ferryboats. Three farmers. One reads English. Three spinners." Most of the households were composed of full-blooded Cherokees, though there were occasional references to "white intermarriage" and individuals "whose degree is not recorded." Most households included at least one farmer and a spinner or weaver, or both. Mechanics (probably blacksmiths and wagon wrights) and ferryboat operators appeared in many others. Many households had an occupant who could read Sequoyah's syllabic Cherokee text; far fewer had members literate in English.[14]

The local Cherokees dwelt in sturdy wooden homes and practiced a mixed agriculture in which they raised corn, beans, and squashes in their gardens; grazed horses and cattle in the pastures; and allowed hogs to roam the woods to fatten on mast from oaks and chestnuts. They left the slopes wooded because they preferred to live and farm in the valley bottoms where the land was rich, flat, and easily worked. Overall, the small population and low-intensity agriculture of the Cherokees had a gentle impact upon the landscape of Ducktown Basin. Twenty years later, the same view would reveal thousands of whites feverishly digging up the basin floor to open copper mines, logging the forests for fuel and lumber, and raising the buildings to house and support them all.[15]

The Cherokees lost the Ducktown Basin and their Southern Appalachian homeland in a three-way battle for sovereignty fought between the Cherokee Nation, the federal government, and the state of Georgia. Sovereignty is defined as the supreme power or authority over a given population and territory. With sovereignty comes the power to control the use of natural resources, whether that power is used wisely or poorly. The Ducktown Basin was the subject of two great Supreme Court battles over sovereignty, each of which had implications bearing on the use of its resources. In the first battle, the Cherokees fought to preserve their tribal sovereignty against the competing sovereign claims of Georgia. Seventy years later, Georgia renewed the battle, this time against the state of Tennessee and its corporate citizens, in order to combat the damaging effects of smelter smoke from the copper works. Both contests were, in essence, struggles to exercise sovereign control over the natural resources of the region.[16]

At the beginning of the nineteenth century, the United States claimed sovereignty over the same lands by virtue of its defeat of the British and the 1783 Treaty of Paris. Georgia's claim rested on its colonial charter from Britain and its status as a sovereign state within the United States. The Cherokee people asserted their sovereign claim on the basis of their ancient occupation and on the terms of their many treaties with the United States, notably the 1791 Holston Treaty, which provided that "the United States solemnly guarantees to the Cherokee nation all their lands not hereby ceded."[17]

The assertion of Cherokee sovereignty met head-on with Georgia's claim of sovereignty over the substantial area of tribal lands within the state's borders. The issue grew as white settlement in the state moved ever west and north toward the tribal lands defined by federal treaties and now administered by a centralized tribal government. Georgia citizens had an insatiable hunger for fresh cotton land and found the existing Cherokee cotton farms especially desirable. The 1828 discovery of gold on tribal lands in the North Georgia Mountains at Dahlonega was another major impetus to Georgia settlers.[18]

Racism provided an additional motive for Indian removal. An earlier generation of Enlightenment thinkers admired the Cherokees. William Bartram praised them in 1791 as "happy in their dispositions," living in "divine simplicity and truth, friendship without guile, hospitality disinterested, native, undefiled, unmodified by artificial refinements." Georgia's governor, George Gilmer, described them in 1855 as "the least worthy of remembrance of any human beings"; their men were slothful except in the hunt or at war, and their women "were the least inviting of their sex." The state's long experience of Indian fighting, most recently in the Creek War of 1813-14, reinforced existing racial attitudes with an argument for frontier safety, even though the defeat of the Creeks ended the threat of serious hostilities.[19]

State sovereignty became the doctrine by which Georgia advanced its material and racial concerns. This was, and would long remain, a core theme of the Georgia body politic. Having won its independence against the British, the state now sought to preserve and advance its prerogatives against the Cherokees. The immediate dispute centered upon an 1802 agreement in which Georgia ceded its claim to western lands (the present states of Alabama and Mississippi) to the federal government in exchange for $1,250,000 and the promise "that the United States, shall at their own expense extinguish for the use of Georgia, as early as the same can be peaceably obtained on reasonable terms the Indian titles to . . . lands within the state of Georgia."[20]

The federal government was thus caught within a tangle created by its civilization program, its recognition of tribal sovereignty in formal treaties, and its 1802 promise to Georgia. Lacking a ready solution, Washington dithered over Cherokee policy for the next thirty years on the fading hope that more voluntary tribal cessions would resolve the matter. In the meantime, Georgia bristled over the delay, which it characterized as an offense to its honor as a sovereign state. The General Assembly also lamented that the civilization program of the federal government attached the Cherokees "to their country and to their homes as almost to destroy the last ray of hope that they would ever consent to part with the Georgia lands."[21]

The state's frustrations escalated with the tribe's adoption of the 1827 Cherokee Constitution and the 1828 Dahlonega gold rush. Choosing to wait no longer, the General Assembly attacked Cherokee sovereignty with a series of laws in 1829 and 1830. It declared all "laws, ordinances, orders and regulations . . . passed or enacted by the Cherokee Indians . . . to be null and void." In 1830 it authorized the governor to take possession of all gold mines, by force if necessary, that lay in "Cherokee country within the chartered limits of Georgia." Another act ignored Cherokee land titles by authorizing the survey of Cherokee territory within the state of Georgia into sections and parcels and calling for a lottery to dispose the land to the white citizens of the state. Yet another act made it illegal for the Cherokees to assemble as a legislative or judicial body upon pain of four years at hard labor. In addition, the act barred whites from residing within the Cherokee Nation without obtaining a license from the state and swearing an oath to "support and defend the Constitution and the laws of the state of Georgia."[22]

Matters came to a crisis when Georgia prosecuted Samuel Worcester for his refusal to obtain the license or to swear the oath. He was not the typical criminal defendant: as an ordained Congregational pastor from Vermont, a federally sponsored agent of the civilization program, a career missionary to the Cherokees, and a prolific translator of scripture and hymns into Sequoyah's syllabary, he could not be more different from the usual array of murderers, thieves, brawlers, and drunkards prosecuted in the Superior Court of the newly formed Gwinnett County. There being no doubt about his refusal to get a license, he based his defense on an argument that the new Georgia laws were an unconstitutional intrusion upon federal and Cherokee law.[23]

The court rejected the argument and sentenced the preacher to hard labor in prison. On appeal to the U.S. Supreme Court, Chief Justice John Marshall ruled in favor of Worcester and the tribe by declaring Georgia's anti-Cherokee laws void. He recognized the competing claims of sovereignty but upheld the superior claim of the Cherokees, based as it was on the tribe's ancient possession and guarantees of federal treaty. Accordingly, he held that "the Cherokee Nation, then, is a distinct community occupying its own territory, with boundaries accurately described, in which the laws of Georgia can have no force." In short, the Cherokees retained their right of self-rule and title to their lands; Georgia had no more sovereignty over them than it did over the land and people of a foreign country such as Canada.[24]

This was a famous legal victory for the Cherokees, but they were soon to learn that the impact of a favorable appellate opinion is contingent upon the political will to enforce it—a lesson that would dramatically shape the course of the Ducktown smoke litigation eighty years later. Georgia chose to ignore the *Worcester* decision on the well-founded confidence that President Andrew Jackson, a son of the Tennessee frontier and a famous Indian fighter, would do likewise. In his 1829 State of the Union address, he announced his sympathies for Georgia and proclaimed to the Indians that "their attempt to establish an independent government would not be countenanced by the Executive of the United States." In the following year, he succeeded in getting Congress to pass the 1830 Indian Removal Act to implement a program exchanging Indian holdings in the East for new lands west of the Mississippi River.[25]

A breakaway group of Cherokee leaders headed by Major Ridge concluded that emigration was inevitable and that it was best to negotiate a voluntary treaty to claim what benefits they could rather than to lose all by a forced ouster. To that end, Ridge and his party signed the 1835 Treaty of New Echota by which they relinquished all Cherokee land east of the Mississippi River and promised to emigrate to the West within two years. Signing the treaty was a desperate act done without tribal authority and in violation of the tribe's anticession law. They also knew that most of the Cherokee people, led by John Ross, were adamantly against the cession. Realizing that the Cherokees' hot anger over the loss of their mountain home would trump a cooler assessment of political realities, Major Ridge announced at the time of the treaty that "I have signed my death warrant." He spoke truly: vengeful opponents executed him, his son, and other members of the Treaty Party.[26]

The Treaty of New Echota brought Cherokee life in the Ducktown Basin to a drastic end. In 1838 soldiers under the command of General Winfield Scott forcibly rounded up almost the entire population of the Cherokee Nation for removal to what would become Oklahoma. Members of the tribe were herded into encampments before being transported by steamboat, by wagon, and on foot. The blockhouse of one of those camps, Fort Marr, still stands at the foot of Little Frog Mountain in Benton, Tennessee. Disease, exposure, and starvation haunted the 17,000 Cherokees sent to the West in the Trail of Tears, causing the death of an estimated 4,000 to 8,000 of them during the journey. Not all the Cherokees were transported. Though the soldiers were thorough, some Cherokees managed to escape removal, including a few from the Ducktown Basin. John Caldwell later hired some of the holdouts to help him build the Copper Road through the Ocoee Gorge in 1851. A few full-blood family groups remained in the basin area, living in enclaves at Turtletown, on Little Frog Mountain, and elsewhere until the 1890s.[27]

As the army harried the Cherokees out of their Appalachian homes, whites poured into the region to take possession, often under the bitter gaze of the former occupants. Adventurers prospected in the mountains for another gold strike like the one at Dahlonega, and farmers coveted the cotton lands in the Ridge and Valley region of Georgia and Tennessee, especially the existing farms developed by the Cherokees under the civilization program. Events moved more slowly in the basin. With its difficult approaches, its lack of discovered gold, and its poor climate for cotton, Ducktown initially remained a backwater for whites just as it had been for the Cherokees. None of the earliest settlers expressed interest in the copper that, in little more than a decade, would transform the basin into one of the richest districts in the South.

Ducktown's low value among the earliest settlers was reflected in prices obtained by the state of Tennessee in the sale of former Cherokee lands within its borders, an area known as the Ocoee Tract. In 1836 the legislature passed a law to dispose of the land by sale at an initial price of $7.50 an acre with stepped reductions in price for unsold land, beginning at $5.00, then falling to $2.00, $1.00, and ending after four more stages at a penny per acre. As in Georgia, cotton land moved quickly, but the highlands of Ducktown failed to draw buyers at the higher prices. Some bought in at $1.00 per acre; most waited until the price fell to the penny. The few white settlers who bought Ducktown acreage engaged in small-scale grazing and farming, using the land as their Cherokee predecessors did, and leaving much the same limited impact on the landscape. During the first few years of white settlement, the basin remained heavily wooded with plentiful game to stir the passions and appetites of hunters.[28]

Indians knew about the copper deposits in Ducktown. The Cherokees and their ancestors used it to fashion decorative plates, beads, armbands, and axes. Spectrochemical tests in the 1970s established that the copper in many of the artifacts found at the Etowah Mounds in Georgia and at other southeastern sites came from Ducktown. Copper from the basin was part of a continent-wide trade in the metal among the tribes of North America, with the largest portion coming from the extraordinary deposits of readily accessible native (naturally pure) copper along Lake Superior. For centuries, Indians had whacked off chunks of metal from the Ontonagon Boulder, a huge mass of native copper in what is now the Upper Peninsula of Michigan. Stories of the boulder told by Indians and fur trappers drew white prospectors who found the legend to be true: even after all the Indian use, the boulder was still a four-foot-long, 3,700-pound copper nugget. Promoters sent the rock to the East Coast, where it enticed miners and capitalists to travel west to a land of long, hard winters. There, beginning in 1843, they created the first major American copper rush of the century. Back in Tennessee, the deposits in Ducktown, though rich, lacked the boulders of native copper that excited attention in Michigan.[29]

Accident, not legendary boulders, led whites to the rediscovery of copper in Ducktown. The story, as recounted by R. O. Currey in his 1857 geological survey of Tennessee, began in 1843 when a Mr. Lemmons was panning for gold in Pumpkin Creek. Lemmons was one of a large number of prospectors who wandered over the Southern Appalachians in the search for gold. He rejoiced to find "large crystals of a deep rich red color" that convinced him he had discovered the precious metal. Lacking a container, he poured the crystals into the sleeve of his coat, secured the ends with a strip of hickory bark, and then found friends and whiskey for an all-night celebration. When he woke from his alcoholic stupor, he discovered that the bright flecks had darkened to a rusty brown. They were iron pyrite, not gold. He was just one more prospector among the many who had been fooled by fool's gold. Others discovered a black oxide of copper without appreciating its worth. Four years later, A. J. Weaver (some sources spell it Webber) showed greater appreciation for the oxide and shipped ninety casks of it to the Revere Smelting Works in Boston. Assayers at the firm, founded by Paul Revere, the patriot and coppersmith of Revolutionary fame, determined the ores to be wonderfully rich, containing 14 to 22 percent copper. Though Weaver gained the satisfaction of a proven discovery, he chose to abandon Ducktown in pursuit of Mexican gold. He later died en route across the Great Plains in a battle with Indians.[30]

Weaver evidently failed to appreciate the money to be made by supplying the large and rapidly growing national market for copper and its two major alloys, brass (a combination of copper and zinc) and bronze (copper and tin). During the colonial era, Americans used copper for coins, kettles, and pots. From brass came buckles, gun parts, sextants and other precision instruments, hardware, and wire for wool cards. Cannons, bells, and statues were made of bronze. Britain supplied colonial needs for the metals from long-established works centered on the copper mines of Cornwall, the smelters in Swansea, and the brass works in Bristol. Americans supplemented the supply with recycled scrap metal and with the output from small copper mines in New England and the Mid-Atlantic colonies. Demand for copper in all forms outstripped supplies soon after the Revolutionary War. A new use for copper arose from the discovery that copper sheathing protected the wooden hulls of naval and merchant fleets from wood-boring teredo worms and barnacles. A brass industry developed in Waterbury, Connecticut, to make brass buttons for uniforms and metal clock works to replace traditional wooden gears. New technology created new product lines, such as parts for steam engines and copper plates for photographers. The few small copper mines in eastern America played out as demand grew, prompting eager investors to develop the new finds in Michigan and Tennessee.[31]

John Caldwell knew this and went to Ducktown in 1849 to make his fortune. As he recounted in an 1855 letter, he was "scouting for copper, and found some five or six tons in a cabin, ten feet square" on property reserved by the state for use as a school. He "found the country unexplored" and noticed that "the school section, a property now worth a million of dollars," was "attracting little or no attention." Having only twenty dollars in his pocket, he "sat down in the woods for three hours to mature a plan to control and open the section." He then gathered the locals to tell them about the money to be made by leasing the tract for mining. Speaking in words calculated to appeal to the growing American spirit of economic boosterism, he told them "their condition would be improved, and that civilization, intelligence, comfort, and wealth would be the inevitable results." He then learned that boosterism was not a trait valued by the many who had settled on penny-per-acre land nobody else wanted. One man rose to say that "a large portion of the inhabitants had come here to get away from civilization, and if it followed them, they would run again." Caldwell's persuasive gifts eventually overcame objections to convince a majority to endorse a bill for the mining lease.[32]

Approval was one major hurdle. Transportation was another. The ninety casks of ore sent by Weaver to Boston in 1847 were hauled by teamsters over a seventy-mile route to Ellijay and then over the Cohutta Mountains to the nearest rail stop in Dalton, Georgia. The later extension of the railroad from Dalton up the Tennessee Valley to Cleveland, Tennessee, brought the tracks thirty miles closer to Ducktown — but on the wrong side of mountains. For the moment, the only direct way between Ducktown and the valley was by the old Cherokee footpath up and over Little Frog Mountain. It was a frustrating situation for Caldwell and like-minded entrepreneurs. The basin had the copper, Caldwell had the lease, investors had the money, and the railroad was tantalizingly near, yet without a viable wagon road to connect the Ducktown mines with the Cleveland depot, the copper industry remained an unrealized hope.[33]

The determined Caldwell solved the dilemma with black powder. He blasted a road through the rocky twenty-five-mile length of the Ocoee River Gorge to make a wagon road where the Cherokees had never made a footpath. To that end, he again applied his persuasive talents to the local crowd, this time at a Methodist camp meeting, to make a pitch for financing and labor. The local Methodists were more attuned to the sacred call of the Almighty than to the profane lure of the almighty dollar; they provided no funding and supplied only three laborers, all of whom quit by the third day of work. Caldwell realized that local opinion "was strong and powerful, being against the enterprise." He instead scrounged for funding from outside investors and hired a dozen of the remaining Cherokees to do the labor. He and the Indians finished the massive task in 1853 after two years of challenging work.[34]

The new road wound through the gorge at a viable grade for wagon travel as it ascended one thousand feet from the valley to the summit. As it did so, it presented a visual spectacle that enchanted many observers. The geologist Currey wrote that "there is presented to the traveler at every turn in the stream, new scenery, and apparently on a grander scale." Another observer said the rocks of the gorge "reminded [him] of the mountains of Switzerland." Investors took pleasure in a different sight: the constant movement of wagons making the four-day round trip between Ducktown and Cleveland, carrying barrels of ore down to the railroad and hauling people, equipment, and supplies back up the gorge to the mines.[35]

The boom could now begin. *Russell's Magazine* reported in 1858 that "a species of contagious insanity broke out; the monomaniacal feature of which was now copper. . . . Coat pockets protruded with the specimens they contained. Everybody talked copper." Speculators from New York, London, and the rich cotton ports of Charleston, Savannah, and New Orleans formed corporations to jump into the scramble for the more promising mining sites. Skilled English copper miners came from Cornwall to provide the technical expertise. Local farmers rejoiced to find a market for their produce and livestock in the mining camps. Any man with a wagon and a team of oxen or mules could join the new workingman's aristocracy of the Copper Road freight haulers. Woodcutters and charcoal burners supplied the need for timber and fuel from the extensive hardwood forests. Certain local advantages aided the mining boom. At Lake Superior, northern ice and snow interrupted work and waterborne transportation for up to six months a year and proved a misery to those whose experience of the lengthy winters was compounded by the great isolation of the Upper Peninsula. In Tennessee, the milder climate allowed for work and travel throughout the year with only brief interruptions for ice and snow, and though the basin was remote by eastern standards, there was no lack of supplies, news, mail, or fresh faces so long as the wagons kept rolling along the Copper Road.[36]

Ducktown mine operators also enjoyed the great advantage that the best copper ores lay at the surface where they could be easily worked. Copper readily bonded with sulfur, iron, and other minerals in its natural state, forming different ores in different combinations. The deposits closest to the surface underwent additional change as exposure to water and atmosphere oxidized or decomposed the ores in a way that increased the percentage of copper. The action of both agents, especially under the influence of Ducktown's high annual rainfall, worked their chemical magic over millennia to create a layer cake of different ores, each readily identifiable by its distinct color. At the surface was the reddish-brown gossan composed of iron pyrite, the stuff that fooled Lemmons, with some native copper nuggets. Below that was an extremely rich layer of decomposed ore known as black ore, though it appeared in different colors: red oxide (cuprite), black oxide, and green carbonate (malachite). The amount of copper in the black ores averaged 43 percent for red cuprite, 24 percent for the black oxide (the best samples contained 60 percent), and 21.5 percent for the green malachite. A layer of rock separated the rich copper oxides from the huge deposits of yellow copper pyrite below. The deep pyrite ores proved to be much leaner, containing only 1 to 5 percent copper, compared to the rich copper oxides above.[37]

The percentages varied from sample to sample and from mine to mine, but all agreed that the red and black oxides were the best. They were cheap to mine because the shallow deposits did not require expensive underground shafts and because "the workman with his pick detach[ed] it with great ease." Miners then stumbled upon an even easier source of copper once they learned how to precipitate it from water pumped out of the mines. They simply pooled the water into reservoirs lined with pieces of rust-free scrap iron. The copper precipitated from the water onto the iron, allowing workers to simply sweep it off with a broom. In 1859 three mines, the Hiwassee, Eureka, and Isabella, obtained more than forty thousand pounds per month in this manner.[38]

Fourteen mines opened between 1850 and 1854, yet for all of the activity, the mining boom did not immediately cause drastic wide-scale changes to the landscape. Frederick Law Olmsted, who would achieve fame for his design of New York's Central Park, traveled up the Copper Road to Ducktown during his 1853–54 travels in the southern highlands. His only mention of the land was that he "approached through the pretty valley of the [Ocoee], where . . . I met with hemlocks and laurels growing in great perfection." The social changes struck him more than the landscape. People at the mines constantly challenged him for being a speculator or a mineralogist. He dissuaded one curious person only after showing that his pouch contained "a pair of gloves, a knife, a corkscrew, a fleam, a tooth brush, a box of tapers, and a ball of twine" instead of ore samples and assaying implements. He noted that the working population consisted of "mostly white North Carolinians" and "several hundred Cornish men." The Cornish men enjoyed high wages but longed for English wheat bread instead of corn bread, and for ale instead of corn whiskey. Yet for all of the people he encountered, he determined that the general population "must be remarkably scattered, for there is nothing like a village."[39]

J. D. Whitney, a federal geologist traveling through Ducktown at about the same time, noted that the area was still heavily timbered with oak and other hardwood. It would not remain so for long because the exhaustion of the best ores led the mine operators to conduct smelting on site in the mid-1850s. Ducktown mine operators initially shipped raw ore by wagon via the Copper Road and then by rail and ship to distant smelters in the northern states and even to Swansea, Wales. Raw Ducktown ore posed a fire hazard in transit. An 1855 entry in *Hunt's Merchant's Magazine* reported that a shipment of copper ore, bound from Savannah to Liverpool aboard the *Albion*, spontaneously ignited because of its high sulfur content, causing damage to a valuable cargo of cotton stowed above it. Captains were warned that the ore was "not fit to be brought across the Atlantic" unless first roasted "to remove all danger from it."[40]

Ducktown miners worried more about the costs of shipping than the fire hazard posed by their ore. Freighters charged by weight, so it was much less profitable for the copper companies to ship raw ore than it would be to ship refined copper ingots. It made little sense to pay for shipping a ton of ore just to transport the small percentage of copper in it. Worse, the cost of shipping effectively increased as the quality of the ore declined because it cost the same to ship five percent ore as it did to ship twenty-five percent ore. Mine operators also realized that the greatest increase in the value of copper occurred at the smelting stage.

The answer to both observations was to refine the ore on site by roasting raw ore on burning heaps and smelting the result within furnaces to increase the copper content. In his 1859 report, Professor Charles Upham Shepard described roasting as the use of fire to produce a form of black ore, "a microcosm of what for thousands of years had been going forward" through the effects of water and atmosphere. The process, developed in Swansea, began by piling eight feet of ore upon stacked logs. Workers ignited the wood to begin chemical combustion, so that "sulfur fumes pass off, and the whole mass begins to heat and ferment." A shed was mounted over each heap to keep the rain from extinguishing the fire and from washing away the copper. The roasting phase lasted for months, during which the expelled sulfur entered directly into the atmosphere in the giant clouds observed by Taliaferro's Skitt. After the initial roasting, the remainder underwent further refining in two different kinds of furnaces interspersed with another round of open roasting. Reverberatory furnaces (essentially two-chamber ovens) calcinated (burned off) additional sulfur, and blast furnaces melted and separated out the iron and slag. The high-quality ingot copper achieved after six stages of roasting and smelting was much cheaper to ship and a much more valuable product to sell. As use of on-site smelting expanded, the amount and quality of Ducktown copper ingots encouraged others to establish a copper rolling mill and wire works in nearby Cleveland, Tennessee.[41]

Local smelting was good for copper companies and terrible for Ducktown forests. Without a railroad to import abundant southern coal into the basin, wood was by necessity the only practical source of fuel. Cordwood fired the roast pits, and the many furnaces consumed wood in the form of charcoal. Crews of charcoal burners followed the loggers as they worked their way through forests of the basin. The process of making charcoal began by stacking timber in a heap and then covering the whole with a layer of clay or soil to slow the rate of combustion. The goal was to achieve a slow burn using little oxygen so that water and volatile compounds were expelled without consuming the carbonized wood. The resulting charcoal had the advantage of burning much hotter and longer than raw wood and was thus of great use to the copper miners at Ducktown and to iron workers around the nation. The great hardwood forests of the Appalachian Mountains fell by the square mile to be reduced to charcoal for iron forges and copper smelting up and down the East Coast.[42]

An 1858 report to the shareholders of the Polk County Copper Company set forth wood requirements in Ducktown: an acre of forest yielded forty cords (a cord is a stack of wood four feet high, four feet wide, and eight feet long); each cord produced thirty-three bushels of charcoal; and a single blast furnace consumed 260 bushels of charcoal each day. Restated, each furnace burned approximately seventy-five acres of forest annually. The company assured its investors that it owned 750 acres of oak forest that would supply one blast furnace for ten years, enough to meet needs until the anticipated arrival of the railroad that would finally bring coal from the Cumberland Mountains.[43]

If it were only a matter of one smelter, the loss of 750 acres (approximately 1.2 square miles) of forest over ten years might have been acceptable given the huge expanses of Appalachian woodlands. And for herdsmen, clear-cutting had the incidental benefit of opening land for cattle grazing to boost local meat supplies. The reality was much worse because of the cumulative fuel requirements for all of the smelters from all of the mines, plus the demand for lumber for building and mine construction, and firewood for home cook stoves and fireplaces. In 1858 *Russell's Magazine* issued a prescient warning of the threat to southern mining from "the consumption of all available timber for fuel, greatly assisted ... by the abominable practice of burning off the cattle ranges" (burning promoted pasturage by preventing the regrowth of trees). The magazine recommended adoption of European forestry practices to ensure sustainable resources of wood and charcoal, but this suggestion would remain unheeded in the Southern Appalachians for a half century until the 1911 Weeks Act authorized the creation of national forest reserves in the East.[44]

Forests could be expected to regenerate in Ducktown's moist, temperate climate except for the impact of sulfur smoke pollution, a problem observed early in the area's mining history. Eugene Gaussoin, a Ducktown mining engineer, applied his professional eye to the phenomenon in 1860, writing "Not a single tree worth sawing is left on the five hundred acres of the Hiwassee property, and on the one hundred and sixty acres of the Cocheco, part of the timber is dead from the sulfurous vapors of the neighboring smelters." That same year, Taliaferro made a similar observation when his Skitt cried out, "See! Their sulphurous smoke has killed most of the timber near them!"[45]

Taliaferro did not intend his Ducktown sketch to be a vision of sylvan apocalypse. It was instead essentially humorous in tone, being one of six "Skitt" pieces he published in the *Literary Messenger* about people and places in the South. The Ducktown article was cast as a framed story involving a dialogue between the visiting sophisticate and several yokels about the boomtown conditions in the basin, of which the smoke was only one aspect. He gave greater space to the rube's tale of how, before the Copper Road and the start of mining, the greatest source of local income was the "witness ticket" or fee given to idle Ducktown folks who hiked the footpath over the mountains to serve in court at the county seat of Benton. Gaussoin, the mining engineer, was even less of an antebellum environmentalist. His concern was the pragmatic determination of whether the company-owned woodlot was large enough to warrant the expense of building a sawmill. As with most others, he considered wood consumption merely as a logistical problem to be solved to keep the smelters burning.[46]

Mine operators solved the wood problem in the near term with resources on their own lands or by purchasing charcoal from local sources. The more serious problem was insufficient capitalization. Many of the several dozen mining companies chartered by the Tennessee legislature either failed to commence at all or spent too much money on acquiring land and not enough on operations. This, and the added costs of building and operating smelters, led to a wave of consolidations in the late 1850s that resulted in three main operations: Union Consolidated Mining Company of Tennessee (the largest), Polk County Copper Company, and Burra-Burra Copper Company of Tennessee. Each of the three managed to grow and prosper right up to the eve of the Civil War, producing together more than three million pounds of copper annually. The mines directly employed about one thousand men and indirectly provided livelihoods to thousands more teamsters, woodcutters, merchants, farmers, and their families.[47]

A German immigrant, Julius Eckhardt Raht, was the guiding figure behind the formation of Union Consolidated Mining Company and became, by virtue of his many skills and indomitable spirit, the leading figure of the first era of the Ducktown copper industry. He was born in 1826 in the German Duchy of Nassau, where his father served as an appellate judge. As a young man, he studied in Bonn and Berlin in the fields of chemistry and mineralogy. The political and religious chaos of the failed German Revolution of 1848 prompted him, and thousands of other Germans, to immigrate to America. He then explored mining operations in several parts of the United States before deciding that Tennessee's new copper industry provided the best opportunity to display his talents and to make his fortune. Others agreed: he was appointed captain of the mines soon after his arrival in 1854. He so dominated the first era of Ducktown's copper industry that the period became known as "back in Raht's time," a phrase that Ducktown historian, R. E. Barclay, captured in the title of his first book.[48]

The outbreak of the Civil War sorely tested Captain Raht's leadership skills. Copper was a critical military resource used by the North and the South to manufacture cannons, impact fuses for artillery shells, bullet casings for the newer breech-loading guns, and percussion caps for the more common muzzle-loading weapons, as well as for miles of telegraph wire. The North had access to abundant copper in the Lake Superior region and elsewhere. For the South, Ducktown was by far the single most important source of the metal within the Confederacy. It was thus a place of great strategic value for both forces, all the more so because it was in a politically unstable part of the Confederacy; many mountaineers remained loyal to the Union, while Polk County voted for secession and formed five companies of Confederate troops, including two from Ducktown.[49]

On the eve of the war, northerners owned most of the mine shares and received most of the mine product. The Confederate government then secured its supply of copper by means of a decree issued in January 1862 that sequestered the shares of Yankee investors for resale to southern sympathizers. Captain Raht was not a southern patriot but chose to remain to supervise mining operations because of his professional and financial stake in the industry. Operations continued until November 1863, when Union raiders destroyed the copper rolling mills and rail facilities at Cleveland, effectively shutting down the mines for the duration of the war. Northern strategists did not need Ducktown copper and were content to leave the mines closed because that satisfied their military objective of denying copper to the South. With the mines dormant, the workers scattered, the idle pumps allowed the mines to flood, and bands of marauders picked over the now empty facilities. Having no more reason to stay, Raht left Ducktown and the South to sit out the war in Cincinnati.[50]

Peace brought an end to the smoking guns and an immediate resumption of smoking copper furnaces. Captain Raht returned from Cincinnati to restore the industry by pumping out the mines, restoring the works, and repairing the Copper Road. Mine workers were no doubt glad to have an income in the economically prostrate former Confederacy. Raht gained the financial backing of reinstated northern owners by selling 200,000 pounds of ingots. Some of the copper had been hidden in the depths of mines to prevent its capture by the Yankee army, and the rest was new copper precipitated out of the water pumped from flooded shafts. Raht's efforts to restore the industry were so successful that it produced more than one million pounds of copper during the first year of peace. It was an industrial miracle in the wreckage of the postwar South, but the spectacular revival could not hide four interrelated problems that threatened a peacetime closure of the mines: lack of direct railroad service to Ducktown, the increasing problem of wood supply, the exhaustion of the black ore, and the postwar decline of copper prices.[51]

The postwar annual reports of the Union Consolidated, Burra-Burra, and Polk County copper companies repeatedly addressed each of the problems. Exhaustion of the rich, easily reached black ores forced miners to incur the expense of digging tunnels to reach the lower-quality yellow ore. Burra-Burra justified the cost of a 250-foot shaft by pointing to the "prospect of soon developing a large and inexhaustible supply of yellow ore." Wood and charcoal presented problems of both cost and supply. Union Consolidated reported in 1866 that it bought a wood lot of 1,800 acres, which was a substantial tract of roughly three square miles, yet the supply of timber within it was insignificant compared to the company's annual wood requirement of 1,200–1,500 acres per year. Together, the copper companies logged out 30,000 acres or 47 square miles of timber in the basin between 1865 and 1878. Plenty of wood remained miles away in the surrounding mountains, but the cost of cutting and transporting it increased with distance and further strained the limited number of men, animals, and wagons in a region that still suffered from the depopulation and destruction of war.[52]

Desperate operators pulled up and burned stumps from older woodlots and turned to the forests at the headwaters of the Toccoa-Ocoee River, where timber from still abundant forests could be logged and floated downstream to the smelters. The state of Georgia did its part to keep the fires burning. It passed a law in 1876 making it a crime "to obstruct, by the erection of fish traps, or otherwise... the main current of Toccoa River, in Fannin County... so as to interrupt or interfere with rafting or floating timber." It also declared the Toccoa "a navigable stream so far as to authorize any person desiring so to do to float timber thereon." Both acts faced opposition from ordinary Georgians who had long asserted their common law riparian rights to fish in the river. For them, fishing was more than a pleasant pastime; it was also a way to supplement the limited corn and pork diets of the mountaineers. The legislature followed a different set of priorities by subordinating the rights and needs of ordinary Georgians to the economic interests of the copper companies across the border.[53]

The efforts involved in mining deeper, leaner ore and hauling wood from ever more distant sources ran up the costs of production at the same time that market prices began to fall. The expected postwar drop in price was compounded by the importation of abundant South American copper to increase overall supply and by the rapid transition from copper-bottomed wooden ships to metal-hulled vessels that acted to reduce demand for American metal. Copper prices fell from thirty cents per pound in 1865 to nineteen cents in 1870. Mine operators struggled to reduce costs by increasing production efficiency in every possible way, but it was not enough. Only the arrival of the railroad in Ducktown could save them by bringing cheap coal for the furnaces and more and larger equipment to boost economies of scale. In their 1869 annual report, the managers of Polk County Copper stated that, "when this road is finished, the cost of producing refined copper... will be reduced forty percent by the substitution of mineral coal for wood and charcoal now used." Coal not only cost less but burned so efficiently that "time now required to covert the raw ore to refined copper, four months, would be shortened two-thirds." The Union Consolidated Mining Company told its investors in 1871 that, "compared with the saving in production of copper this railroad would effect for us, every other improvement sinks into insignificance." Captain Raht stated the matter in more ominous terms, warning that, without a railroad, "nothing can be done at Ducktown until the Resurrection takes place."[54]

The point was not lost on others. As Barclay related in his *The Railroad Comes to Ducktown* (1973), more than a dozen competing, usually underfunded, rail ventures attempted to reach Ducktown from the east over the mountains from Asheville, North Carolina, from the south through the Blue Ridge from Atlanta, Georgia, and from the west up the Ocoee Gorge from Cleveland, Tennessee. Each state saw the Ducktown copper industry as the justification for a railroad that would serve the larger purpose of boosting all forms of commerce in the Southern Appalachians by creating a link to the national rail network. Year after postwar year, the mining companies repeatedly upheld the prospect of tracks running through Ducktown from "the Atlantic seaboard on the one hand, to the Great West on the other" in their annual pleas for financial support and patience from their shareholders.[55]

Hope failed on both ends of the line. One railroad company after another collapsed under the financial and technical burdens of laying track in the mountains. As one company failed, another rose to continue the work. Yet overall, progress failed to occur at a pace that could justify continued operation of the mines under conditions of low copper prices, local wood shortages, exhaustion of the richer ores, and the high cost of wagon transportation on the forty-mile Copper Road to the nearest tracks in Cleveland, Tennessee.[56]

The local mining industry was dying. Captain Raht, the hero of its postwar recovery, was now the target of shareholders in the United Consolidated Mining Company. They could not understand how he remained one of the richest men in Tennessee, while their dividends stopped and share prices declined. As the only person or entity in Ducktown that still had a deep pocket, he inevitably came under fire in a shareholder lawsuit alleging various financial misdeeds during his stewardship of company operations. The truth was much simpler. He made his fortune legally by running a commissary business to supply food and sundries to the miners, and then by making shrewd investments of his profits throughout Ducktown and eastern Tennessee.[57]

He won his case, but nothing short of rail access could save the local copper industry. One by one, insolvent mining companies allowed the smelter fires to die, the last of them in 1878. Captain Raht then died in 1879. It was the end of an era. Workers dispersed once more, mines again flooded, and investors counted their losses. And, as anyone could have seen from the mountain peaks surrounding the basin, "those huge columns of smoke" no longer covered "the heavens with a smoky pall."

2 THE REVIVAL OF DUCKTOWN MINING AND THE FIRST SMOKE SUITS, 1890–1903

The railroad eventually reached Ducktown. On a bright summer day in 1890, a train full of businessmen, politicians, and reporters left Atlanta for the first rail trip to Knoxville by way of the mining district. Proceeding over tracks laid by wage-earning white mountaineers and shackled, mostly black, convict laborers, it chugged northward ever higher through the foothills in Cherokee and Pickens counties to pierce the Blue Ridge Mountains in Gilmer County at the old Indian town of Ellijay. It continued up the Ellijay Valley into Fannin County to the town of Blue Ridge, and from there made the descent into the Ducktown Basin by following the Toccoa River. Waiting at the McCays Depot (later renamed Copperhill) were several hundred people and a brass band to honor the momentous occasion. All eyes turned up river to see the first puffs of smoke from the train's smokestack; all ears strained to hear the locomotive's whistle and bell. When the train arrived, the brakes squealed, the band played, and then a hush came over the crowd as local dignitaries read and delivered a proclamation to the crowd. In the grandiloquent phrasings of the times, the proclamation stated, "We hail this event as the beginning of an era of matchless growth and prosperity to this, one of the richest sections of the South," made possible because "our inexhaustible deposits of copper . . . are now accessible to the outside world."[1]

This was one of those rare moments when booster rhetoric rested firmly on present realities rather than airy hopes. Ducktown's copper resources were a proven fact, as was the demand for the metal to supply America's growing telephone, telegraph, and electrical industries. Completion of rail connections to Atlanta and Knoxville provided the last needed element for the revival of mining in the district. Capitalists on both sides of the Atlantic were alert to the new opportunities. London investors, operating as the Ducktown Sulphur, Copper & Iron Company, Ltd. (DSC&I), monitored the course of railroad construction, acquired the holdings of the old Union Consolidated Mining Company, and spent heavily to reopen the mines as the final lengths of track were laid. In 1891, another group of investors from Pittsburgh leased the School Property Mine and Polk County Mines and operated them under the name of the Pittsburgh & Tennessee Copper Company. The third, and eventually, dominant entity in Ducktown's restored industry was the Tennessee Copper Company (TCC). Created by New York City capitalists in 1899, it obtained the Burra Burra mine and other dormant mines from the heirs of Captain Julius Eckhardt Raht and then acquired the assets of the Pittsburgh & Tennessee Copper Company when it experienced financial difficulties. With new capital, engineers and workers rebuilt, refurbished, and expanded every aspect of the mine works. The expensive efforts to revive the mining industry could now be realized: with a locomotive puffing away at the depot, the basin's economic prospects were as solid as the rails on which the local trains now rolled.[2]

Rail access secured Ducktown's future and ended the twelve years of local economic depression that followed the closing of the mines in 1878. The closing had struck the district with terrible force because a generation of Ducktown residents had grown up with the copper industry. The industry began with the discovery of copper in the 1840s and, in 1851, grew when the completion of the Copper Road made wagon travel possible from Ducktown through the Ocoee Gorge to the nearest rail stop in Cleveland, Tennessee. During Captain Raht's time, Ducktown achieved a level of sustained industrialization that most areas of the mountain South would not experience until the end of the nineteenth century, when advancing railroads allowed large-scale coal mining to penetrate higher into the Appalachians.[3]

The first era of Ducktown mining ended when the lack of direct rail access to the basin, falling copper prices, exhaustion of local timber resources, and the playing out of the rich copper oxide deposits combined to make further mining operations unprofitable. When the clouds of smelter smoke drifted away, the local economy drifted away with it. The population collapsed as jobless miners and their families left the Ducktown Basin to pursue gainful labor elsewhere. Most of the loggers and teamsters also lost their livelihoods. Many local farmers followed their former customers out of the basin. Perhaps four or five thousand people lived in the basin before the closure. Ten years later, Carl Henrich, a mining engineer for the Pittsburgh & Tennessee Copper Company, wrote that "less than five hundred tried very hard to eke out a scanty subsistence, hoping for better times from the revival of the once flourishing copper mines."[4]

The farmers who remained lamented the loss of the market for the grain, vegetables, and fruit they formerly sold to the mining community. This was more than an inconvenience; little of what they grew could be profitably hauled over the mountains to distant markets. As Georgia geographer Laurence LaForge wrote a half century later in 1925, "The country is so rough and so stony and muddy in bad weather that communication and development are hindered," and the long distances to markets over miserable roads "have rendered farming unprofitable." The farmers, he said, "have won a living, but not much more," and "the profits are insufficient to attract any but hardy people whose needs are simple." They could supply most of their food needs from the crops and hogs they raised, but they needed to purchase or barter for coffee, salt, powder, shot, and other items that could not be made or produced on the farm. They also needed cash for property taxes. Some farmers made a tedious one- or two-week journey to lowland towns to sell white beans, sorghum molasses, and perhaps ginseng and other wild products. Others raised cattle on the cutover woodlands of the basin. As had been true since colonial times, it was often easier and more profitable to drive livestock on the hoof out of the Appalachians than it was to carry produce by wagon over the terrible roads.[5]

Though the mining-dependent populace suffered, the suspension of operations allowed the Ducktown landscape to begin a limited recovery. A quarter century of mining had transformed much of the heavily wooded landscape that existed at the time of the Cherokee Removal in 1838. Most of the nearby forests were leveled. Mine buildings, stores, and dwellings stood where only scattered farms could be found earlier. Open pits and mine shafts marked the areas where miners dug to reach the rich oxide ores near the surface and the sulfurets deeper down. Some areas of barren ground remained in the vicinity of the roasting yards and smelters along Potato Creek, where extensive exposure to toxic fumes denuded the soil and left it exposed to erosion under the heavy mountain rains. Even so, the scale of operations had been small enough during the first era of mining that most of the land retained sufficient fertility for some form of vegetation in the moist and now smoke-free climate. Second-growth forest rose in many areas of the cutover land where the great hardwood stands once grew.[6]

In other areas, cutover woodlands developed into pasture as the open ground attracted herds and herdsmen to Ducktown from throughout the region. The local law of open range, which prevailed at that time, allowed everyone to graze their livestock, without regard to property titles, except where landowners took the trouble to fence livestock out. Farmers had the burden of protecting their crops from free-ranging animals by raising fences around their cornfields, vegetable gardens, and orchards. (The later passage of fence laws eventually reversed the burden by requiring livestock owners to fence in their animals.) The teeth and hooves of cattle and sheep kept trees from reclaiming the pasturelands. Herders enhanced the process by lighting fires every year to burn away young trees and renew the grass. The basin's hardwood forests had been removed by the combined impact of logging and sulfur smoke from the roast heaps and smelters. Now, while the industry was dormant, livestock grazing and the deliberate use of fire continued and extended the changes over much of the basin.[7]

The few who stubbornly remained in the basin during the hard times did so by clinging to the hope, and not without reason, that the railroad would eventually come to revive the copper industry and the fortunes of all who depended upon it. Even in the depths of the basin's economic doldrums, the financial logic for a rail line to Ducktown remained strong enough to inspire continued investment and construction even after the mines had closed in 1878. Ducktown's copper resources and rapid industrial growth in the American economy made it likely that mining companies would return to the basin when the transportation problem was solved. The hopes of those who longed for the return of the industry gained strength with the advance of the Marietta & North Georgia Railroad (M&NGRR) north from Atlanta and the Knoxville Southern Railway southward from Knoxville.

The M&NGRR was the older of the two efforts, being chartered by the Georgia General Assembly in 1854. The burdens of the Civil War, the dearth of capital during Reconstruction, the distractions of business litigation, and the technical challenges of building railroads in the mountains caused it to flounder for decades. An infusion of northern capital in 1880 and the eager contributions from the mountain counties enabled it to extend northward to Jasper in 1883, Ellijay in 1884, and Blue Ridge soon thereafter.[8]

Notoriety surrounding the use of convict labor cast a shadow over the progress of construction. The state of Georgia leased black prisoners, often convicted on false charges, to private contractors at rates much lower than the going wages for labor. Contractors eagerly snapped up underpriced labor. The going rate for free black male labor was $120 per annum, and yet the state leased each man to the railroad for $11 per annum. From the perspective of the state, the practice accomplished several purposes: it relieved the cash-strapped postwar government of the expense for housing prisoners while ostensibly providing revenue from the leasing; it provided a pool of labor for rebuilding the state's war-torn infrastructure, especially the railroads; and it provided a powerful means to reassert white control over blacks that had been lost with emancipation.[9]

The system was rife with racism and brutality. The *Atlanta Constitution* considered the practice on the M&NGRR and elsewhere in an article under the provocative headline, "The Biped Zebra" (a reference to striped clothing worn by the prisoners). The paper noted that so many black men served as convict laborers that "at this rate it appears that the Negro is firmly and surely going back to a slavery that is worse than the one from which he was rescued, viz. a penal servitude." The author then added a sardonic observation that "there is not a court in Georgia in which the Negro is not given precisely the same justice that is meted to the white man, and yet nineteen of every twenty convicts are Negroes." News of abuses to black prisoners could be suppressed; but when a young white man was whipped to death for escaping from an M&NGRR chain gang, the incident led to press coverage and an investigation.[10]

Construction on the Atlanta & Knoxville line gained notice on a more positive note: the great engineering challenge required to bring the railroad up from the Tennessee Valley through the gorge of the Hiwassee River to reach the heights of Ducktown. Earlier plans to reach Ducktown by rail through the Ocoee Gorge, along the Copper Road, had never succeeded. Now, a later set of visionaries with greater capital, including money from a bond issued by the city of Knoxville, sought to reach Ducktown by way of the even more tortuous Hiwassee Gorge. Track workers clawed their way up the Hiwassee to a point where they were confronted with a 425-foot escarpment between the riverbed and the Ducktown plateau above. They first created a series of switchbacks, carving the side of the mountain wall with a series of giant zigzags of track. This required a train to proceed forward up one leg and then to back up the next leg in successive fashion until gaining the top. Though it worked, it was slow and it necessarily reduced the length of a train to three or four cars to accommodate the shortest leg of the zigzags. Engineers soon devised a better approach by running a climbing spiral of track around Bald Mountain to its top where they built a trestle to cross over the chasm to the plateau. The resulting Hiwassee Loop circled the mountain one and a half times so that the front end of a long train crossed over its tail as it climbed. The boldness of the solution testified to the lure of Ducktown copper.[11]

The mighty feats of railroad engineering, the injection of foreign capital into the mines, and the realistic prospects of renewed prosperity gave the crowd at the McCays Depot plenty to cheer about as it met the train from Atlanta. Everyone on the station platform knew that the distinct puffs of locomotive smoke they saw gently wafting over the river would soon be followed by massive clouds of dense, odorous sulfur smoke from the reopened roast heaps and smelters—the sight and smell of money. Miners and farmers rejoiced together, but few if any of them anticipated that the industry's revival would lead to a quarter century of smoke litigation.

The railroad immediately produced economies of scale for Ducktown miners. One mining engineer commented that before rail access, the industry relied on wood charcoal at the cost of ten cents a bushel (twenty pounds), and afterward trains brought coal by the carload from Kentucky at three dollars per ton. Other forms of technology improved production and lowered costs. Miners once used sledgehammers and hand drills to make holes for explosive charges to rip the ore from the earth; now they used diamond-tipped steam powered drills. Dynamite replaced black powder. A steam plant powered the mine hoists that lifted ore from the depths and the crushers and stamping mills that broke it into bits. Locomotives, instead of mules, hauled ore cars from the mines over a network of local rail lines to centralized roasting yards and smelting furnaces. Blast furnaces further refined roasted ores by forcing air through molten metal. Many of the advances were first employed by Captain Raht's miners toward the end of Ducktown's first era of mining; the railroad and abundant capital now made their use widespread.[12]

New technology employed by Ducktown Sulphur, Copper & Iron Company and by the Tennessee Copper Company boosted production rates and reduced costs for almost every aspect of copper mining, except for the continued use of open heap roasting. For centuries, this had been the state of the miner's art for the initial smelting of pyritic (sulfurous) ores and remained in use despite the production inefficiencies it created. Roasting remained a slow process that could not keep up with the much faster secondary smelting performed in the furnaces. Roasting required one to three months for each mound of ore. Blast furnaces could do the secondary smelting of the same amount of ore in a few days. The difference created a production bottleneck that could be solved only by increasing the number of roasting heaps—or by finding a means to accomplish primary smelting without roasting. The copper companies chose to expand the number of roast heaps. DSC&I established two huge roast yards in the early 1890s, and then TCC built another when it began production in 1901. The three major roast yards, each up to half a mile long, consisted of hundreds of burning heaps. Each heap started with a layer of cordwood, then coke, then hundreds of tons of ore in a pile five feet wide, ten feet long, and five feet high. Open sheds covered the heaps to prevent rain from slowing combustion and from leaching out the copper.[13]

Though they continued roasting, it was not for lack of trying alternatives.

Powerful economic incentives spurred efforts by Ducktown Sulphur, Copper & Iron Company to create a viable method for roasting sulfurous ores within a furnace, a process called pyritic smelting. If successful, the anticipated speed and efficiency of pyritic smelting would reduce the production bottleneck. Enclosed furnaces would retain heat to cook the sulfur out of the ore more efficiently than open heap roasting while reducing the amount of fuel needed for smelting. Pyritic smelting would also be a major step toward the recovery of the sulfur because the sulfur-laden smoke could be then drawn into the furnace chimney and redirected for further processing. Sulfur was a valuable commodity, and its use, especially in the form of sulfuric acid, was important to almost every aspect of industrial chemistry. It was also essential to convert phosphate rock into the superphosphate fertilizer so badly needed to replenish the worn-out cotton fields of the South. London investors proclaimed the importance of sulfur by the priority they gave to it in the company's name: the Ducktown *Sulphur*, Copper & Iron Company, Ltd.[14]

With these incentives, the race was on during the 1890s to develop pyritic smelting. The key problem was the determination of the correct charge or mixture of ore, coke, and quartz for the furnace. Mining engineers achieved some success in experiments in Colorado, Montana, and Tasmania, but attempts to duplicate the results in Ducktown failed because of the different composition of the local ore. The pyritic ores mined during Ducktown's revival contained only 1 to 2 percent copper, about 25 to 30 percent sulfur, and up to 40 percent iron, a combination that frustrated repeated efforts by DSC&I to determine the correct charge. Each time, the furnace fire began properly and then died prematurely, leaving a partially processed slag that had to be dug out of the furnace. The company would eventually solve the problem, but in the meantime both it and the later-arriving Tennessee Copper Company were forced to continue open heap roasting into the first years of the new century if they were to remain in the copper business.[15]

The roast yards filled the mountain-rimmed basin with astonishing amounts of smoke. The fires burned every hour of every day, rain or shine, creating a pall of smoke so thick that it hindered visibility even in the daytime. Windy days might disperse it, but on the frequent days of calm air and damp weather, it settled close to the ground to create a sulfurous fog. Even without the smoke, fog was a common feature of weather in the southern mountains. The geographer LaForge wrote, "The fogs of the valleys ... gather during the summer months to a depth of 100 to 200 feet where the valleys are surrounded by mountains." It would often "fill the entire area of the intermountain plateaus, so that the mountains and hills stand above them like islands in a sea." Natural fog became an unnatural nuisance when mixed with sulfur fumes from the mine works. James Smallshaw, a Tennessee Valley Authority scientist, interviewed longtime Ducktown residents in the late 1930s about the smoke experienced at the turn of the century. E. R. Wallace said it was so thick that wagoners put bells on their teams so that the jingling sound would serve as a warning to oncoming travelers. Octavus Hankivell related that at night the smoke made it impossible for him to follow wagon tracks on horseback. He had to dismount so he could walk the tracks with a lantern in one hand while leading the horse with the other. The horse was partly to blame, though; it was reported that "after getting himself and the horse lost several times...he bought another horse with more sense and had no further trouble."[16]

The greater volume of smoke soon became an intolerable burden to many local farmers. G. W. Prince claimed that "the smoke settles down on my place deeply unless the swift wind drives it away, and remains until nine and eleven o'clock AM so thick and dense as to deprive me and [my] family from the sunlight." A. J. Bell told how the smoke "settles on vegetation and when the sun strikes, it turns white and it dries it up. It also settles on iron tools and causes the rust to eat them up." Their testimony would be echoed by many other witnesses over the next twenty years in hundreds of smoke cases and described the effects of what would later be called acid rain. Sulfur from the roast yards entered the air as sulfur dioxide (SO_2) that either precipitated in dry form or fell in wet form when it combined with water to form sulfuric acid (H_2SO_4). Both forms were troublesome because frequent temperature inversions trapped the gas within the basin, and the prevailing humid, rainy weather encouraged acid formation. The consequences that the farmers understood experientially would eventually be considered at length from a scientific perspective in the smoke litigation to come.[17]

Smoked-out farmers had only three options: tolerate the smoke for as long as possible, abandon farming altogether, or file suit against the copper companies. If they litigated, their only recourse was under the law of nuisances, whereby they could seek either monetary damages for their crop losses or an injunction against the smoke. If they filed suit, it would be on their initiative and at their time and expense. There were no regulatory agencies to act on their behalf because there were no federal or state regulations to control the smoke. None would be enacted until well into the twentieth century when California passed the first statewide air pollution law in 1947, followed by the federal Clean Air Act of 1963, and Air Quality Act of 1967.[18]

At the municipal level, antismoke initiatives arose in some major cities during the late nineteenth and early twentieth centuries to combat smoke from thousands of industrial furnaces, steam locomotives, coal cook stoves, and home hearths that coated every surface with layers of sooty filth. Progressive citizen groups, often acting at the instigation of middle- and upper-class women, urged passage of smoke ordinances to advance what David Stradling described as the "Victorian notions of cleanliness, health, and aesthetics" associated with the City Beautiful movement. The regulations were a start but failed to accomplish much. They were often voided on legal grounds, especially when cities attempted to regulate interstate polluters such as railroads. Realistically, city action suffered the impracticability attendant to the local regulation of a regional problem. The legendary King Canute could not stop the tide, and New York City could not stop the prevailing winds from carrying New Jersey smoke into Manhattan.[19]

Regulation was even rarer in one-industry towns because citizens often lacked the political will and resources to force regulations that might kill their golden goose. The citizens of Butte, Montana, proved to be the exception when they forced passage of a municipal smoke ordinance in 1890 that forced Anaconda Copper to move its smelting operations out of town to rural Deer Lodge Valley. Issues of health, not cleanliness, prompted the initiative when a local physician noticed rising death rates during periods of intense smelter smoke. He and others soon realized that Anaconda's smoke was poisonous because the local copper ores contained significant amounts of arsenic that smelting released into the atmosphere. The removal of the smelters improved the health of townsfolk but proved lethal to livestock downwind of relocated works because the arsenic then precipitated onto pastures, causing horses, cattle, and sheep to drop dead after eating the toxic grass. Unlike the assembled citizens of Butte, the farmers and ranchers of the Deer Lodge Valley lacked the political clout to force a regulatory cure and instead turned to the courts.[20]

Ducktown's ore, and hence its smoke, was blessedly free of arsenic, and its absence was a major reason why there was no public demand for health-oriented antismoke ordinances in basin communities. Yet, the high-sulfur content of Ducktown smelter smoke had a toxic impact on crops, orchards, and woodlots that was plain enough to local farmers. The amount of smoke increased as the revived industry hit its stride, leading a group of ten farmers to file the first wave of suits in 1896 and 1897, later consolidated for appeal under the caption, *Ducktown Sulphur, Copper & Iron Co. v. Barnes*.[21]

The *Barnes* lawsuits, and several hundred others to follow, provide a description of the smoke suitors, their farms, and their motives. They were men and women (Mrs. Margaret Madison filed many suits as a *femme sole*, a woman authorized to file suit in her own name, without the intervention of a husband or male guardian). And their names, such as Barnes, Bell, Carter, Fortner, Johnson, Madison, Runnion, Stuart, and Thomas, reflected their cultural roots among the Scots-Irish and English stock that settled and farmed the southern mountains. Italians, Poles, Slavs, and other immigrants from a dozen other countries came to the Ducktown Basin in the early years of the revival, mostly to find work in mine construction, but few of them remained in the area for long and fewer became farmers with a claimant's interest in smoke litigation. R. E. Barclay, a lifelong Ducktown resident and its leading historian, observed that "the foreign element disappeared after seven or eight years."[22]

All of the suitors were whites. Blacks lived in pockets throughout the Southern Appalachians, though their numbers as a percentage of the population were much lower than in the lowlands. More than four hundred black slaves once toiled in the Tennessee Valley portion of Polk County, west of the mountains that walled in Ducktown. Many of them remained in the county after emancipation; but they were not welcomed in the highlands of the Ducktown Basin, where whites were determined to preserve steady jobs in the copper industry for themselves, by force if necessary. In 1894 a group of armed whites raided a camp of black laborers brought in to lay railroad track at the Mary Mine. The raiders came out of the night, fired an estimated fifty to a hundred shots, and tossed a few sticks of dynamite to roust the blacks out of their bunks and out of the basin. Nobody was hurt but the point had been made. Barclay noted in 1934 that "the Negro has always been banned from the basin, except one employed privately," and added that "there has never been a Negro home here."[23]

The suitors were all landowners for the reason that only a person holding property rights could maintain an action in nuisance. Most owned their farms in fee simple, partly because the rate of farm tenancy was much lower in the highlands than in the cotton regions of the South, and more because tenants lacked the incentive and time necessary to litigate cases that might last for years. The ownership requirement shut out the squatters who built little farmsteads on the land of absentee timber owners in the surrounding mountains. The squatters nearest the roast pits and smelters must have suffered smoke damage, but without land titles, they could not make a legal claim for damages and were, in any event, loath to risk exposure to the authorities. George Peabody Wetmore, a New England aristocrat with extensive timber holdings in Polk County, spared no expense to prosecute squatters. He filed suit, through his local counsel, Charles Seymour, to eject them whenever found, lest they establish title by adverse possession. He and Seymour took one such case all the way to the U.S. Supreme Court in 1898.[24]

The smoke suitors came from all three counties of the basin and thus involved the citizens of three different states. Of the ten *Barnes* claimants, six were residents of Polk County, Tennessee; three owned land in Fannin County, Georgia; and the farm of the tenth lay in Cherokee County, North Carolina. In later years, as the zone of smoke damage expanded, the greatest number of claimants would be Georgians, followed by the North Carolinians, but it bears noting that Polk County farmers were among the first to litigate.[25]

They practiced a thoroughly mixed form of agriculture on farms of 40 to 250 acres. A few farms approached 1,000 acres along Hot House Creek where some of the best tracts in the district were located. Good bottomland could also be found along other major streams, including the Toccoa-Ocoee River, Wolf Creek, and Fightingtown Creek. It is no accident that most of the early smoke suits arose from those areas. Sulfur dioxide smoke was heavier than air, especially when saturated with moisture, so it tended to descend into stream valleys and to flow along their courses. Many observers commented how clouds of smelter smoke could be seen flowing along the Toccoa-Ocoee Valley before branching off into tributary valleys, passing over the prized bottomlands as it did so.[26]

The hilly topography of the basin made the size of farms deceiving. This was not a land of broad level farms like those in the Midwest, where row crops could be planted from border to border; instead, mountain farms usually consisted of fertile bottomlands, the less fertile but still workable gentler slopes, and the steepest sections that remained forested. The varied topography and highland climate encouraged a variety of crops. The fields were planted with corn (maize), oats, rye, and wheat. Every farm had a garden patch for Irish and sweet potatoes, cabbages, peas, beans, squashes, and assorted greens. The orchards contained apple trees, peach trees, and hives of bees to pollinate them. Thousands of cattle, sheep, and swine provided meat, milk, leather, and wool. They grew some burley tobacco but no cotton.[27]

Mixed agriculture led to the rational expectation that basin farmers could provide most of their food needs from their own farms; yet they were not independent of the market economy, nor did they want to be. The farmer litigants of the 1890s lived adjacent to one of the great industrial complexes of the South in a district that now enjoyed daily rail service, received daily newspapers from Atlanta, Nashville, and Knoxville, and had access to telegraphic communication with New York and London. They were not the isolated subsistence farmers of hillbilly legend.[28]

By comparison, the area's earliest farmers in the 1840s did live in isolation behind the mountain barriers and practiced subsistence agriculture to a considerable degree. Those days ended with the arrival of the mining industry and the Copper Road, though they reappeared for a decade when the mines closed in 1878. When the mines reopened in 1890, Ducktown farming and the copper industry advanced side by side. As miners repopulated the basin, each farmer expected to produce surplus meat and produce for sale in the mining communities in order to buy a wide variety of goods from merchants. A journalist observed in 1876 that "nearly all the grain raised in Polk County finds a market at Ducktown," and it sold for a higher price there than in the Tennessee Valley. The same was "equally true of meal, hay and all other supplies." Another journalist noted, "If it were not for the minerals, especially copper," the area "would be little else than a huge pile of mountains, or a pile of huge mountains, whichever you choose." Happily, Ducktown was "about the best produce market in East Tennessee."[29]

Ducktown farmers agreed with the journalists, and for that reason they entered suit against the copper companies with some reluctance. They valued the mining community as their best and nearest market and longed for a mutually beneficial economic relationship. They were not Luddite reactionaries against technology, nor were they environmental activists in the modern sense. Nuisance law in the industrial context inevitably touched on the greatest issues of environmental policy, but the Ducktown litigants did not see it that way. No environmental or conservationist organizations stood behind them when they fired the first shots in the smoke wars. Instead, the farmers sued the copper companies because smoke hindered their ability to participate in the thriving economy created by Ducktown's mining industry. The sulfur smoke killed their crops, and thus their stock in trade. Crop losses translated into lost income and lost purchasing power. Ultimately, smoke put their agricultural livelihoods at risk. A. J. Bell, one of the *Barnes* litigants, stated the problem succinctly when he complained that he was "amply able to support himself and family from his farm by raising good crops of grain ... and all kinds of fruits and vegetables, and bee culture ... and could carry to market some of the crops, etc. before the smoke, but now he can sell nothing."[30]

Bell and his fellow *Barnes* litigants gave expression to their market frustrations by filing suits in the circuit court at Benton, the seat of Polk County,

Tennessee, against Ducktown Sulphur, Copper & Iron Co. At the time of filing in 1897, the Ducktown Company was the only major copper firm in the basin; the Pittsburgh & Tennessee Copper Company was in decline, and its successor, the Tennessee Copper Company, did not begin smelting operations until 1901. The gravamen of their claims was that DSC&I created a nuisance "by sending forth from its roast piles, smelters, and furnaces volumes of sulfur smoke, fumes, and poisonous gases, thereby poisoning the air," killing their crops, orchards, and timber, and "shutting them off from the sunlight and rendering their homes unpleasant and unhealthy, ruining all their edged tools, killing their bees." The litany of wrongs ended, in a biblical turn of phrase, with the allegation that the smoke made "the whole country a barren plain." For such wrongs, each claimant sought damages in the range of $1,000 to $2,000.[31]

The law of nuisances is part of the great heritage of English common law that serves as the foundation for American law. Nuisance law involves offenses caused by one party that impairs another party's use and enjoyment of property. The term derives from Middle English, through Old French, and ultimately from the Latin, *nocere*, to injure or harm. Typical actions giving rise to nuisance suits include the diversion of water onto another person's land and the creation of offensive odors from tanneries and slaughterhouses that wafted onto the property of another. For the *Barnes* claimants, the offensive agent was smelter smoke. Whatever the offense, the law of nuisances sought to enforce the maxim, *sic utere tuo ut alienum non laedas* (so use your property as not to injure your neighbor's). It should be noted here that American lawyers also use the term "nuisance" in another sense, meaning a frivolous lawsuit, particularly a claim for damages arising from a minor or questionable injury. The Ducktown smoke litigants used the term in its traditional sense, in reference to the infringement of property rights.[32]

Mrs. Margaret Madison's complaint in nuisance was typical of the other *Barnes* suitors. She first set forth allegations of her ownership, her enjoyment of the property, and the offensive interruption to her enjoyment caused by the defendant's smelter fumes. In phrasings cast in an Arcadian or perhaps Old Testament air, she proclaimed ownership of valuable farmland that produced "wheat, corn, rye, oats, barley, Irish and sweet potatoes, and vegetables in goodly quantity," and on which "the apple and peach, the plum and quince, the cherry and the grape, and the berry family generally grew to perfection." Her land normally provided "a generous return for labors expended in their care and cultivation," and "by that reason the small farmers around were contented and happy." Then came the allegation of harm, where she stated that her once healthy gardens, orchards, and fields "have disappeared under the blighting poisons of the smoke, vapors, and gases emitted from the works" of the defendant copper company. Specifically, the nuisance was "smoke and noxious vapor charged with sulfuric acid, which is varied by air currents across the Tennessee line into Georgia and which destroys" the vegetation. Claiming the mantle of the humble yeoman farmer, she denied a desire for personal gain and instead wanted "simply to make her bread on her farm, and eat it in peace and comfort."[33]

Because Ducktown Sulphur, Copper & Iron Company was the sole defendant in the *Barnes* actions, it fell to the firm's counsel, Mayfield & Son, to frame the legal strategy later adopted by its soon-to-be rival, Tennessee Copper Company. The dense clouds of fumes and the barren lands around the DSC&I works at Isabella made it pointless to deny the presence or toxicity of the smoke, though the degree of damage on a given farm could be argued. Rather than litigate damages, the Mayfields adopted a policy of doing everything possible to keep cases from going to jury trials. James G. Parks, who succeeded the Mayfields as DSC&I counsel, explained the policy in a letter to management about rising court costs. The costs for repeated continuances and appeals were "due to the long time the cases have been pending and the policy of delay that has been followed from the beginning." He justified the expense saying, "It has undoubtedly given the 'smoke suitors' a decided distaste for lawing, and has deterred many from suing who would have done so, if they could have 'railroaded' their suits through the courts."[34]

This was sound defense policy. As attorneys, they were advocates for their clients, not dispassionate framers of public policy about industrial smoke. As advocates, their job was to protect company coffers by defeating smoke claims. The safest way to do this was by use of procedural maneuvers to keep cases away from juries. Underlying the strategy was the bedrock truth that Polk County juries would likely award damages to claimants if given the chance.

It would at first seem counterintuitive that Polk citizens would punish the only major employers and by far the greatest taxpayers in the county, but the very wealth and power of the copper companies made them targets in a context of strained local relations. Labor tensions placed the DSC&I in a negative light, especially because many farm families had members who worked at least seasonally in the mines. The troubles escalated in the spring and summer of 1899 with a series of incidents, some of them bloody. One worker, W. A. Curtis, was charged with carrying a concealed weapon and then firing it near a powder magazine on DSC&I property. Another was charged with assault with intent to kill when he swung a shovel at a deputy sheriff who was moonlighting as a night watchman at the smelting furnaces. L. Y. Henson was charged in a third incident when he shot a man deep underground in the Mary Mine. W. H. Freeland, the general manager of DSC&I, wrote corporate counsel P. B. Mayfield and insisted upon prosecuting all three men for what he considered a "prearranged conspiracy." Matters worsened a few days later when a DSC&I superintendent, S. M. Reese, was ambushed and shot to death. (Bloodhounds soon tracked the assailants into the mountains and treed them.) Notwithstanding the violence of the recent events, Mayfield soon learned what Polk County jurors thought of the company. He reported to Freeland that "the Grand Jury ignored the bill against Curtis and possibly no action was taken against Henson."[35]

Labor tensions led to formal union activity later that year. In September the American Federation of Labor (AFL) established an affiliate among workers frustrated over the requirement that they consult a company doctor for work injuries instead of seeing their personal doctor. They did not trust a company physician to be fair when evaluating and treating injuries. DSC&I refused to recognize the union and discharged its members, but the halt in production soon forced the company to relent on the physician issue. That success soon led to a demand for better wages and hours. The workers asked for a reduction of the daily shift from ten hours to eight and a rate of pay equivalent to nine and a half hours at the old rate. A strike ensued when management refused, and tensions increased when the company attempted to replace striking union members with non-union "scab" labor. The union held the upper hand because the AFL threatened a sympathy strike by union engineers and firemen on the Atlanta & Knoxville Railroad, which handled freight for the company. The threat that Ducktown's rail lifeline would become a noose then moved DSC&I to agree to an eight-hour shift and a 10 percent pay increase.[36]

A much less violent matter, tax appraisals, also stirred local citizens against the copper companies. Polk County school and government leaders periodically accused the copper firms of failing to pay their fair share of property taxes to support education and public improvements, notwithstanding the truth that the companies were by far the greatest source of tax revenue in Polk County. The argument was appealing to local farmers—and potential jurors—burdened with property taxes to be paid in years of good harvest and bad. Leaders in the Ducktown mining community saw things differently; the local quip was that "Polk County's got a big cow. We feed it up here, and they milk it in Benton." When the copper companies fought a tax appraisal suit in 1900, they both lost. DSC&I's valuation quadrupled from $55,000 to $200,000, and TCC's more than tripled from $75,000 to $250,000.[37]

Absentee ownership also worked against the companies. Most Polk County jurors struggled to make a living, whether on the farm or in the mines, and they readily identified with hardscrabble smoked-out farmers against the interests of remote copper plutocrats. The Tennessee Copper Company headquarters were on Broadway in New York City, and its major owners were the Lewisohns, a German Jewish mercantile family. The firm began when Samuel Lewisohn sent his sons, Leonard and Adolph, to America during the Civil War to import horsehair (used for stuffing mattress and upholstery) and ostrich feathers (for decorating hats). Upon arrival, the brothers began to export copper back to Germany and soon became major figures in American copper with mine holdings at Lake Superior and Butte, Montana, along with a large copper refinery in Raritan, New Jersey. Their financial peers were the Guggenheims, Bigelows, and other prominent mining families, not Scots-Irish southern mountaineers. The London owners of the Ducktown Sulphur, Copper & Iron Company, Ltd. were even more remote geographically, but present tensions with the English firm stirred old cultural memories of the American Revolution. Margaret Madison, owner of 160 acres in the Wolf Creek area of Georgia's Fannin County, counted herself among the "small farmers in the rural district [who] were contented and happy" until the recent smoke damage; according to her own declaration, she was unwilling "to see her home wrecked, her patrimony destroyed, and . . . in her old age be driven out by her lordly neighbors from Great Britain."[38]

Margaret Madison's Jeffersonian yeoman rhetoric could be expected to play well with Polk County's rural jurors. Southern farmers struggled to maintain personal economic independence in a world increasingly shaped by national and international market forces. Hundreds of thousands of cotton farmers, including many in the Tennessee Valley lowlands of western Polk County, like their counterparts throughout the cotton South, blamed their slide from ownership to tenancy on the actions of cotton brokers, industrialists, banks, and railroads that kept cotton prices low and growing costs high. The farmers high up in the basin struggled against visible smoke, rather than Adam Smith's "Invisible Hand," but shared in the problem of keeping their farms viable. Agriculture was a losing proposition for most small southern growers, and for that reason the smoke litigants could count on a sympathetic hearing from jurors drawn from among their own kind.[39]

To prevent that from happening, company attorneys exploited every procedural ground for delay by means of motion, injunction, and appeal. Much of the strategy turned upon the sharp distinction that Tennessee jurisprudence made (and continues to make) between law and equity. The circuit courts exercised jurisdiction over law cases such as criminal matters, contract disputes, and personal injury claims, whereas the chancery courts exercised jurisdiction over equitable matters such as wills, estates, and land titles. The distinction was an accident of Anglo-American legal history but had great practical importance for Ducktown smoke litigation. Nuisance law allowed smoke suitors a choice of remedies: monetary damages to compensate for crop and timber losses, or an injunction requiring the copper companies to abate or eliminate the smoke. The choice of remedy dictated the choice of court because a circuit court could not order injunctive relief, and a chancery court could not award unliquidated damages (a monetary award not readily fixed or calculable by contract) unless it first gained jurisdiction on an equitable issue. Only circuit courts allowed for trial by jury and the chance to have the amount of damages set by sympathetic local citizens. Each type of court met infrequently in rural counties such as Polk, with sessions scheduled only three or four times a year, and with each session lasting just a week or two. Lengthy delays usually followed the transfer of a case between the two courts.[40]

The Mayfield firm first deployed its stalling tactics against the ten *Barnes* cases that the farmers filed in circuit court. The firm responded to the nuisance allegations with several arguments. It asserted that it should not be held liable in nuisance for conducting a lawful business when the use of open heap roasting was the best practical smelting technique. Next, the "incidental advantages" of mining operations to the local economy should be considered when awarding damages. Third, it argued that the community acquiesced to the smoke by welcoming the return of mining. It then sought dismissal of several of the suits upon allegations that lawyers for the plaintiffs committed champerty. The charge referred to the then illegal contingent-fee contract whereby a lawyer agreed to take a percentage of the recovery as a fee instead of charging an hourly or flat rate for time and services. Though illegal at the time of the *Barnes* cases, the contingent-fee contract eventually became widespread after Tennessee and other jurisdictions abolished champerty laws. The arrangement is now commonly advertised by personal injury lawyers with their promise of "no fee unless you recover."[41]

Jurisdiction was the last prong of the defense. DSC&I contended that citizens of Georgia and North Carolina had no right to sue in Tennessee courts for smoke damage occurring to lands situated outside that state. The argument was that the alleged offense was local in nature because it impacted real estate and thus had to be tried in courts having jurisdiction over the affected land. It followed that damage to Georgia land should be tried in Georgia courts.

The argument was coupled with a trick: Georgia courts had no jurisdiction over DSC&I under the narrow understanding of jurisdiction applicable at the time. Broadly speaking, a court's jurisdiction over a defendant rested upon the presence of the defendant's person (jurisdiction *in personam*) or property (jurisdiction *in rem*) within the geographic bounds of the court's authority. The copper company, as a corporation, was deemed to be a legal person with a residence where it was chartered (London), where it conducted its operations (Polk County, Tennessee), or in such other states where it registered with a state to do business (which it had not done in Georgia). In the eyes of the law, the company had no legal presence in the state of Georgia, never mind how much smelter smoke it allowed to drift south over the state line to trouble the farmers of the Peach State.[42]

The beauty of the argument for the DSC&I lawyers was that if it succeeded, Georgia smoke suitors would face the near impossibility of suing the company in either state. They would be barred from the Tennessee courts because an action for damage to real estate was local to Georgia. They would also be unable to litigate in their own state because Georgia courts had no jurisdiction over the company. It was the sort of legal logic that befuddled first-year law students but became a fine-edged tool in the hands of a skillful attorney. It was also an argument that was more plausible in the 1890s than it would be in the future when notions of jurisdiction broadened. Long after the era of Ducktown smoke litigation ended, the "minimum contacts" test announced by the Supreme Court in *International Shoe Co. v. Washington* (1945), as well as the enactment of "long-arm" statutes, would greatly expand the ability of courts to assert jurisdiction over nonresident defendants in a wide variety of cases.[43]

Underneath all of the defenses was the question of why the suits were filed at all, given the long and generally cooperative relationship between Ducktown farmers and Ducktown miners. The mines had been roasting raw or "green" copper ore in open heaps and then smelting the roasted ores in furnaces ever since the early 1850s. The only times when the skies over the Ducktown Basin had been smoke-free were during the dreadful last years of the Civil War and the twelve-year suspension of mining from 1878 to 1890—events that few cared to see repeated. So the question arose, Why did smoke suddenly become an issue at the turn of the new century? Why then, in the late 1890s, and not fifty years earlier?

The company raised the point in the *Barnes* litigation when it argued that its methods were no different from those employed by Captain Raht's Union Consolidated Mining Company back in Ducktown's first era. Operations after resumption of mining in 1890 "caused smoke and the gasses to be emitted as in a manner had always been the case." Those who complained now of smoke produced by the same methods used in the past were "malcontents fraudulently confederated and combined to vex, harass, and annoy" the company. This put the onus upon the farmers to differentiate post-1890 smoke from that of Captain Raht's time.[44]

Their efforts to do so ranged from specious to credible. William Madison, Margaret's son and fellow claimant, pointed to the recent use of coked coal in lieu of the former wood charcoal as the fuel for the roast heaps and furnaces. He was correct about the change of fuel but offered no support for the proposition that one fuel caused more smoke damage than the other. Charcoal and coke were analogous products, both being organic fuel carbonized in airless fires. Coke began as coal dug from the nearby Cumberland and Alleghany Mountains. It was then baked, as wood charcoal was, in airless ovens at high heat to drive out sulfur and other volatile components, to produce a high-carbon smelter fuel that burned cleaner and hotter than ordinary coal. Madison nonetheless argued that use of wood charcoal was better because it "gave out no vapor or poisonous smoke of consequence." If anything, the long-term reliance on wood had been even more destructive than the smoke from the fires it fueled. More than fifty square miles of forest fell to the ax and saw to fuel the roast heaps and furnaces, and the deforestation would prove to be a major element in the creation of the badlands of the man-made Ducktown Desert. Coke imported by rail from outside the basin had the great benefit that it ended the total reliance upon local timber for smelting.[45]

William Madison then argued that he suffered no damage during the Raht years because the miners then "carried the ores through from five to six processes" of roasting and smelting "whereas the present company merely roasts the ores one time and smelts, thereby throwing a greater quantity of sulfur into the overhead air in a given time." The assertion carried little explanatory force because again, whether done in two steps or six, the goal of smelting was to expel sulfur from the copper ore from whence it entered the atmosphere as sulfur dioxide. His mother, Margaret Madison, came closer to the mark with her allegation that "current operations are worse than in the past because of use of high-sulfur ore."[46]

It was true that the first era of mining employed the copper oxides found at and near the surface, ores that were rich in copper (often 25 percent or more) and low in sulfur. When mining resumed in the 1890s, the industry processed leaner sulfite ores containing only 2 to 5 percent copper and 25 percent or more sulfur. This meant that more ore had to be smelted to attain the same amount of copper, and that each ton of ore roasted and smelted released up to a quarter ton of sulfur into the atmosphere. John Quintell, a Ducktown miner for forty-five years, provided the most insightful response to the problem in his deposition. Yes, Captain Raht used richer ore that contained less sulfur, but his daily production was also a fraction of that in the modern era.[47]

Quintell had it right. Though the type of ore had changed, the root of the problem was the rapid expansion of every phase of mining made possible by the railroad, electricity, dynamite, modern drills, and larger furnaces. Production figures told the tale. During the thirteen-year period from 1866 to 1878, the combined production from all of the local mines was 24 million pounds of copper, resulting in a total annual rate of about 1.8 million pounds per year, and most of it was from rich, low-sulfur ores. In the modern era, total copper production dwarfed earlier levels despite the use of lean, high-sulfur ores. By 1903, DSC&I and TCC were producing 14 million pounds of copper annually, seven times the production in the former era. It was not a matter of whether the roast heaps were fired with coke or wood charcoal. The new companies simply processed far more ore containing far more sulfur. Both factors combined to fill the skies of the Ducktown Basin to a degree never witnessed in the days of Captain Raht.[48]

All of the legal defenses raised by DSC&I could have been heard in circuit court, but this would have exposed the company to the risk of jury trials, where farmer jurors would listen sympathetically to the woes of farmer smoke litigants. Instead, the company filed a bill in the Polk County Chancery Court in 1897 to enjoin the circuit cases of the *Barnes* claimants on grounds that, together, they constituted a "multiplicity of vexatious suits" and a "fraudulent confederation and conspiracy" against the company. If the company's bill of injunction succeeded, it would eliminate the risk of jury trials in circuit court, and the company could still pursue most of its legal defenses in chancery court.[49]

Legal authority for the company's strategy was murky. The basic rule was that when a defendant was burdened with multiple lawsuits in the circuit court arising from the same circumstances, here being the claims of crop and timber damage from smelter smoke, the defendant company could file a bill in chancery court to be delivered from the "consequent harassment and vexation" of handling a series of separate though similar cases. The chancellor could enjoin further proceedings in the circuit court and then combine the separate actions into one matter for disposition in chancery court. Some authorities questioned whether this would allow courts of equity to effectively usurp actions for damages from the circuit court. In any event, the smoke suitors failed to effectively argue against the point, so the chancery court granted the DSC&I bill and removed the cases from circuit court. There would be no jury trials for the *Barnes* claimants. Yet, to the surprise of DSC&I, the chancellor referred the cases to the clerk and master for determination of damages. Although the chancery court, being a court of equity, normally could not award unliquidated damages, an exception to the rule provided that when the chancery court properly gained jurisdiction on equitable grounds, as here with DSC&I's assertion about the multiplicity of actions, chancery then gained the power to completely resolve the case, including the award of damages.[50]

The company's strategy had backfired, and it was now ordered to pay damages to at least some of the smoke suitors. Even though it had invoked chancery jurisdiction, the company now chose to appeal the awards to the Tennessee Court of Chancery Appeals on the argument that chancery had no power to award unliquidated damages. The appellate court rejected the point, saying "the complainant cannot be heard to seriously advance this contention." The company's arguments on nuisance and jurisdiction also failed. First, the court stated the basic rule of nuisance, that when a business "is carried on in such a manner as to interfere with the reasonable and comfortable enjoyment by another of his property," an action in nuisance for damages will lie. Liability for the nuisance was strict; therefore, "it is no defense" that the business is "a lawful activity" conducted "at a suitable location," that it is "one useful to the public," and that "the best and most approved appliances and methods are used." Second, the damages could not be offset by whatever "incidental benefits" the copper works created by offering a local market for farm products. Third, past mining activities of Ducktown's first era did not give the company "a prescriptive right to flood the adjacent lands with smoke and sulfur gases, destructive to timber and vegetables." And, finally, on the jurisdiction issue, the court ruled that smoke damage to vegetation was considered personal, not local, and thus claimants from other states could claim jurisdiction in Tennessee courts for a nuisance originating in Tennessee. Smoke suitors from Georgia and North Carolina now had the court's approval to file smoke damage cases in the Tennessee courts.[51]

DSC&I appealed once again to the Tennessee Supreme Court, but in *Ducktown Sulphur, Copper & Iron Co. v. Barnes* (1900), the high court adopted the opinion of the Tennessee Court of Chancery Appeals with only minor changes about the correct rule for determining damages to growing crops. The cases were then returned to the Polk County Chancery Court to redetermine the damages according to the proper formula. The farmers had won — or so it seemed.[52]

The company and its lawyers in the Mayfield firm had suffered defeat on every substantive issue of nuisance law and jurisdiction but did accomplish the primary goal of preventing the *Barnes* cases from reaching a jury in the circuit court. Of the ten claimants, six became worn out by the litigation and settled for token amounts. The seventh suffered dismissal for an illegal attorney fee contract. The Polk County Chancery Court entered awards to the three remaining claimants in 1901. The amounts were disappointing and hardly worth the bother of five years of litigation in multiple courts: J. A. Fortner won $66.66, Margaret Madison a like amount, and her son, William Madison, $100.00. The company then caused another two years of delay by appealing the modest awards to the court of chancery appeals and the supreme court. Fortner's award was increased to $92.50, William Madison's remained unchanged, and his mother's fell to $1.00. The three punch-drunk litigants did not receive payment until January 1903. The legal frustrations of Ducktown farmers would become even worse as the litigation that started with the ten *Barnes* cases escalated into a full-scale legal war over Ducktown smoke.[53]

3 THE FARMERS AND THE COPPER COMPANIES WAGE BATTLE IN THE TENNESSEE COURTS

Attorney P. B. Mayfield bragged that his aggressive defense strategy in *Ducktown Sulphur, Copper & Iron Co. v. Barnes* (1900) "resulted in all the suits, but three, being dismissed, and doubtless other suits contemplated, were delayed and abandoned." It was not an empty boast. Margaret Madison, her son William, and J. A. Fortner were the only claimants of the ten to recover damages, and the sums they received were so small and so delayed as to make a mockery of their seven years of smoke litigation in the Tennessee courts. They had won every point of law, yet if success is measured by the money awarded, especially considering the great expense of time and trouble to obtain it, the mountaineer farmers had to admit that they lost the first round of the smoke wars to the copper industry.[1]

Their cause had seemed so simple at the outset, a straightforward matter of seeking compensation for crop damage caused by the smelter smoke produced by the copper companies. The laws they employed were not new and were, at first glance, a well-traveled path defined by centuries of English and American judges, a set of principles clearly set forth in the legal hornbooks used by their lawyers. The farmers then learned the hard way that almost nothing in the law is that simple. They had instead entered into a bewildering tangle of law and procedure comparable to a common feature of their mountain landscape: the laurel hell.[2]

Hell is an apt description for the great thickets of mountain laurel (*kalmia latifolia*) and rosebay rhododendron (*rhododendron maximum*) found on the slopes of the Ducktown Basin and throughout the Southern Appalachians. Both species are evergreen plants that grow closely together so that their woody, twisting branches are thoroughly interlocked. Being too tangled to allow walking, a person must either crawl or chop a way through them. There are many accounts of hunters and other travelers who came to appreciate their hellish features. Horace Kephart, a famed if often criticized author who wrote of the southern mountains during the time of the Ducktown smoke wars, recalled a story of "an old hunter and trapper ... born and bred in the mountains" who became lost in the laurel "for three days, although the maze was not more than a mile square." George Ellison, a modern-day Appalachian writer, recalled his own experience of trying to navigate a similar thicket in winter near Clingman's Dome in the Great Smoky Mountains. He and his companions found that "belly-sliding around in the snow, trying to get under the hellish thickets of rhododendron was bad enough," but the situation was made worse "by the intermittent growls of a family of black bears who resided therein."[3]

Like Ellison's episode in the laurel hell, the legal tangles experienced by the smoke suitors in the *Barnes* cases were bad enough and would soon get worse with the filing of twenty-one new smoke suits, collectively known as the *Fain* cases, against Ducktown Sulphur, Copper & Iron Company (DSC&I). More suits followed when the Tennessee Copper Company (TCC) added to the smoke problem by commencing operations in 1901. Overall, claimants filed more than 250 actions against one copper company or the other, or both, during a twenty-five-year period from 1895 to 1920. The actual number is difficult to determine because the Polk County Circuit Court suffered the loss of its records to fire in the 1930s. Fortunately, surviving records from the Tennessee appellate courts, as well as local docket sheets and attorney correspondence archived at the Ducktown Basin Museum, allow for a reasonable estimate.[4]

Most of the cases were actions for monetary damages. An individual claim by the owner of a modest mountain farm posed little economic threat to either of the copper companies, but collectively the scores and hundreds of such claims put corporate profits and shareholder dividends at serious risk for TCC and DSC&I. The threat grew when other, high-dollar claims were filed by the owners of the largest and finest local farms, particularly those along Hot House Creek, and by the owners of valuable timber in the mountains rimming the Ducktown Basin.

The suits for monetary damages posed one threat to the copper industry. Other suits seeking permanent injunctions against the smelter smoke posed an even greater hazard because, if granted, an injunction could force the companies to close and risk the livelihoods of the thousands of people dependent upon the industry. Some farmers considered the smoke problem so bad, so injurious to their agricultural way of life, that they were willing to pursue the drastic remedy on the principle of *fiat justitia, ruat coelum* (let justice be done, though the heavens fall), even if the price to be paid might fall upon the newly revived Ducktown economy.

With the stakes this high, lawyers and litigants fought in the courts for every possible advantage in the law. Each legal tactic provoked a counter-tactic, then another, resulting in a legal thicket every bit as forbidding as a laurel hell. Judges at every level of the Tennessee courts struggled to find a path through it all. At the same time, the litigants carried the battle to the Tennessee General Assembly, with each side working to add, alter, or preserve those features of the statutory law most beneficial to its cause. In the courts and in the legislature, farmers and industrialists thrust and parried over the law of nuisances in a struggle that brought about its permanent transformation in the Volunteer State.

The battles in the legislature began even as the *Barnes* litigation continued along its tangled path through the courts. The conservative opinions rendered by the state supreme court in favor of the smoke suitors convinced the copper companies of the need for statutory change to the common law of nuisance. Tennessee, like most jurisdictions, rooted the law of nuisance upon the maxim *sic utere tuo ut alienum non laedas* (so use your property as not to injure your neighbor's). This principle made a nuisance case a matter of strict liability that could be assessed in simple if-then terms. If the defendant's activity caused an objectionable consequence—whether smoke, flood, stench, noise, or otherwise—that impaired a person's enjoyment of his land, then the "neighboring owner is entitled to recover all the damages he sustained."[5]

Courts in some of the other states had already begun to modify traditional nuisance law out of a growing concern that the principles framed in England centuries before the Industrial Revolution were not responsive to the realities of modern industrial civilization. Traditional nuisance law originally developed as means for resolving small-scale disputes between neighbors. If a soap maker offended his neighbor with the stench of boiling fats, matters could be readily adjusted by paying damages, or if need be, by issuing an injunction requiring the offender to take his smelly business elsewhere. Either way, the limited size of preindustrial businesses allowed a court to order appropriate relief between the neighbors without causing wide-scale economic upheaval that might reshape the fortunes of thousands of other people.[6]

The huge scale of modern industry presented the biblical problem of pouring new wine into old wineskins; just as a fermenting new wine would cause inflexible old skins to burst, the size and scope of industrial mining and manufacturing operations threatened to tear apart the rationale of nuisance law framed in simpler times. It was one thing to enjoin a local butcher shop when blood, guts, and flies frayed the sensibilities of others. It would be quite another to force relocation of a huge integrated enterprise like Chicago's Union Stockyards and its associated packing plants when that would impact jobs and fortunes in an entire city or region.[7]

Smoke from modern mining, railroad, and manufacturing operations presented just such a problem. Some courts resolved it by refusing to consider industrial and railroad smoke to be a nuisance at all. Others abandoned strict liability and instead adopted a balancing test, especially when asked to impose injunctions to abate or terminate the nuisance. The test attempted to weigh the impact of the offense upon the neighbor against the economic and social consequences of an injunction. Elements of the test varied among jurisdictions, but on the whole, little attention was given to the ecological and aesthetic values that shape modern environmental law. In practice, the balance weighed the damage to the neighbor's land against the utility of the industry for its products and the jobs it generated. It was a very hard test for individual farmers to win against great mining concerns with large payrolls.[8]

The national trend toward adoption of the balancing test reflected a shifting conception of what constituted the public good. Traditional nuisance law rested upon the belief that the public good was best served by protecting the private enjoyment of property free of the intrusive activity of others. The standard of strict liability imposed by the maxim *sic utere tuo ut alienum non laedas* elevated private property rights above general economic considerations. The balancing test reversed the analysis by allowing local and regional economic concerns to rise above personal property rights. This occurred in Massachusetts when traditional rights of citizens to use and fish local streams were subordinated to the industrial use of the Merrimack River system. The state legislature and courts modified traditional uses of the river (the law of riparian rights) by allowing the Boston Associates to erect a series of dams to provide waterpower for textile mills in Lawrence, Lowell, and other cities. Private owners saw their property flooded, and fisherman lamented how the dams prevented salmon, shad, and other migratory species from swimming up the river for spawning. As historian Ted Steinberg noted, "The notion of public good had been further refined to favor the industrial use of the river's water."[9]

The reaffirmation of traditional nuisance law by the Tennessee Supreme Court in the *Barnes* case prompted lawyers and managers for the Tennessee Copper Company and the Ducktown Sulphur, Copper & Iron Company to seek a legislative redefinition of the public good to favor their industrial use of the atmosphere. If the legislature agreed, the change would improve the legal posture of the copper industry, and as stated by John Allen, the London solicitor for DSC&I, it would "render nugatory the smoke suits with which we have so long had to contend." To that end, P. B. Mayfield, local counsel for DSC&I, offered several bills in the Tennessee legislature in 1899 and again in 1901. The first bill sought to redefine the meaning of nuisance per se to exclude smoke from industrial plants in rural districts—a circuitous way of describing smelter smoke in Polk County. The bill would end the smoke suits simply by declaring that such smoke was no longer recognized as a nuisance under the law. Attorney Mayfield saw the measure as "manifestly just" and one that "eminently stands to the protection of all mining and manufacturing enterprises in the state."[10]

Mayfield drafted the second measure, the "incidental benefits" bill, to modify the strict liability standard of traditional nuisance law. It would direct chancery courts to exercise discretion when ruling on requests for injunctions. Specifically, "the Court exercising sound discretion *may* immediately, upon petition of plaintiff, order *or* decline to order the nuisance to be abated." In effect, the measure was intended to give chancery courts the authority to apply the balancing-of-interests test that had been adopted in other jurisdictions. The bill also sought to allow proof of economic benefits received by the plaintiff from the industry. The idea was that the presence of the industry made crops and timber more valuable because of the local market it created and that many tracts might also increase in worth, especially if they had value for mining or mine-related activities. In practice, the bill would allow a court to offset the amount of crop and timber losses by the supposed increase in land values.[11]

The third bill was a procedural measure that sought to limit out-of-state smoke suits by allowing only Tennessee citizens to file lawsuits *in forma pauperis* (literally in the form of a pauper) in the Tennessee courts. Normally, a claimant initiating a lawsuit had to pay filing fees and post a bond guaranteeing payment of court costs if the suit failed. The requirements were burdensome to the poor but could be avoided by filing *in forma pauperis* and swearing an oath of poverty. Though the procedure was a practical necessity for cash-poor litigants, the very idea of having to swear an oath of poverty chafed mountaineer pride. Margaret Madison protested that she was an old woman without political influence and found it "passing strange that an English Corporation, backed by influence that money alone can command, should for a moment consult the wishes or interests of the poor who it derisively classes as paupers."[12]

The copper companies resented the procedure because it made it easier for humble farmers to file suit, but tolerated it when exercised by Tennessee citizens. Their tolerance ended when citizens from other states exercised the same privilege in Tennessee courts. P. B. Mayfield, the DSC&I attorney, wrote Tully Cornick, counsel for TCC, to complain "that the body of the suits pending" were "brought by citizens of North Carolina and Georgia, and all under the pauper oath." The proposed bill would discourage such suits by forcing out-of-state plaintiffs to pay court costs and to post bond up front. This was more than a matter of border pride because the number of out-of-state smoke suits continued to grow as copper production increased. The DSC&I works at Isabella were only four miles from Georgia and even closer to North Carolina. The TCC works at Copperhill were less than a mile from the Georgia line. The combined flow of sulfur smoke from their roast yards and furnaces were sure to impact more and more farmers across the border.[13]

The legislative package fared poorly in the 1899 session because of anticorporate fervor in general and anticopper hostility in particular. Tennessee had growing industrial centers in Knoxville, Chattanooga, Nashville, and Memphis, but populist themes, if not the Populist Party, remained potent within a state that was still predominantly rural and agricultural. Given that context, DSC&I made a strategic mistake by offering the measures through R. Meigs Copeland, a local attorney and legislator identified as an ally of the company. P. B. Mayfield grumbled that the measure would have passed "but for demagoguery in the House," but he learned the tactical lesson: the 1899 legislative session taught him that "open and public advocacy" in the legislature by known DSC&I interests "served to abort rather than to accomplish the end had in view."[14]

The newly arrived Tennessee Copper Company joined the Ducktown Company in devising legislative strategy for the 1901 General Assembly. Tully Cornick, of the Knoxville law firm Cornick & Cornick and local counsel for TCC, wrote Mayfield to encourage a joint effort before the legislature. Mayfield responded with a proposal to disguise copper company sponsorship because "I have slight confidence that open or known advocacy by any of us here would facilitate the passage of any of the measures." The two companies agreed to introduce the bills through a Memphis lawyer whose district was more than three hundred miles west of the smelters. The trick worked. The antipauper bill passed as proposed. The incidental benefits measure also passed, though in modified form. The new law encouraged chancery courts to employ the balancing-of-interests test when deciding upon injunctions, but it excluded the use of the incidental benefits test in circuit court when assessing damages. Only the third measure, the desired exclusion of rural industrial smoke from the definition of nuisance per se, failed to pass in any form.[15]

A few months after the successful 1901 legislative campaign, the Ducktown Company faced a batch of twenty-one new circuit court smoke cases, collectively known as *Ducktown Sulphur, Copper & Iron Co. v. Fain*. The corporate response to the new cases invoked a contest of dueling injunctions, some filed by the copper companies against the farmers, and others filed by the farmers against the copper companies. Before they were done, both sides managed to clog the Polk County Chancery Court, the Tennessee Court of Chancery Appeals, and the Tennessee Supreme Court with a series of appeals that went repeatedly up and down the appellate ladder.

The Mayfields responded to the twenty-one *Fain* cases with a repetition of the *Barnes* strategy. They filed another chancery bill for injunction on the grounds that the suitors acted unlawfully in combination to "vex, harass, and annoy complainant with a multiplicity of suits." In truth, the smoke suitors had no choice but to file separate claims for damages because, at the time, there was no provision for class actions in the Tennessee circuit courts. The modern law of class actions, set forth in rule 23 of the Federal Rules of Civil Procedure and analogous state law, allows for the combination of separate claims into a single action "where there are questions of law or fact common to the class." The procedure is frequently used in airline crashes, product liability cases, and consumer actions that generate multiple claims. In Tennessee, chancery courts had had a similar procedure to combine separate equity bills with a common basis, but the concept of class actions did not extend to the circuit courts until the state adopted its version of the Federal Rules in 1971. Thus, smoke suitors could combine their cases for injunctive relief in chancery court, but they had to maintain separate actions for damages in circuit court.[16]

The point was not lost on the smoke suitors. One week after DSC&I filed to enjoin the *Fain* circuit court cases, a group of farmers led by William Madison responded with a chancery bill of their own, *Madison v. Ducktown, Sulphur, Copper & Iron Co.*, seeking both preliminary and permanent injunctions to "inhibit and restrain" DSC&I "from destroying any other timber, growing crops or other vegetation, or creating any other, or more noxious, foul, offensive and disagreeable odors and smells, poisoned by the sulfur fumes." The farmers cast the issue as a fight to preserve hearth and home from the power of industry. They alleged that because of smoke they would "be compelled to leave their homes and farms and sacrifice their lands which they had bought and paid for, upon which they were able to live in peace and plenty and enjoy the liberty of Americans, prior to the invasion of their rights by defendant." A tone of bitterness, rooted in the growing sense of being trapped and manipulated, seeped through the language of the chancery bill. The petitioners had offered to sell their lands to the company but were refused because the company "could not afford to buy all the lands within the territory covered by the smoke and fumes from its works." Fourteen nonresident claimants who failed to file suit ahead of the new antipauper law also complained of a double injury: the smoke damage reduced them to poverty, yet they had no access to the courts "for the fraud of defendant in sending its attorneys to Nashville ... and procuring the passage of Senate Bill number 307."[17]

DSC&I responded in its demurrer, or answer, by reciting the "universal delight and encouragement" it received when mining resumed in 1890. Most people in the district understood that copper was the economic engine that provided income to miners, farmers, loggers, and draymen. When the industry collapsed, so did the economy and individual livelihoods. The claimants attempted to argue that they had regional markets apart from the mining industry, but this was not so. Ducktown remained an isolated section of the Appalachians except for the railroad built to serve the copper mines. Local roads were terrible, especially in bad weather, and the old Copper Road through the Ocoee Gorge had fallen into ruin. Squeezed as it was through a chasm with cliffs on one shoulder and a torrent on the other, it required constant expensive upkeep to repair damage from falling rocks and washouts. The collapse of the copper industry in 1878 removed both the funding and the motive to maintain it, and the arrival of rail access to Ducktown in 1890 removed the need for the road, at least until the future when the coming automobile age required it. Its loss was hardly noticed because the new railroad served both the mining industry and the farmers. By 1902 the Brownlee Brothers store in Fannin County's Mineral Bluff sent "large quantities of chickens, butter and eggs to Atlanta every day" on behalf of local farmers. It was unlikely that the railroad would continue to serve the farmers without also receiving the much more lucrative business of the copper companies.[18]

After asserting the industry's economic importance, DSC&I's lawyers adopted an impolitic, even vituperative tone in other portions of its response. Most corporate lawyers prefer the velvet-covered fist to the poisoned pen out of justifiable fear that inflammatory prose only goads the opponent to further action. Lawyers for the Ducktown Company chose otherwise. They ignored composer Stephen Foster's popular sentiment, "Be it ever so humble, there is no place like home," and instead belittled the farms of the claimants as "poor barren mountain soil, almost wholly unfit for agricultural purposes and upon which only a scant existence can be eked out." They also mocked the farmers' cries of looming impoverishment from smoke-damaged crops as being "the staple demagogical cry that they were being put at the mercy of a rich and merciless private corporation." Populist themes were not to be taken seriously for they merely afforded "so much pabulum to the average plebian of dwarfed mentality."[19]

In August 1901 newspapers from cities at both ends of the Ducktown rail line considered the Madison injunction bill a threat to the industry that called for something more than sarcastic pleadings. As the *Atlanta Constitution* noted, "There are 10,000 people who are dependent for sustenance upon the works' successful operation." The *Knoxville Journal and Tribune* wrote that an injunction "would have confiscated or nullified three million dollars worth of property in Polk county and pauperized thousands of her men, women, and children." Chancellor T. M. M'Connell of the Polk County Chancery Court had these thoughts in mind when he refused to issue a preliminary injunction to the farmers. The petitioners then exercised the well-worn tactic of forum shopping by traveling away from Polk County to present their case before a friendlier court in middle Tennessee. There, Chancellor J. S. Gribble issued an injunction that threw DSC&I, TCC, and the newspapers into a state of alarm. The case returned to Polk County before Chancellor M'Connell, who then dissolved the injunction on August 20, 1901, pending a permanent ruling. The companies were safe for the moment but now realized that smoked-out farmers had learned to balance the scales of power with injunctions of their own.[20]

As the opponents swung away at each other with their mighty injunction cudgels, the number of smoke cases rapidly increased. DSC&I management grew so concerned that it hired James G. Parks away from the Polk County Circuit Court bench to serve as lead counsel over the Mayfield firm. General Manager W. H. Freeland explained to the Mayfields that "the growing magnitude and critical condition of our litigation calls for the almost exclusive attention of someone familiar, in detail, with our operations, the district and its people." This was an evasive way of saying that the company wanted someone it perceived to have greater legal clout. The Mayfields had designed and implemented the injunction strategy, but Parks was a master of circuit court procedure and wrote the leading Tennessee treatise on the subject, *A Manual of the Law of Pleading*. Thus, he literally wrote the book on strategies for legal delay. The Mayfields felt the change as an insult, refused to work under Parks, and instead switched sides; P. B. Mayfield's son, J. E. Mayfield, became a prominent attorney for the smoke claimants.[21]

Parks would face the younger Mayfield time and again in smoke suits and would suffer many jury verdicts against DSC&I when the circuit court logjam finally broke in 1905. In the meantime, he worked his procedural magic toward the goal he described to his TCC counterpart, Howard Cornick, as "keeping a blocked docket." Like his predecessor, P. B. Mayfield, Parks understood that the best way of preventing the award of expensive monetary damages was to keep the cases from ever reaching the jury. Any method that would accomplish that goal was useful to the defendant copper companies, whether by filing a motion to challenge the legitimacy of a claim of *in forma pauperis*, exploiting a procedural defect in a plaintiff's suit, filing a petition in chancery court to bar prosecution of a case in the chancery court, or pursuing endless interim appeals.[22]

Defeat in the appellate courts only furthered the strategy of delay. The court of chancery appeals dissolved the circuit court injunction won earlier by DSC&I in the twenty-one *Fain* cases, and in September 1902 the Tennessee Supreme Court affirmed. In *Fain*, unlike *Barnes*, the court rejected the multiplicity of suits argument. It held that the concept applied only to traditional areas of equity jurisdiction, such as land titles and estates. Chancery could consolidate separate cases pertaining to a given tract of land or a decedent's property but could not do so in actions for damages arising in nuisance or other torts. Accordingly, the cases were returned to the circuit court for jury trials.[23]

Once again, DSC&I lost on a substantive issue but accomplished the real goal of forcing more delay. Now that the *Fain* cases were back on the circuit court docket, Parks informed DSC&I that "we shall interpose a motion to de-pauperize the plaintiffs, and thus get a continuance, unless they should be ready to make bond, which is improbable." Judge George L. Burke, the new circuit court judge, responded favorably to his predecessor by granting the motion and thereby delaying the cases until another term.[24]

Dozens of claimants tried to circumvent the delaying tactics employed in circuit court by instead filing suit before local justices of the peace, a venue that functioned as a small claims court. This required claimants to reduce their demands to $500 in order to come within the lower jurisdictional authority authorized for justices of the peace, but they willingly did this in the hope for faster action on their cases. Parks and Cornick responded with yet another injunction, *Ducktown Sulphur, Copper & Iron Co. Ltd v. Crofts*. All of the new suits before the justice of the peace were now stayed from further action. In 1903 Parks gloated to DSC&I general manager, W. H. Freeland, that "we are now in good general shape with all our litigation, so far as delay is concerned."[25]

More opportunities for delay arose when claimants began to file damage suits naming both Tennessee Copper Company and Ducktown Sulphur, Copper & Iron Company as co-defendants, on the theory that each was jointly and severally liable for the harmful smoke. This made sense from the claimant's perspective because both companies now ran at full capacity using the same open-heap roasting process, and each contributed its respective portion of smoke to the massive cloud over the Ducktown Basin. It was usually impossible to determine which company's smoke ruined that year's apple crop. An action against both companies avoided the difficulty of allocating fault between them and saved time and court costs.

Parks and Cornick realized this also and collaborated to frustrate the new strategy. The refusal was in part based on the reluctance of the two copper companies to submit themselves to proportional damages. TCC was by now the larger company, and most of the locals attributed two-thirds of the recent smoke damage to it and a third to DSC&I. The formula was convenient to claimants and perhaps accomplished a rough sort of justice, but it also reinforced the idea of collective liability for smoke damage. The greater problem in the eyes of defense counsel was that litigation of suits with co-defendants was too convenient and expeditious for the plaintiffs. Delay was the goal, not judicial economy or justice.

After sharing legal research on the problem, the company lawyers each selected a two-defendant suit to challenge on the grounds that the companies were improperly joined as co-defendants. Judge Parks's circuit court successor, Judge George L. Burke, ruled in favor of the copper companies in both cases, and the Tennessee Supreme Court affirmed in *Swain v. Tennessee Copper Co.* (1903). It was a straightforward decision according to the weight of precedents, holding that if two or more people commit independent acts of nuisance, then "each is liable for damages which result from his individual conduct only." The rationale was that "if the law was otherwise, the one who did the least might be made liable for the damages of others far exceeding the amount for which he really was chargeable, without any means to enforce contribution or to adjust the amount among the different parties."[26]

Once again, the Tennessee Supreme Court had issued a conservative ruling in keeping with traditional principles of nuisance law designed to address small-scale disputes in a preindustrial age. Most of the cases it cited involved water disputes in which two or more parties acted separately to interfere with a stream in a manner that combined to flood the land of a neighbor. Other cases arose from quaint matters such as the combined stench from several urban pigpens, and the loss of sheep from attack by the unrestrained dogs of several owners. By comparison, the combined discharge of sulfur dioxide smoke from the two copper companies was commingled by wind and weather within the mountain walls of the Ducktown Basin to impact miles of farmland and forests owned by scores of separate owners. The Tennessee Supreme Court essentially forced each smoke suitor to divide the smoke cloud by filing separate lawsuits against each copper company.

The *Swain* opinion forced every claimant to dismiss his or her co-defendant action and then refile as two separate actions, one against TCC and the other against DSC&I. The local actions before the justice of the peace were also dismissed and refiled, this time in circuit court. Thus, the work of the corporate lawyers in *Swain* caused the already terrible circuit court backlog to instantly double. The number of pending cases continued to grow for other reasons as well. The passage of time and ever-increasing rates of production expanded the zone of smoke destruction to encompass more farms and woodlots. On the older, already afflicted properties, the ongoing nature of the nuisance continued to generate new claims with every growing season, and each new claim had to be filed within the three-year statute of limitation lest it be time barred, and because of the *Swain* decision, every new claim had to be filed twice. Many farmers filed three, four, and even more claims in separate suits against the Ducktown Company and then against Tennessee Copper.[27]

Another spur to litigation occurred in 1899 when the Tennessee General Assembly repealed the champerty law, thereby legalizing contingent attorney fee contracts. A farmer could now hire a lawyer with little financial risk because the lawyer agreed to be paid only if the smoke suit resulted in a recovery by settlement or judgment. If the suit failed, the lawyer received nothing. Although the arrangement switched the financial risk from client to lawyer, a cluster of country lawyers in the Polk County seat of Benton and others in Georgia and North Carolina sought to build portfolios of contingent-fee cases by hustling for smoke suits. Claimant's lawyers found smoke suits appealing because they met the three criteria of a good contingent-fee case: clear liability, provable damages, and a defendant with deep pockets. The *Barnes* decision resolved the liability issue by declaring copper smoke to be an actionable nuisance. Damages were relatively easy to prove so long as farmers took turns testifying for each other about crop losses. And best of all, DSC&I and TCC had the deepest pockets in East Tennessee, something claimants' lawyers already knew from their lucrative tort actions for worker deaths and injuries. It certainly beat trying to make a living by doing legal piecework, such as wills, deeds, and divorces for cash-poor farmers.[28]

Defense attorneys expressed irritation at the rise of contingent-fee smoke suits. Before they switched sides, the Mayfields offered deposition testimony from N. B. Graham, a newspaper publisher and the local clerk of court, that "when these parties began to import lawyers into the country from Marietta and Morganton, Georgia, Cleveland [Tennessee], and other points for the purpose of bringing suits for smoke damages it looked like the industry in Ducktown had been and would be menaced." The new contingent suits added to the already large backlog in the Polk County Circuit Court. When the smoke suitors asked the legislature to add additional terms of courts, Parks opposed the bill, arguing that new terms of court would only work to accommodate "a lot of conscienceless scoundrels [who] have been riding the Georgia and North Carolina territory working up 'smoke' claims." He added that the lawyers lured clients "into their hands with the understanding that they will try to get something for them, and if not, they will lose nothing." By his estimation, such lawyers were "shysters" and "a lot of blackmailers ... who are trying to hold us up for blood-money."[29]

The cynicism was not total; corporate counsel often acknowledged the reality of smoke damage, if not the dollar values claimants placed upon it. Parks recommended settlement with Billy Humphrey, a farmer and preacher "who was forced to leave his little farm with his old wife and no property" after living "across the Ocoee [River] opposite the mines as long as he could stand it." Owners in the Hot House Creek district had some of the best farmland in the basin. Parks acknowledged that they were men of "higher standing" and that "there can be no dispute that these farmers have been pretty badly injured."[30]

Parks and his allies were more skeptical of other claims. Their attitude toward individual cases bore some relation to the distance of claimants' farms from the roast heaps and smelters. Parks mentioned "the law of the diffusion of gases" in a letter to co-counsel, W. B. Miller, adding that whenever a claimant's property lay closer to the TCC works than to the DSC&I smelters in Isabella, "that fact ought to weigh very greatly in our favor." Actually, patterns of smoke damage failed to correspond solely to distance. Sulfur dioxide smoke was heavier than air and tended on still days to collect and flow in the lowest portions of the basin, along the Toccoa-Ocoee River and its tributaries. That is why the early *Barnes* suitors were clustered along Wolf Creek and also why Parks worried about cases filed by the farmers of Hot House Creek.[31]

On windy days, clouds of smoke blew over higher ground and often reached into the mountains that walled the Ducktown Basin. J. K. Haywood, a chemist with the U.S. Department of Agriculture, noticed a similar pattern in his 1905 study of smelter smoke at the Mountain Copper operations in the Mount Shasta region of California. He observed that northerly and westerly winds drove smoke up the gulches because "the fumes have a tendency to keep together and drift for long distances up these natural chimneys." Southerly winds sent smoke "drifting over one high hill" and then "down the valley of the Sacramento." Nonetheless, the law of diffusion of gases provided a plausible basis to question claims arising from the more remote property owners of the Ducktown Basin. The suspicion grew with distance that litigating was more lucrative than mountain farming. J. V. Kisselburg, a farmer near Mineral Bluff in Fannin County asserted, "Some of my neighbors kicks sometimes, but they say they are going to get damage, as they claim that it is easier to get money out of the copper co. by lawing them than it is to work for it."[32]

Company attorneys also noticed a link between mining employment and the filing of claims. Many farmers worked for the mines on a seasonal basis, either as wage laborers or by doing contract work as suppliers of wood and quartz rock (a flux for smelting) or as teamsters. The argument arose that some claimants filed only in retaliation for the loss of employment. The Ducktown Company alleged that it "found inexpedient and impracticable to give to all who so desired official position . . . and out of this as a consequence, dissatisfaction and disaffection arose." J. H. Barnes, whose farm was only two miles from the DSC&I roast heaps, admitted that he filed his lawsuit only after the company ended his contract for furnishing wood: "If I had worked on for the company, perhaps I would not have sued." The reverse was also true: legitimate smoke claims were occasionally settled in exchange for an offer of work. J. E. Mayfield, now representing claimants, wrote Parks to suggest that Avery McGhee's case could be settled if the company would receive and pay for rock he delivered. DSC&I agreed to allow smoke suitors in the Wolf Creek area to resume hauling rock for the company if they would settle for only payment of the court costs.[33]

Mixed employment was a common practice among mountaineer farmers living near mining and logging operations. Highland farming was never easy and was rarely a totally self-sufficient livelihood. Opportunities for supplementary income were welcomed, especially if the work coincided with slack times on the farm. Farmers preferred both sources of support if possible, but if smoke rendered farming unprofitable or impossible, the importance of nonfarm work became all the greater. They considered it only right that suits should be filed if the copper companies simultaneously ruined farms and denied employment. Conversely, cases could be settled cheaply if the opportunity to resume mine employment made it possible to hold onto farms a little longer. In Ducktown, the fight to preserve home and livelihood through agricultural or public work, or both, remained an expression of the southern populist spirit that sustained people in the struggle against the great economic forces of market and industry.[34]

With the courts being so crowded, claimants were caught in a three-way problem: their ability to make a living at farming was impaired or ruined by smoke, their opportunities for wage and contract work were often barred at the mines, and the successful strategy of "keeping a blocked docket" prevented relief in the courts. Avery McGhee complained that "all of the efforts of complainants to obtain some compensation for the loss of their homes, crops, and timber have so far been in vain ... on account of the delays defendants have been able to obtain." Each time they filed in circuit court, the companies "waited until a few days before the term convened at which the suit stood for trial and ... then obtained an injunction from the chancellor."[35]

Despite their frustration over delay tactics, most claimants continued to file their cases in the Tennessee state courts for lack of practical alternatives. Georgians and North Carolinians could not litigate in their own state courts because of the inability to establish jurisdiction over the copper companies under the narrow jurisdiction standards of the day. Some tried to litigate in the federal courts of Tennessee but found it to be an expensive and disappointing effort. Article 3 of the Constitution and a series of federal judiciary acts allowed for diversity jurisdiction, meaning that federal courts could assert jurisdiction in suits between citizens of different or diverse states. Wealthy lumber owners from New York, Rhode Island, and Georgia, all of whom held title to valuable timber in Fannin County, Georgia, filed suits against the Tennessee Copper Company in the U.S. Circuit Court for the Eastern District of Tennessee (the pre-1911 federal circuit courts are not to be confused with the later federal circuit courts of appeal created in 1911). The circuit court dismissed the suits because the legal residence of the corporation was in New Jersey, where it was incorporated, not in Polk County, Tennessee, where it conducted its operations. In short, only the federal circuit court of New Jersey had jurisdiction over TCC under the rules then applicable. The U.S. Supreme Court affirmed both dismissals in *Ladew v. Tennessee Copper Co.* (1910) and *Wetmore v. Tennessee Copper Co.* (1910). The significance of the decisions was not lost on local plaintiffs' counsel. When asked to file a smoke suit on behalf of ginseng harvesters in Union County, Georgia, attorney J. E. Mayfield noted the decisions and advised, "If the parties could arrange in any way to bring their suits in the U.S. Court, we are satisfied it would be to their interest, but we can see no way for this to be done."[36]

Few of the smoke claimants had the resources to litigate in New Jersey, and most of them would not have been able to sustain the costs and travel needed to litigate in the federal courts of Tennessee if it was legally possible to do so. Many claimants were barred from federal court by the "amount in controversy" requirement established by Congress in diversity jurisdiction cases. Its purpose was to prevent the federal courts from being burdened with financially small matters. The amount in controversy requirement in force at the outset of the Ducktown cases was $2,000, as set by Congress in 1877. It was increased to $3,000 in 1911, midway through the sweep of private smoke suits. The requirement posed no difficulty to the several timber barons who filed large claims in federal court but was an effective barrier to most of the smaller farmers. Experience in the Polk County Circuit Court showed that most verdicts involving small mountain farms were considerably under $500, which explained why the *Crofts* claimants were willing to set their claims at that amount in the attempt to have them expedited for trial before a justice of the peace. It was hard for all except the most prosperous farmers to make a plausible claim for damages in excess of $2,000.[37]

Distance was another barrier to federal court. The federal courts in Chattanooga and Knoxville—never mind New Jersey—were considerably further than the Polk County courts in Benton. Train travel was the only practical way out of the Ducktown Basin to Benton and points beyond. A rail trip to Benton was shorter and less expensive than travel to Chattanooga or Knoxville. Travel by automobile was not yet a realistic alternative. Ducktown smoke litigation occurred at a time when automobiles were a novelty (the first automobile appeared in the basin in 1906), and good roads were almost nonexistent in the Southern Appalachians. The roads were so bad that when L. H. Abernathy bought a car in Atlanta in 1907, it took him fifteen days to find a viable route to his home in Copperhill. Ducktown would not have a direct automobile highway over the Unakas to Benton until construction of the Kimsey Highway in 1920.[38]

The state courts in Benton remained the best of courthouse destinations for all but the wealthiest smoke suitors. Even then, some suitors were so poor that they could not afford a local train ticket and were compelled to travel to the Benton courthouse on foot or by horse via the now ruined Copper Road.

E. M. Harbison reported to the *Knoxville Journal and Tribune* about a November 1905 session of court where he observed "women and children in attendance" who had "been forced to travel thirty-five miles through the rain and over one of the roughest roads that mortals ever had to travel." He also noted men who had suffered so badly from exposure from rain and cold on their way through the gorge that "they never attended another earthly court." The travel situation improved in 1911 when the Tennessee legislature authorized sessions of district court in Ducktown, though that increased the risk that the jury pool would contain members sympathetic to the mining industry.[39]

For all of these reasons, claimants continued to pursue their claims locally, even though their choice subjected them to the array of delay tactics in the state courts. In 1902 the farmers fought back with three more anticompany injunction bills in addition to the pending Madison bill. This time the companies aggressively countered the populist threat of injunction by loudly proclaiming the workingman's disaster that would ensue. When P. J. Farner filed an injunction bill in January 1902, local papers and even the distant *Washington Post* reported that "when the injunction is served upon him [Tennessee Copper's president] will shut and close down the entire works and mines and pay off and discharge every man." In a front-page article captioned "Disaster Threatens Ducktown," the *Knoxville Journal and Tribune* feared a repetition of the mine closure in 1878, when men too poor to pay for transportation "were forced to put their wives and little children in the public road and walk away, carrying as much of their personal effects as possible upon their persons." TCC counsel Tully Cornick admitted to the papers "that there is no doubt but that the abutting land owners suffer some injury by reasons of the smoke" but avowed that the company was willing to settle on reasonable terms by paying damages or by purchasing "the right or easement to flood the land in question with the smoke generated on their roasting yards." Settlements failed to occur because "lands assessed at taxation for sixty cents per acre are claimed to have been damaged not less than $10 per acre." Cornick did not mention the delay strategy that he and Parks used to frustrate farmers in the local courts.[40]

Two more injunction bills followed, one by Farner's kinsman, Isaac Farner, in the following November, and the other by Avery McGhee in September 1903, making a total of four pending anticompany injunction bills. It was here that the tactical wisdom of the copper companies in their 1901 legislative campaign came into play. The new incidental benefits act imposed the duty of "exercising sound discretion" when chancery courts considered injunctive relief. With power came responsibility; the genius of the act was that elected chancellors (the title of judges in chancery courts) answered to local voters for their decisions about whether to close the mines. They legitimately framed their rulings in terms of balancing costs and benefits but would be less than human if they failed to consider the number of voters impacted by smoke compared to the number that would suffer from unemployment or conditions of general economic decline. When Chancellor N. Q. Allen granted an injunction by fiat to P. J. Farner, he then took only seven days to reverse it, saying, "I decline to be the instrument either lawful or otherwise of stopping an industry for even a short time in mid-winter which furnishes food and fires to a laboring population of over 5,000 people." The claimants, he ruled, had adequate remedy at law by pursuing claims for monetary damages, notwithstanding the role that the chancery played in creating the appalling backlog in circuit court.[41]

Chancery issued similar rulings in the other three injunction bills, thus putting the burden of appeal on the smoke suitors. P. J. Farner abandoned his case, but the others took theirs to the court of chancery appeals. There, in a series of two-to-one rulings, the court ruled in favor of the farmers. Writing for the majority in the *McGhee* injunction case, Judge S. F. Wilson framed the issue as "whether these complainants shall be permitted to live in the homes that they have occupied and on which they have supported their families, or shall they be compelled to leave their homes and seek residences elsewhere?" He then asked, "What right has any enterprise, however beneficial in its results to the public generally, to say that he must vacate his home?" He concluded that he knew of "no principle of law and of no decision that authorizes works, manufacturing establishments, or other enterprises, however extensive their operations, or however beneficial to the public in general, by the methods of their business, however necessary, to drive a citizen from his home."[42]

The majority opinions in *McGhee*, *Madison*, and *Farner* were not well reasoned. The sweeping terms of the rulings elevated the status of private dwellings to a degree of absolute right and ownership never contemplated in Anglo-American law and without apparent regard for the consequences the rulings caused to others. In his dissent, R. M. Barton Jr. argued that a person's right to life and property had never been absolute but was subject to limitations such as eminent domain. He opined that the majority, in its pursuit of absolute right, had abandoned the exercise of discretion that lay at the heart of equity jurisprudence. Specifically, the grant of extraordinary relief by injunction "was always a matter of grace or discretion and not of fixed right."

In its willingness to destroy a mighty industry with a payroll of thousands "without regard to consequences" for the sake of modest farms appraised for taxes at $100 (Madison) and $83.00 (F. M. Carter), the majority acted according to the maxim *fiat justitia, ruat coelum* (let justice be done, though the heavens fall) instead of the equitable principle *salus populi est suprema lex* (let the welfare of the people be the supreme law). Barton then recited national precedents where courts of equity employed the balancing test in industrial cases. Oddly, he failed to cite the 1901 Tennessee incidental benefits act to the same end. He concluded with a warning that the rulings in the Ducktown injunction cases would put "the entire business interest of the state ... at the mercy of those who are willing to set in motion such destructive agencies."[43]

Copper officials and railroad executives agreed and worked together on the legal and political fronts to reverse the three rulings of the court of chancery appeals. The legal staff of the Louisville & Nashville Railroad offered the assistance of its corporate counsel because of the railroad's growing exposure to nuisance claims and because of its direct financial stake in Ducktown as the new owner of the railroad that served the mines. TCC counsel Howard Cornick advised DSC&I's James G. Parks that "we have the Governor interested in the matter and think he will exert such influence in our behalf as he may be able to exert within the strict bounds of propriety." Parks, a Republican, ran for office as state representative "to be there to try to get needed legislation in case the injunctions were granted." To that end, he asked and received the tacit backing of Cornick's Democratic law firm (which had grown to become Cornick, Wright & Frantz). Parks also insisted on the support of DSC&I wage employees. He complained to management when one DSC&I mine worker said "he would not vote for me because of ... my stand for the company in the smoke suits while I was Judge."[44]

Legal precedent, rather than political pressure, led the Tennessee Supreme Court to reverse the court of chancery appeals in each of the three injunction cases. In its opinion, issued on November 26, 1904, the court agreed with Judge Barton's argument that the enjoyment of one's home was not an absolute right, and then it employed the 1901 incidental benefits statute to balance the interests of the parties. The court determined that it could do justice without letting the heavens fall upon "two great mining and manufacturing enterprises that are engaged in work of very great importance, not only to their owners, but to the State, and to the whole country as well." The claimants had an adequate remedy for damages. Thus, when weighing their respective interests, the Tennessee Supreme Court refused to destroy the copper industry by injunction in order to protect "thin mountain lands of little agricultural value" and an aggregate tax value of less than $1,000.[45]

Corporate relief was immediate. Ducktown's general manager, W. H. Freeland, fired off a coded cablegram to the company's managing directors in London advising that "the injunction has been refused in every case by supreme court." The directors responded with another cable stating that the ruling "has given us the greatest satisfaction." Freeland then proclaimed that the ruling "will practically mean that so far as our own state courts are concerned, no future action for injunction on the grounds of nuisance will stand." This proved to be true.[46]

The 1904 *Madison v. Ducktown* decision ended the debate over injunctive relief in the Tennessee courts by taking the weapon away from farmers. At the same time, the companies were losing their own injunction club as they ran out of procedural grounds to enjoin circuit court cases. Having learned from their mistakes in earlier smoke suits, claimants' attorneys now avoided the procedural defects that the copper attorneys had so frequently exploited for purposes of delay. By 1903, Parks lamented, "the delays hitherto have not been the result of continuances from term to term, but by injunction. It is plain that we have about exhausted this means." Local juries could now dip into company coffers by making damage awards to their fellow citizens.[47]

Trials of the *Fain* cases against DSC&I began in late in 1903 and continued for several more years. Most cases resulted in verdicts for the plaintiffs with awards ranging from $46 to more than $500, and most falling in the $100–$200 range. The amounts were not huge but were not insignificant either, given that the daily wage for miners was around a dollar per day. By 1907 the newer claims, filed separately against each copper company, also began to move through the circuit courts. Awards for the newer cases tended to be higher, with claimants recovering totals of $300 to $500 for each pair of cases. Some of the prominent farmers of the Hot House Creek community won combined amounts exceeding $1,000.

The copper companies could no longer enjoin the circuit court trials but continued other maneuvers in the attempt to stem the increasing number of plaintiffs' verdicts and to delay payment of judgments. They routinely appealed verdicts to the Tennessee Supreme Court even though the grounds for doing so were slim, often amounting only to a blanket assertion that the awards were excessive or not supported by the evidence. The supreme court just as routinely affirmed the verdicts, though the exercise did accomplish the goal of delaying payment by another year or so. No sanctions were imposed for what would now be considered frivolous appeals filed only for purposes of delay.

The parties also continued their legislative wrangling. Smoke suitors tried and failed to win repeal of the 1901 antipauper and incidental benefits acts. The companies tried and failed that same year to block addition of another week of circuit court intended to expedite the backlog. Parks complained that his judicial successor, Judge Burke, declined to oppose the bill. In 1909 the companies did manage to block an act authorizing special terms of courts with imported judges, after a letter-writing campaign in which Parks solemnly assured legislators that "we have not been trying to delay the trial of these cases."[48]

The great number of cases passing through circuit court created a new problem for the copper companies: witnesses. The fears of corporate counsel concerning local bias against the copper companies proved all too true. Farmers easily obtained testimony from fellow farmers about crop losses. The copper companies had a much harder time trying to find effective rebuttal witnesses. Parks complained to Cornick that "it is out of the question to depend upon local witnesses" because "neighbors having no suits expect to have at some time; besides they all sympathize with one another." Credible and local procompany witnesses were hard to obtain because communities were tight and grudges long lasting: after all, "you can hardly prevail upon our best citizens to . . . make voluntary witnesses." As for bringing in outside witnesses, "to send a lot of scalawags would be worse than sending nobody."[49]

Parks proposed legislation in 1903 to provide another procedural fix. Instead of relying on local witnesses and local juries, he proposed a new system called jury of view. If enacted, the court would select a panel of five witnesses from a wide area, perhaps even outside of Polk County, and charge them to view the allegedly damaged farms and then set the amount of compensation. This would bypass a regular jury and the need for rebuttal witnesses. Parks listed other advantages in a long letter to Cornick. First, experience with a similar system in eminent domain condemnation cases showed him that "a jury of view rarely gives one-fifth what an ordinary jury in the court will allow." This was because his "good men" would be more willing to evaluate damage if summoned by the court instead of a company attorney. Second, the jury-of-view process "would give us more delay than is possible under present laws."[50]

This time, the Tennessee Copper Company refused to make common cause with Ducktown Sulphur, Copper & Iron. Despite repeated letters and meetings, TCC refused to back the bill, largely for fear that it would backfire on the company as "an evident admission of liability for damages." Parks privately asserted his greater procedural knowledge compared to the young attorneys in Cornick's firm, who, though "above the average for their years," have yet "to acquire that calm judicial poise and level-headedness that constitutes the chief element of value of a lawyer to his client." Still, he admitted it was pointless to introduce the measure because it could not overcome anticorporate populist resistance in the legislature if only one of the companies backed it. The jury-of-view idea never again resurfaced. The circuit court, now working apace with added sessions, would continue to handle cases with local witnesses and traditional juries for the many remaining years of Ducktown smoke litigation.[51]

The copper companies survived the farmer smoke suits. So did the livelihoods of the thousands of mine workers and their dependents. Their success in the legislature made the balancing-of-interests test a key part of Tennessee nuisance law, and their victory in *Ducktown Sulfur, Copper & Iron Co. v. Madison* effectively ended the threat that a Tennessee court would impose an antismoke injunction. They failed in their attempt to exclude the fumes from rural smokestack industry from the statutory definition of nuisance per se, but did, for a number of years, stifle the flow of monetary claims through their skillful manipulation of law and procedure. Their success in holding back the damages suits would eventually prove to be Pyrrhic: the frustration and anger experienced by Georgia farmers under the industry strategy of "keeping a blocked docket" would lead the state of Georgia to file suit on its own behalf in the U.S. Supreme Court. The copper industry would be forced to continue the Ducktown smoke wars against a much more powerful opponent in a court far removed from the familiar political environment in Tennessee.

Results for the farmers, their farms, and the principles of rural populism were also mixed. Many farmers suffered lasting damage to their lands and livelihood that money could not replace. They lost their power of injunctive relief in the Tennessee courts. Even so, rural populists and their supporters in the Tennessee legislature achieved major accomplishments. They added more weeks of court to restore viability to a circuit court system that almost collapsed under corporate delay tactics. They defeated the attempt to redefine nuisance per se in a way that would have barred nuisance actions against rural smokestack industries. They prevented the imposition of the incidental-benefits offset in the calculation of smoke damages. The right to jury trial in nuisance actions and the right to call one's own valuation witnesses remained intact. Local farmers continued to serve as witnesses and jurors. Farmers received verdicts set by their peers.

Neither side achieved a complete victory; yet, by the vigor and persistence of their struggle, they transformed the Tennessee law of nuisance in a way that achieved a rough accommodation between the populist concerns of the farmers and the economic realities of the industrial age. As events would prove, Georgia's suit before the U.S. Supreme Court would extend the transformation by converting nuisance law from a private remedy into a state tool for responding to interstate smoke pollution—an outcome that more accurately reflected the scale of industrial nuisance at the turn of the century, if not the personal harm suffered by property owners.[52]

GEORGIA ENTERS THE FRAY 4

In 1903 Georgia farmers in the Ducktown Basin were beside themselves with frustration. Each growing season brought another round of dismay as acid from the smoky clouds killed the fruits of their labors on the stalk, the vine, and the branch. It was bad enough to suffer the damage. It was worse to know that they could have profitably sold all of their produce to the miners—if only they could grow it. Added to this was the realization that their efforts to fight the smoke problem in the Tennessee courts were, for the moment, going nowhere.

Ambrose Bierce, a contemporary journalist, self-described cynic, and author of *The Devil's Dictionary* (1911), defined litigation as "a machine which you go into as a pig and come out of as a sausage." After five or six years of litigation in Tennessee, it was obvious to the smoke suitors that they were being ground into sausage and that the company lawyers were turning the crank. James G. Parks, counsel for the Ducktown Sulphur, Copper & Iron Company (DSC&I), and Howard Cornick, his counterpart for the Tennessee Copper Company (TCC) had outmaneuvered them at almost every turn by applying their considerable legal skills, bolstered by the resources and power available to them as corporate attorneys. In the Polk County Circuit Court, their strategy of "keeping a blocked docket" had prevented all but a few cases for monetary compensation from reaching juries. In the Polk County Chancery Court, they blocked the farmers' repeated attempts to abate the smelter smoke. Each of the petitions for injunction against the copper companies languished in seemingly endless rounds of appeal.[1]

The copper companies had also won key legislative battles in Nashville. With the active support of corporate counsel and lobbyists from the railroads, they changed the common law of nuisance in favor of industry. The new law required chancery courts to employ the balancing-of-interests test when ruling upon petitions to enjoin smokestack industries. Now, the dollar impact of industrial smoke upon the operations of typically modest mountain farms had to be weighed against the economic consequences of an injunction—the millions of dollars invested in plant and equipment, the number of employees and households dependent upon industry wages, and the general economic impact of a closure. Another statute hindered cash-poor

Georgians and North Carolinians from initiating suits in the Tennessee courts by prohibiting out-of-state litigants from filing *in forma pauperis*. They now had to spend scarce dollars to post a bond for costs at the commencement of suit.

Parks and Cornick defended their clients almost too well. Georgia litigants concluded that they had little or no effective recourse in the Tennessee judicial system and instead voiced their complaints to more sympathetic ears in their home state. If individual Georgia mountaineers were outmaneuvered by corporate powers in Tennessee, the obvious solution was to cajole the powers in Atlanta to continue the fight in a more favorable venue with legal resources that only the state could wield. Or, to return to Bierce's definition, Georgians wanted a different sausage grinder, one that was capable of grinding up corporations, with the strong right arm of their home state at the crank.[2]

It was not clear, as a matter of law, what Georgia could do about cross-border smoke pollution. Given the lack of federal and state antipollution legislation at the time, the issue appeared to be governed by the common law of nuisances, yet it remained an open question whether the ancient principles encompassed an action by one state to enjoin pollution generated by individual or corporate citizens in another state. It was a problem of historical scope. English law lords developed the principles of nuisance law in a nation where sovereign power rested ultimately in the Crown, as tempered by the powers of Parliament. Unified sovereignty over England and Wales favored the consistent formulation and administration of nuisance law throughout the realm from Kent to Cumberland.

Ducktown smoke blew over a different legal landscape, encompassing three sovereign states, each with its own judicial system and legislature, and each jealous to maintain its prerogatives against neighboring states. A state's judicial and legislative power was nominally commensurate with its geographic borders. Yet the cross-border realities of commerce, crime, and now corporate smoke pollution confused the clarity promised by lines on a map. Smoke clouds, and the winds that moved them, did not respect state lines. Issues of interstate jurisdiction were further clouded by national sovereignty whereby each state shared sovereign authority with a federal government that operated its own courts and legislature. The confused interplay of federal and state jurisdiction created a complicated constitutional struggle marked by endless litigation, even a civil war, to determine the respective powers of the states versus Washington, and the states vis-à-vis each other.

Divided sovereignty gave rise to a number of fundamental questions: Did Georgia have the power to compel corporate citizens in Tennessee to abate the production of poisonous fumes that drifted over the common border? Could the matter be litigated in Georgia courts? If not, did the federal courts have the power to provide an effective remedy? What role could or would the U.S. Supreme Court play in the matter? The answers to the questions involved arcane issues of civil procedure—the law of how and where cases are to be tried—concerning such topics as jurisdiction, venue, justiciability, standing, pleadings, and motions. They were the sort of issues that a law professor might pose to befuddled law students. There were also serious questions about the central claim—what lawyers call substantive law, as in the substance of the case—that needed to be answered. The most important question was whether Georgia had a legally recognizable interest in the fight over Ducktown smoke. Nuisance law turned upon the violation of property rights, but the state of Georgia long ago disposed its northern holdings to private citizens in the 1832 lottery of Cherokee lands. Did it retain sufficient property interests to have a legally recognized stake in a nuisance case? If not, could Georgia obtain relief under some other theory? All of the questions, both procedural and substantive, would eventually be addressed at length by lawyers and judges during the course of Ducktown smoke litigation.³

In the meantime, the suffering farmers and timber owners in Georgia were neither lawyers nor law students, and their patience with abstruse legal principles had long since been exhausted in their as yet futile private cases in Tennessee. They demanded a political solution. To that end, citizens from the mountain counties circulated petitions to Governor Joseph M. Terrell and goaded their delegation in the General Assembly to action. Mercer Ledford, a state senator from Union County, located just east of the heavily damaged Fannin County, heeded the call. Lacking a solution to all of the bedeviling legal issues, he employed a time-honored maneuver that gave a positive response to his constituents without encumbering the legislature with any real responsibility to solve the problem: he proposed a resolution for a study commission. Fellow legislators quickly grasped the wisdom of this approach. The measure easily passed in both the House and the Senate and was enacted on August 17, 1903, as Resolution No. 47.⁴

The document began with two recitals (the "whereas" clauses), the first noting that "it has been represented that great and irreparable damage has been, and is being done to the timber, fruits, and agricultural interests in the counties of Murray, Gilmer, Fannin, Union, and Towns" by the "smoke and fumes produced by the smeltering of copper ores" in Ducktown. The second recital declared that "some steps should be taken looking toward the suppression of this evil." The text continued with two resolutions. One called for the creation of a five-member commission consisting of the commissioner of agriculture, the state chemist, the state geologist, and two citizens to be named by the governor. They were to investigate "the damage already done and the damage likely to be done" and then to report their findings to the governor. The other resolution charged the governor, upon receipt of the report, "to take such steps as shall be deemed proper and necessary to correct this evil and to prevent future damage."[5]

Governor Terrell embraced the resolution. He was young for a governor, entering office at age forty-one, full of the vigor and zeal that enabled him to earlier serve ten successful years as attorney general. In that office, he proved his skills as a litigator on criminal and civil matters. He won all twelve of the cases he argued before the U.S. Supreme Court in a series of death penalty appeals, business tax disputes, and railroad regulatory matters. As governor, he demonstrated a progressive bent with enactment of a pure food and drug law, the creation of the state court of appeals, and laws to limit child labor and to curb agricultural speculation. The Ducktown resolution provided him with an opportunity to lead the state in another legal battle—this time on a progressive crusade against smoke damage.[6]

Not one to dally, Governor Terrell named the commission members only two days after enactment of the resolution. It was a well-balanced team designed to be responsive to investigative and political needs. Commissioner of Agriculture O. B. Stevens held his post by popular election in a state where the farm population still outnumbered the urban citizenry. His presence on the commission gave it political heft. The other designated members, State Chemist John M. McCandless and State Geologist W. S. Yeats, were trained specialists who gave the group scientific credibility. For the at-large members, Governor Terrell named two North Georgians, J. H. Witzel, the ordinary (probate judge) of Fannin County, and W. E. "Buck" Candler, a prominent citizen of Union County, to represent local interests.[7]

Altogether, the terms of the resolution and the composition of the commission demonstrated the understanding that Ducktown smoke posed its greatest threat to the vitality of the agricultural and timber economy in the mountains. The political machinery of the state had thus moved to protect livelihoods and the economically valuable natural resources of wood, water, and soil rather than natural beauty, wildlife populations, or what later generations would describe with terms such as the environment, ecology, and ecosystems. In 1903 Georgia's mountain farmers and loggers were the endangered species deserving the aid of the state, not the wild creatures of the woods and waters.

Despite its weighty and urgent charge, the commission moved with less dispatch than the governor, allowing two months to pass before making its investigative tour of the mountain counties. The growing season was over and the leaves had fallen by the time they arrived in late October. If they failed to observe growing crops and leafy trees, they saw enough to conclude that the copper works "are hourly, daily damaging and destroying vegetable life . . . for miles within the boundaries of this state." They built a lengthy record based on personal inspection of smoke-damaged farms, a view of the mine works, and meetings with afflicted farmers. On that basis, they warned that unless current practices stopped, "a large area of farming and timber country within the limits of the State will certainly be destroyed and lost to the present owners." The observations, though dramatic, were to be expected given the well-publicized scale of the smoke problem. The more surprising language came in the conclusion, where the commissioners opined that "it is better that this industry should be entirely annihilated than that the present intolerable conditions should continue." This was the first official declaration that Georgia's entry into the Ducktown smoke wars threatened the continuity of Ducktown's copper industry.[8]

Wilmon Newell, the state entomologist, was not a member of the commission but toured the region that same month to provide another scientific appraisal at the request of Governor Terrell. Upon arrival, Newell traveled the basin with Judge Witzel and three local farmers, A. B. Dickey, J. W. Anderson, and J. Wilson. The farmers, all from the Hot House Creek area that extended from Fannin County, Georgia, into North Carolina's Cherokee County, were locally prominent men who each had private smoke cases languishing in the Tennessee courts. Dickey had emerged as spokesman for the Hot House community through his frequent demands and explanatory letters to Judge Parks, the trial counsel for the Ducktown Sulphur, Copper & Iron Company. Together, Witzel and the farmers made sure Newell saw their own damaged lands and guided him to the blighted farmsteads of their neighbors.[9]

The tour allowed the entomologist to note the concentric zones of destruction extending from the DSC&I works in Isabella and the Tennessee Copper Company works at Copperhill. In the first zone, with a radius of two miles, sulfurous acid "has removed from the country all traces of vegetation, save an occasional patch of Bermuda grass." Within four miles of the works, "all forest trees . . . are destroyed and the earth devoid of practically all vegetation except broom-sedge." In the next zone, extending to five or six miles, "50 to 75 per cent of all timber is dead," and beyond that, areas of smoke damage could be detected ten to twelve miles from the refinery. Worse, "the area of devastation is steadily increasing." The zones of destruction correlated to the presence or absence of agriculture. Newell reported that "in the area already denuded of forest growth no attempt is made to grow crops"; in fact, it "is totally abandoned except by the employees of the copper-mining companies."[10]

Newell then applied his entomological training to the question of whether the trees died from insect infestation or diseases as opposed to damage from sulfur fumes. He cut down and examined trees, both dying and dead, to examine the insect populations within them. There was little evidence of insect infestation. Instead, the insects he observed "are of species which feed upon *dead* vegetable matter and are therefore not responsible for the death of the trees." Newell determined that "possibly the death of one tree out of every five hundred dead is chargeable directly or indirectly to insect work," a rate well within normal occurrence in healthy smoke-free forests. If insects did not kill the trees, neither did fungal diseases. No signs of fungal growth appeared upon examination of the roots. Observations pointed instead to death from sulfur fumes. Healthy trees could withstand insect infestation for extended periods and often put out suckers from the trunk and roots. The smoke-damaged trees in Ducktown did not put out suckers, and instead the "root, trunk, and branches are killed at almost the same time." The cambium layer (the cell forming tissue) of smoke-damaged trees "becomes blackened before the tree has succumbed to the poison."[11]

His findings represented the first scientific examination of Ducktown smoke damage. Before his study, litigants and judges in the Tennessee courts had conceded to the damaging effect of smoke upon crops and timber. Just as they interpret the weather, people could experience and react to the smoke without understanding the related science. Everyone could see and smell the smoke. They saw it blot out the sun on frequent occasions. All could observe the expanding zones of denudation. These factors matured into a point of law three years earlier when the Tennessee Supreme Court declared smoke to be an actionable nuisance in *Ducktown Sulphur, Copper & Iron Co., Ltd. v. Barnes* (1900), a decision reached with a notable lack of scientific testimony from either side. Claimants and copper lawyers therefore gave little attention to issues of causation in the course of the private smoke suits and instead debated the extent of damage upon individual properties.[12]

Governor Terrell and Attorney General John C. Hart were both experienced lawyers who appreciated that Georgia's impending lawsuit against the copper companies would recast smoke litigation upon a much greater scale, one that required a solid scientific foundation. The higher stakes of a state-initiated lawsuit required a factual basis that rested on more than anecdotal descriptions of smoke damage. The governor's appointment of Newell was the first step toward that end. The entomologist was a scientist familiar with the methods of biological field study and equipped by training and intellect to compare his findings against known biological data in the fields of entomology and botany. After employing his skills to rule out insect infestation and fungal diseases, he concluded that "many square miles are being denuded of all forest growth and the devastation can be traced to no other agency than the sulfur fumes from the copper refineries at Ducktown and Isabella."[13]

Newell then broadened his attention from a close study of bark and roots to survey the larger landscape. He added that "within a radius of ten miles ... all merchantable timber, i.e. timber suitable for lumber, has already been destroyed" and warned that "a continuance of present conditions will convert this territory into a barren desert." His prescient comments provided an accurate description of present conditions and future results. This was one of the first recorded applications of the term "desert" to Ducktown.[14]

He did not use the term carelessly. As a life scientist, he presumably understood the idea of a desert in biological terms. His use of the term in the present context stemmed from an experiential, perhaps even atavistic, appreciation of the shocking expanse of empty, barren terrain in the midst of what used to be lush Appalachian forest and farmland. In this way, he anticipated later observers who would describe the area with phrases such as "a small but extreme man-made desert" and as "the only bona-fide desert east of the Mississippi, the handiwork not of nature but of man."[15]

Newell was correct about the impact of smelter fumes. A parade of scientific foresters would later testify to their toxic effect, but forest loss in the Ducktown Basin could not be attributed only to the smoke. He failed to address the impact of more than a half century of intense logging in the basin. The ax, not sulfur smoke, caused most of Ducktown's first great wave of deforestation. During the three decades of the industry's prerailroad years, fifty square miles of standing timber were converted to charcoal for consumption in the roast heaps and furnaces or were sawn into lumber for mine and building construction. Logging created an area of deforestation much greater than the zones of smoke damage around the small-scale roast heaps and furnaces of the first era. Circumstances changed by 1903, when the revived industry hit its full stride. Modern mining ran on coal and coke imported by rail rather than on locally produced wood charcoal. That alone might have encouraged areas of forest regeneration except for the vastly greater amounts of sulfur smoke caused by the refining of high-sulfide copper ores and the much larger scale of operations. The area of denudation was expanding, but the question of how much of that was caused in the first era versus the second would remain for other investigators to determine.[16]

Newell returned to firmer analytical ground when he considered the consequences of Ducktown deforestation. Forest preservation was a "matter of moment" to local inhabitants and to the state at large. Stable forests "supplied the people with wood, one of the prime necessities." Healthy forests prevented erosion and floods by absorbing runoff. Shady forests and their leaf litter preserved watersheds by keeping groundwater from drying out. The converse was also true. Newell warned that "removal of forests means rapid soil erosion" making "the land utterly useless for agriculture." Tree loss "conduces to extremes of flood and drought."[17]

Each of his comments marked him as a student well schooled in the principles of watershed conservation articulated forty years earlier by George Perkins Marsh in his landmark work, *Man and Nature* (1864). Marsh, a Dartmouth-educated Vermonter, was a man of remarkably wide-ranging talents and interests. Trained as a lawyer, he became a noted philologist fluent in twenty languages with a specialty in Icelandic. He was also an experienced diplomat, serving as ambassador to Turkey and Italy. Observations of barren and arid lands made during his extensive travels in the Mediterranean basin led to his insight that human conduct, specifically deforestation, could lead to disastrous consequences such as erosion, flood, drought, loss of arable land, and even permanent climate change. The roots and humus of forests, he concluded, regulated stream flow by retaining water from heavy rains. The shady forest canopy slowed evaporation within the watershed, thereby releasing water more gradually to keep streams flowing during drier periods. Marsh was a man ahead of his times; but his book soon became a foundational text for the nascent conservation movement and inspired the state of New York to create the Adirondack State Park to protect the Hudson River watershed.[18]

His stream-flow theories would eventually undergo serious challenge in 1908 by H. M. Chittenden, an officer in the Army Corps of Engineers, but that was in the future. For now, Newell applied Marsh's doctrines with a disciple's devotion in the hope of reversing the damage to the Ducktown mining district. Reclamation of the devastated areas was urgently needed, but he advised this "is possible only by reforestation, which under present conditions is impossible." The clouds of sulfur smoke had to be stopped, and if that required the closing of the mines, then so be it, for "the forest interests of Fannin County, Georgia, are of far more value and importance to the commonwealth than any revenue to be derived directly or indirectly from copper mining and refining." These were bold words for a man whose primary duties involved the detection and eradication of insect pests to cotton, corn, vegetables, and fruit. The once relatively anonymous civil servant, a person inevitably called "the Bug Man," now rose to the status of a prophet for natural resource conservation; his pronouncements were quoted verbatim in the *Atlanta Constitution* and the *New York Times*.[19]

The 1903 reports by Newell and by the legislative commission received generous press coverage in Georgia under headlines such as "Sulfur Fumes Destroy Vegetation in Fannin," "Copper Fumes Killing Trees," and a seven-column illustrated article captioned "Georgia Forests Withered by Sulfur Fumes." In another article, published a month before the Wright brothers made their historic first powered airplane flight, an unnamed reporter imagined a flight over the blighted Ducktown Basin: "If a man were up in a balloon, hovering over the boundary line between Georgia and Tennessee, and were to take a look far down in the direction of this Ducktown, he would see a cloud of smoke that obscured the landscape, as though ... a giant had laid down an enormous cigar, still burning." Underneath that cloud "the smoke settled on trees, leaving an acid deposit like a blight and the trees shriveled and died." It was as if the countryside "had contracted some terrible leprosy of vegetation that ate its sad way farther and farther from the point of first infection."[20]

Coverage like this could have occurred only with the cooperation and encouragement of the governor's office. The *Atlanta Constitution* returned the favor, showing Governor Terrell and Attorney General John C. Hart at their best and busiest on the Ducktown matter. The paper gave daily mention in captioned articles, and in its political column, "Gossip at the Capitol," of the many meetings held by the two men as they weighed the state's legal options in response to the two investigative reports. The possibilities included the state's participation in the farmers' suits for damages and injunction, or perhaps an action by the state in the Tennessee courts against the copper companies. The third and most dramatic alternative was a direct action in the U.S. Supreme Court "against the state of Tennessee for a mandamus to compel the latter to stop the destruction of property in Georgia." The paper noted that if the latter course was adopted, it would present "an entirely new question ... arising out of one of the most important provisions of the federal constitution—that dealing with the relations of one state to another." Such a course "will be unique, and will be fraught with interest to the legal profession all over the country; indeed, it will be of such peculiar nature as to interest lawyers in all parts of the world, for there is no other nation on the globe in which a similar action at law could be heard." Together, the published contents of the reports and news of the evident determination of Terrell and Hart to act upon them in a constitutionally novel way combined to stir excited expectations of an interstate conflict of historic proportions.[21]

James G. Parks and Howard Cornick, lawyers for the two copper companies, kept abreast of news from Atlanta by means of the Georgia newspapers brought to them on the daily trains and as reprinted in Tennessee journals. Parks sent a copy of the commission report to W. H. Freeland, the DSC&I general manager, along with a clipping from the *Ducktown Gazette*. He noted that the newspaper "has gotten the bug man's report mixed up with that of the other members of the Commission." In the same letter, he mentioned that verdicts were rendered in two of the local smoke suits, one for $170 and the other for $100, dismissing them as "little suits." Parks showed greater concern about the progress of the three injunction suits filed by William Madison, Isaac Farner, and Avery McGhee that were now in the Tennessee appellate courts. In October 1903 the Tennessee Court of Chancery Appeals ruled against the copper companies in the same month that Newell and the commission were tramping the basin landscape in the company of smoked-out farmers. For the moment, it appeared that Madison, Farner, and McGhee might succeed in their quest to shut down the mines by injunction. They were the enemies at the gate who posed a threat far more pressing than the maneuverings of a potentially dangerous but still distant foe on the other side of the mountains in Atlanta. Parks faced the nearer threat by collaborating with Cornick and corporate lawyers volunteered by the Louisville & Nashville Railroad to prepare what would be a successful appeal of the three injunction cases to the Tennessee Supreme Court.[22]

The threat from Atlanta appeared even more remote at the Broadway offices of the Tennessee Copper Company in New York City. At the same time that anticopper press coverage rose to a high pitch in Atlanta, TCC officers informed the *Wall Street Journal* of the company's profitable year and its intention to double production. The company paid a dividend of $1.25 per share in July 1903 and anticipated a like amount at the close of the year. Present earnings and future prospects were so good that it planned to install more and larger furnaces to increase production from the present rate of one million pounds of refined copper per month (twelve million per annum) to twenty-five million pounds per annum. The company was flush with cash, allowing it to finance the expansion from earnings without the need to borrow funds or to float a stock offering. Sulfur smoke from the company's smelters exposed it to lawsuits by the score, constant bad press, and now the risk of state action, yet the firm's response was to increase operations. Expansion would inevitably increase the amount of acidic fumes belched into the air and thus provoke more smoke litigation. So be it: Tennessee Copper officials were content to let farmers, journalists, and politicians yammer about smelter smoke so long as there was money to be made in the southern mountains.[23]

Governor Terrell and Attorney General Hart could follow the New York financial press just as easily as the copper companies monitored the Georgia papers. The published reports of expansion plans added yet another element of urgency to the demand for state action on the problem of Ducktown smoke. Inaction is a frequently used and occasionally wise tool of public policy, but it was not a viable option for the smoke problem. The cry for relief shouted out by North Georgia mountaineers matured to a matter of state policy when adopted as a joint resolution of the General Assembly. Newell and the commissioners provided empirical support with evidence of the widening zones of crop and timber damage extending from the mine works across the border into Georgia. The press repeated the findings in a series of articles that elevated a once-local nuisance, in both the legal and the popular senses of the term, into a popular cause. With those factors in mind, Terrell and Hart took the next step by drafting and sending a formal demand letter to Governor James B. Frazier of Tennessee.[24]

Some might have considered the mere sending of a letter to be a weak, tentative opening to a legal war. It could be done more dramatically by filing suit at the courthouse without warning and then having a uniformed sheriff or marshal appear at the doorstep of the unsuspecting defendant to serve the summons and process. Despite the perhaps malicious pleasure of litigation by ambush, demand letters served several important functions that led to their use in most civil cases. If the recipient was agreeable, the letter provided an opportunity to resolve disputes without the costs, delays, and risks of a lawsuit. Even if it was refused or ignored, the effort of making it tended to put the plaintiff in a better light during the ensuing trial by allowing him or her to argue that "we tried to resolve this without wasting the court's time but the defendant would not listen to reason." This was even more important in the present case because Georgia intended to invoke a court's equitable power to grant an injunction to abate the smoke. Every court of equity in America sought to apply the ancient maxim "he who seeks equity must do equity." This was a fundamental rule of fairness that meant, at the very least, that the attempt to seek a peaceful resolution by the use of a demand letter was encouraged and might even be rewarded by the court.[25]

Demand letters—whether by a creditor seeking payment from a debtor on a delinquent account, by an accident victim seeking compensation from the negligent party, or here by the governor of a sovereign state seeking an end to cross-border smoke pollution—all have the same three basic elements: the statement of grievance, a request that the recipient resolve the grievance in a specified manner, and the threat, whether expressed or implied, that litigation will ensue if the recipient fails to act as requested. Governor Terrell's one-page letter of November 25, 1903, contained all three elements. He first briefly recited background events, beginning with mention of the joint resolution and the formation of the commission in response to complaints from mountain counties "that the timber, fruit and agricultural interests ... had suffered great and irreparable damage on account of the fumes produced by the smeltering of copper ores." This much was not news. Farmers and timber owners in the tristate area had been litigating smoke damage claims against the copper companies for the better part of a decade. That made it necessary for Governor Terrell to explain how a long-running private dispute had now become a controversy between two sovereign states. For Georgia's part, the state became "interested in the question because of the great damage done to a large area of the public domain and the threatened total destruction of all vegetation within thirty or forty miles" of the copper works.[26]

Turning to the second element of a demand letter, Terrell asserted that Tennessee had a duty to intervene to abate the smoke damage because it was caused by corporations, "which I understand are created by the laws of Tennessee." Specifically, he requested that his counterpart in Tennessee "take steps to prevent a continuation of the methods now used for smeltering ores by these corporations." The third aspect of a demand letter, the threat of lawsuit, remained implicit owing to the dignified tone required in a letter between the leaders of two sovereign states. The bullying tone, often used in a demand by a landlord against a delinquent tenant, had no place in a communication between parties of this stature.[27]

Governor Terrell's letter, though admirably brief, clearly written, and dignified in approach, was nonetheless flawed. He asserted a claim of damage to Georgia's public domain that begged the question of whether the state actually suffered harm to its property interests independent of the property rights of its citizens. The state had little if any remaining public domain in the affected part of the state. Except state-owned buildings and highways, all of the land in North Georgia was in private hands. There were no state parks or wildlife refuges at the time. The letter also revealed that the state's hope for a technological cure to the smoke problem was more a matter of wish than reality. The demand for an end to "methods now used for smeltering ores" was, more specifically, a demand to end the roasting of ores on open heaps, in the expectation that the long-sought pyritic method of performing primary smelting in enclosed furnaces would somehow produce less sulfur smoke. Why that would be so was a mystery because whether green ore was roasted on heaps or burned in a furnace, the end remained the same: the expulsion of sulfur from copper ore and its ejection into the atmosphere.

Tennessee's Governor Frazier jumped on the third major problem of the demand letter in his response of December 14, 1903. After first expressing regret over any possible injury to citizens of another state, he wrote, "I know of no power vested in me as Governor of Tennessee to interfere and prevent such injury." The statement, to Georgia ears, was on a par with the classic response of a defendant in a dog bite case: "I am sorry, but that's not my dog." It was nevertheless correct. The Georgia demand letter implied that Tennessee was vicariously liable for the acts of corporations that it had created by grant of a charter or, in the case of foreign corporations, had authorized to do business by action of the secretary of state. The argument carried little weight because the very heart of corporate law was the creation of a new, artificial entity that shared some of the attributes of a natural flesh-and-blood person, including the capacity to sue and be sued in court. A corporation enjoyed the great benefit of limited shareholder liability upon condition that it could be sued and held liable for its conduct. That is why Governor Frazier insisted that "if the citizens of Georgia are injured by the operation of these works *by their owners* . . . the courts of Tennessee and possibly of the United States are the only tribunals I know of in which redress can be sought."[28]

Governor Frazier would have done well to end his response on that point. Instead, he added another page of text that drew heavily upon the arguments and attitudes of the copper companies, leading to the reasonable inference that copper lawyers helped to draft the response. He observed that, "from information obtained from Ducktown, the extent of the injury done by these industries has been very greatly exaggerated" and then slightly distanced himself from the remark by adding, "but as to this I have no opinion." He repeated the contentions made by DSC&I and TCC in the pending Madison, Farner, and McGhee injunction cases. Yes, the smoke was an inconvenience and possibly caused injury, but the mines gave employment to many. And the benefits of the industry outweighed the damage to the surrounding lands: "I am informed that the lands both in Tennessee and in Georgia, lying near to these works, would, in the absence of the copper mines, not be of very great value."[29]

As a matter of law, the arguments were essential to application of the newly enacted balancing-of-interests test then under consideration in the Tennessee's appellate courts regarding the three injunction cases. It would have been better for Tennessee and the copper companies if the arguments had been confined to court and out of the reply to the demand letter. Georgia mountaineers and their political leaders inevitably heard Governor Frazier's denigration of their lands, and the suggestion of exaggerated damages, as a form of mockery that became fighting words when quoted in Georgia newspapers. The *Atlanta Constitution* repeated the comments under the headline "Frazier Says He Cannot Act" and ended the article with the observation that Governor Terrell "at once turned the letter over to Attorney General Hart" for further action. "It is confidently expected that a suit at law will be the outcome." Popular expectations were one spur to action. A greater spur was the knowledge that that state of Georgia would appear a toothless opponent unless it filed suit soon after receiving Governor Frazier's response. Acting on Governor Terrell's instructions, Attorney General Hart simultaneously met public expectations and gave effect to the implied threat of the demand letter by filing suit in the U.S. Supreme Court on January 20, 1904.[30]

The owners and officers of Ducktown Sulphur, Copper, and Iron and of the Tennessee Copper Company recognized the suit as a dramatic and threatening change in Georgia's policy toward the Ducktown mining district. It ended a half century of close cooperation in which the state reached across its northern border to embrace the copper industry. Though the mines, and hence their taxable property, lay in Tennessee, Georgia had long valued the industry as a powerful economic magnet that could attract a railroad from Atlanta up and over the Blue Ridge Mountains to unite North Georgia to the rest of the state. The northern counties could not develop economically without a railroad to transport their resources to distant markets, and nothing was more likely to spur railroad investors to build it than the prospects of handling freight for a well-established copper industry.

Georgia legislators understood this from the earliest days of Ducktown mining. The 1853 Georgia General Assembly authorized construction of a line upon the finding that "there are large bodies of copper ore now raised and raising . . . on both sides of the state line between Georgia and Tennessee which cannot be shipped to places of manufacture without the aid of a railroad." More than a dozen other acts and resolutions followed in succeeding decades to promote the Ducktown line. Georgia provided convict labor to build it, loaned state funds to finance it, and even forgave the loans when construction faltered. The General Assembly also encouraged extensive logging in the Toccoa River watershed for the benefit of the copper industry. It passed laws declaring the Toccoa to be a navigable stream and forbade dams, fish traps, and other impediments that interfered with the rafting of logs from Georgia forests down the river to Ducktown to fuel the roast heaps and smelters.[31]

The *Atlanta Constitution* had long supported the copper industry and the railroad that was intended to reach it. It reported in 1870 that the people of North Georgia "are all alive for the project" and declared that "the railroad is the great instrumentality of progress and development" that would allow the section to "blossom like the rose." When Governor James M. Smith traveled the proposed route of the line to Ducktown in 1874, the paper said "the people of these sequestered counties are enthusiastic over the prospect of a connection with the rest of mankind." It noted the governor's "deep interest in the early development of the great mineral resources of the State" and his concern that he "did not see how this could be done without the aid of railroads."[32]

The *Constitution* lamented the closing of the mines in 1878, asserting that "the suspension of work there at any time, and especially at the present, would be a great drawback upon all this section of country." It worried when North Carolina's Governor Thomas Jarvis boasted that a competing rail link from North Carolina would reach Ducktown first and that "Chattanooga and Atlanta will be completely flanked and their importance as strategic railroad points will disappear." When the Georgia line won the race, allowing the mines to reopen in 1890, the paper proclaimed, "The completion of this road opens up a limitless territory rich in agricultural resources, minerals, timber, and water powers . . . and well may Tennessee and Georgia rejoice over its completion."[33]

Editorial policy and legislative action toward Ducktown were an expression of the New South ideology that prevailed in Georgia during the last quarter of the nineteenth century. The term "New South" is historically elastic and has been applied to every era of southern history from the end of the Civil War up to and beyond the modern civil rights era of the 1950s and 1960s. In its narrowest sense, as articulated by the *Constitution*'s nationally famous editor, Henry Grady, it was a campaign to revive the war-torn region with a combination of agricultural diversification (to reduce the near total reliance upon cotton as a money crop), promotion of industry (with northern capital and expertise), and white supremacy (to reverse civil rights advances for African Americans imposed by the North during Reconstruction). Diversification largely failed; southern agriculture remained fixed upon cotton until the boll weevil and other factors forced reconsideration after the turn of the century. White supremacy prevailed through the institution of Jim Crow segregation and the elimination of black voting strength through intimidation and eventual disfranchisement.[34]

Results on industrialization were mixed. Drawn by the attractions of the South's large pool of unskilled labor, low wages, and low taxes, investment money poured into the South to build railroads and factories and to exploit the region's mineral and timber resources. Southern transportation, utilities, and industries increased markedly, but at the price of perpetuating a colonial economy that exported raw materials and partially finished goods to the North and then purchased higher-value finished goods made by better-paid workers in northern factories. Over time, the South became, in the phrase of historian Gavin Wright, "a low wage region in a high wage nation." Even so, throngs of sharecroppers were eager to leave cotton farming, with its uncertainties of weather and crop prices, in the hope of a somewhat steadier life of wage labor in the mills. Many mountaineers from areas where they had never grown cotton were just as eager to leave their hilly farms for work in the mines.[35]

The revival of copper mining at Ducktown fit well within the pattern of New South industrialization. The basin provided an array of attractive incentives for investors from New York and London. The *Wall Street Journal* reported that "with low wages, cheap fuel, a mountain of low grade copper ore, and a modern smelter, the Tennessee Copper Company has attracted considerable attention in the copper world." The same prospects enticed DSC&I investors in London to enter the district a decade earlier in 1889. The copper interests delivered on their end of the New South bargain. Their faith in the viability of the Ducktown mines spurred railroad investors to finish the line through North Georgia, to the great economic advantage of every mountain county along the way. The county seats of Canton, Jasper, Ellijay, and Blue Ridge were all now securely tied to the regional and national rail network. Timber from Gilmer County, marble from Pickens County, minerals and farm products from Fannin County all found ready markets in Atlanta and beyond. Hundreds, maybe thousands, of Georgians in the Ducktown

Basin earned their living in whole or in part from mining. Having done their part to fulfill the New South agenda, the two great firms, Ducktown Sulphur, Copper & Iron Co., Ltd., and the Tennessee Copper Company, might have expected the long embrace with Georgia to continue with even greater fervor.[36]

Smoke had always been a part of the relationship between the state and the industry. Ducktown smelter smoke had drifted across the border for more than half a century before it became a matter of official state concern in Atlanta. Georgia had long accepted smelter smoke as the expected price of industrial and economic progress. It was not a problem worthy of state attention until the volume of smoke—and consequent damage to local crops and vegetation—reached an unprecedented scale with the rapid expansion of the industry and the switch from high-grade oxide ores, rich in copper and low in sulfur, to the current use of sulfide ores, containing roughly 2 percent copper and 25 percent or more sulfur. With those factors in mind, the 14 million pounds of copper produced in 1903 revealed the scope of the problem: the copper came from 350,000 tons of ore that also yielded almost 90,000 tons (or 180 million pounds) of sulfur. Some of the sulfur was consumed during smelting. The rest ascended from the roast heaps and furnaces to enter the air over the basin where it remained trapped by the mountains on all but the more windy days.[37]

The growing volume of sulfur smoke inevitably left its mark on the landscape with the expanding zones of smoke seen and documented by the legislative commission and the state entomologist, Wilmon Newell. It also ended the rhetoric of shared prosperity and replaced it with the rhetoric of war. Mountaineers first raised the cry. The General Assembly authorized combat by its legal champion. Attorney General Hart, in his formal challenge before the Supreme Court, alleged that sulfur smoke from the mines was "a hostile invasion on the part of Tennessee and its citizens" upon the state of Georgia. The *Atlanta Constitution* echoed the phrase with the headline, "Georgia Seeks to Repel Invasion by Tennessee."[38]

Invasion was a strong word with important legal and historical connotations. The legal significance would surface later in the Supreme Court litigation; in the meantime, the term struck a powerful chord in the hearts and souls of Georgians who shared a visceral hatred for an earlier invader, William Tecumseh Sherman. Thousands of living Georgians retained vivid personal memories of the sufferings they endured when Sherman's Army of the Tennessee marched, fought, and burned its way through the entire length of the state. After burning Atlanta, his army left a three-hundred-mile trail of deliberate destruction to cities, towns, factories, mills, gins, farms, and homes. Attorney General Hart was ten years old when Union troops plundered his native Greene County on the way to the sea. Those Georgians born after the event were raised on a diet of biscuits, gravy, and tales (both real and apocryphal) of Sherman's depredations.[39]

The state's political and cultural leaders carefully nurtured memories of the event as part of a larger campaign to enhance white identity and supremacy with stories of the glories and sufferings of the Lost Cause of the Confederacy. In 1874 the General Assembly set aside April 26, the day of General Joseph E. Johnston's surrender to Sherman, as Confederate Memorial Day. Atlanta's official seal depicted a phoenix rising from the ashes in defiance of Sherman's burning of the city.[40]

Characterization of Ducktown smelter fumes as an invasion was more than a figure of speech. The smoke confronted the senses in a way that gave weight to the rhetoric of invasion. It was not a hidden industrial menace like lead in house paint. It had the sensed immediacy of Sherman's army. His troops could be seen as they marched out of Chattanooga for Georgia. The endless train of its wagons could be heard at a distance, and the sound of its guns yet further. And, if one came too close, the odor of thousands of unwashed men, their cooking fires, and their poorly sanitized camps could be smelled. Ducktown smoke shared some of those attributes. It carried the acrid stench of rotten eggs. It was a visible mass that blocked the sun, and if viewed from the surrounding heights, it appeared to flow over the landscape as if it was an advancing army, leaving a trail of destruction behind it.

Invading smoke made a refugee of J. H. Verner and his family. He testified in 1905 that the smelter smoke "killed all my fruit trees, timber and crops, and I had to leave and follow something else for a living." He abandoned the farm in 1899 and then failed to get work at the mines. He then moved to the town of Ducktown, where he supported his family as a teamster "hauling wood for citizens, goods for merchants, and any one I could get." Scores of other Georgia farmers had their own complaints about smoke damage. The problem was getting worse. The invasion needed to be repelled.[41]

Georgia's fight against Ducktown smelter smoke was a specific response to a geographically isolated threat. It was not part of a general state campaign against air pollution. The smoke suit was a high-profile exception to a longstanding policy of legislative and judicial tolerance of industrial smoke. The greatest human source of sulfur dioxide in Georgia and the rest of the nation was the burning of coal to heat homes and to fire up steam boilers to power locomotives and factory equipment. Coal generated about 100,000 tons of sulfur dioxide pollution in Georgia in 1900, and that amount rose to 200,000 tons by 1920. The amount of pollutants, though great, did not lead to a general outcry for relief. Georgia's cities and industries were growing, but did not approach the smoke-producing magnitude of New York's population or Pittsburgh's steel industry. Unlike Pittsburgh or Ducktown, the state's open topography south of the mountains allowed coal pollutants to disperse through the atmosphere before causing the visible damage so evident in the basin. That being so, the General Assembly did not feel compelled to act against coal pollution until 1967, when it joined the modern campaign against air pollution by enacting its first comprehensive air quality act.[42]

Georgia courts also tolerated smoke as an acceptable consequence of progressive civilization. This posture first surfaced in opinions involving nuisance claims against railroads, especially concerning the operation of trains in urban settings. In a set of cases from Augusta, neighboring property owners, long accustomed to the urine, manure, and noises of horse-drawn wagons, complained that steam locomotives were "shaking the houses, thereby breaking the plastering and filling the houses with dust and smoke." The engines spooked horses, causing them to run away pell-mell to the risk of others, and exposed homes to the risk of fire from flying cinders. Trains and their tracks impeded traffic. The trial court refused to enjoin rail operations. The Georgia Supreme Court then affirmed, saying, "in these days of modern improvements, we admit the legality" of permitting trains on city streets, though it did allow individual actions for damages to proceed. The right to damages was restricted in a later case involving a railroad built through the village of Stone Mountain to haul granite quarried from the mountain of the same name. As in Augusta, the villagers sued for damages, but this time the court applied the incidental-benefits rule to affirm a zero verdict for claimants. It held that the complaining neighbors were not entitled to damages if the presence and improvements brought about by the railroad increased the value of their properties in an amount greater than the amount of the damages caused by the trains. On this point, the Georgia Supreme Court was even more favorable to industry than its Tennessee counterpart, because the latter court refused to apply the incidental-benefits rule on behalf of the copper industry in *Ducktown Sulphur, Copper & Iron Co. Ltd. v. Barnes* (1900).[43]

Georgia's Supreme Court showed the same favorable disposition toward steam-powered factories. In 1856 it affirmed judgment against homeowners when they attempted to enjoin a steam-powered carpentry shop in Columbus because of noise. Chief Justice Joseph H. Lumpkin first dismissed the noise issue by observing that "we know of no sound, however discordant, that may not, by habit, be converted into a lullaby, except the braying of an ass or the tongue of a scold." He then declared that one may "as well attempt to stop up the mouth of Vesuvius as to arrest the application of steam to machinery at this day." Judicial tolerance had its limits as demonstrated in a case by the owners of an office building against a coal-fired steam laundry in downtown Athens, a few blocks from the University of Georgia School of Law. The landlord contended that the laundry's use of soft coal created a dense cloud of dark smoke that forced its professional and business tenants to keep their windows closed on hot days to their great discomfort (this occurred long before the advent of air conditioning). The court recognized that the use of coal was necessary in homes, businesses, and factories, and that "as population thickens, the impurities thrown into the air are increased." It added that "pollution of the air" is "actually necessary to the reasonable enjoyment of life and indispensable to the progress of society." As an inescapable component of modern life, smoke pollution did not give rise to an actionable case for nuisance unless it was generated "in an unreasonable manner so as to inflict injury." Whether the laundry's use of soft coal in lieu of cleaner-burning hard coal was unreasonable under the circumstances was a question for the jury.[44]

Neither the courts nor the legislature had occasion to consider smelter smoke generated by operations within the state. The major Georgia mining operations—marble, stone, kaolin clay, gold, coal, and crushed stone—did not involve the smelting of sulfurous ores. The government was largely tolerant of mining's other impacts. Gold mining drew white settlers into Georgia's Blue Ridge Mountains in the 1830s and remained a favored endeavor of the state for the remainder of the century. The General Assembly did all it could to promote gold mining with little regard to the consequences to the surrounding lands and waters. It repeatedly authorized hydraulic mining, the technique of blasting high-pressure water shot from hoses to wash gold-bearing deposits out of the hillsides. The water jetted from the nozzles with such force that, as a consequence of "accidentally striking a man or animal, [that man or animal] would be killed in an instant," and it carved up the landscape just as easily. The muddy runoff then passed through sluice flumes that isolated the gold while allowing enormous loads of dirt and gravel to wash into the streams. The method eroded slopes, ruined watercourses with loads of silt, and flooded downstream lands wherever it was employed.[45]

Nevertheless, the General Assembly granted hydraulic miners a general right of way to impound and divert streams for their operations on the argument "that by this process alone can the mineral wealth of large portions of our mining districts be ever fully developed." It also issued corporate charters to numerous companies with unambiguous names describing the streams they intended to despoil, such as the Yahoola River and Cane Creek Hydraulic Hose Mining Company.[46]

In California, where there was a much larger hydraulic gold industry, farmers fought and prevailed in a nuisance action against mining interests in 1884. Nothing similar appeared among the reported cases in Georgia. Georgia's Supreme Court did enjoin two small mining operations near the piedmont town of Cartersville when neighbors complained that operations diverted and polluted local streams, but neither case involved hydraulic mining. Nor did the cases do anything to quell statewide excitement over the revival of gold mining when the Dahlonega Consolidated Gold Mining Company began operations at about the same time that copper mining revived in Ducktown.[47]

Georgians were also excited about the possibility of copper mining of their own. A report by Walter Harvey Weed, published by the United States Geological Survey in 1904, renewed hope of creating a viable copper industry in several parts of Georgia, especially where the Ducktown ores extended over the border from Tennessee into Fannin County. There was no reason to expect that Georgia ore was any less sulfurous than ore from Tennessee. Nonetheless, the *Constitution*'s editors cited the report and insisted "the time may come when the new battleship *Georgia* will be sheathed with Georgia copper." Sporadic copper mining in Fannin County proved disappointing with the brief exception of the No. 20 Mining Company that produced 86,000 tons of ore over a three-year run from 1916 to 1918. All of it was carried by rail across the border for smelting at the works of the Tennessee Copper Company.[48]

Aldous Huxley wrote, "Consistency is contrary to nature, contrary to life. The only completely consistent people are the dead." Even so, the *Constitution*'s booster rhetoric for copper on the Georgia side of the border was jarring when read alongside the war rhetoric of its ongoing articles against the operations of Tennessee Copper and Ducktown Sulphur, Copper & Iron, located less than five miles away on the other side of the border. The difference turned on the geographic consequences of the state line and the local topographical features that made that line problematic. Topography concentrated the smoke problem, allowing it to flow over the border into the

Georgia portion of the basin. The same mountains contained the problem in ways that minimized the political and economic consequences of Georgia's response to it.[49]

The Ducktown Basin is roughly oval in shape, narrow in the north and widening to the south.[50] The northern fourth of the pan lies on the Tennessee side, and the southern three-fourths are in Georgia. Only a tiny portion of the basin extends into North Carolina, a major reason why that state declined to join suit against the copper companies. Ligon Johnson, an attorney who would soon play a major role in the smoke litigation on behalf of Georgia, wrote that the fumes from the roast heaps "did not float off in the atmosphere but hugged the ground, rolling along in front of the wind in constant volume until absorbed by the soil and vegetation." He added, "There was little diffusion and the radius of fume influence gradually widened with the destruction of each successive barrier of vegetation." Instead of rising, the smoke flowed over the cultivated bottomlands and climbed the slopes into the orchards and woodlots. Instead of dispersing into the atmosphere, it remained in a highly concentrated state, causing damage too severe and too pervasive for Georgia authorities to ignore. A different result might have occurred if smelting occurred in a flat, windy landscape that allowed smoke to disperse over a wider region.[51]

The same mountains also made the damage a geographically isolated problem. Dora Galloway, daughter of a TCC copper miner, noticed this effect when she traveled out of the basin to visit her grandparents in Farner, a village a few miles up the track beyond where Stansbury Mountain guarded the basin's northern rim. In a trip of only a dozen miles, she traveled from a manmade desert back into the Appalachian hardwood forest: "We had barely settled in our red plush seats when the train slowly moved away from the station; and hardly no time until we left the barren hills . . . with trees looking as though they were flying by the windows." The basin's drainage patterns also worked in Georgia's favor: all the local streams flowed into the Toccoa-Ocoee River, which then drained into Tennessee. Silt from denuded slopes and waterborne mineral toxins from the mines flowed away from Georgia's portion of the basin without troubling the rest of the state.[52]

Favorable drainage patterns and protective mountain walls made it possible for Georgia to fight smelter smoke with a targeted lawsuit, without altering its pro-industry New South policies for the rest of state. The problem was specific to the Ducktown Basin. TCC and DSC&I were the only offenders of note, and they were on the other side of the border. The fight against their smoke did not require enactment of a general pollution law at the risk of irritating thousands of voting polluters, as would be the situation in the event of a statewide effort to control sulfur dioxide from coal and other sources.

The geography of the state line also shaped Georgia's response in other ways. The border controlled tax benefits, but it did not control the smoke. Georgia industry generated Georgia taxes, and so the state might have tolerated the smoke if the smelters were on its own side of the border. They were not. Tennessee enjoyed almost all of the direct tax benefit of the copper industry in the Ducktown Basin. Property taxes assessed against the two great companies made Polk County one of the wealthiest counties per capita in Tennessee. Georgia could not tax property across the state line. Nor did it have an income tax or sales tax at the time that would have allowed for a relatively direct link to the industry by taxing earnings paid to Georgia mine workers and sales to the mining community from Georgia merchants. Smelter emissions troubled both states, but Tennessee had all the taxable benefit. Georgia just got the smoke. Those simple facts made DSC&I and TCC vulnerable to the demands of Georgia smoke suitors for action by their government in Atlanta.[53]

Georgia's lack of direct tax benefit in the Ducktown mining industry served the political cause of the smoke victims. The citizenry of the North Georgia mountains did not, as a rule, carry much weight in the legislature because the region was geographically smaller and far less populated than the broad piedmont and coastal plains. The mountains added relatively little to an economy based on cotton and urban industrialization. Cotton grew poorly north of the Blue Ridge, and the industrial centers were concentrated in Atlanta and the fall line cities of Augusta, Macon, and Columbus. Nonetheless, the smoke resolution passed in both houses with near unanimous votes for the simple reason that the copper companies were on the other side of the border. Action against Tennessee Copper and Ducktown Sulphur, Copper & Iron would not pit Georgia farmers against Georgia industry. Nor would it cost the state anything in lost tax revenue. No state is eager to kill a taxable golden goose, but the copper industry was Tennessee's goose, not Georgia's. Tennessee protected its economic interests by amending the law of nuisance in favor of the copper industry. Georgia could afford to heed the political demands of smoked-out farmers and loggers by authorizing an investigation and a lawsuit.

In January 1904 Hart headed to the Supreme Court in Washington, D.C., for the opening round of Georgia's legal war against Ducktown Sulphur, Copper & Iron Co. and the Tennessee Copper Company. With him was Congressman Farish Carter Tate, whose district embraced the Georgia portion of the basin. Tate might have had mixed sympathies as he weighed the factors for and against his visible support of the smoke suit. The Tate family owed more than a little to the copper industry, yet Tate's district embraced the areas most damaged by the smoke.

His political stature rested, in part, upon his family's New South industrial success. That success began with the good luck of his grandfather, Sam Tate, when he acquired land lot number 147, in what is now Pickens County, during one of the lotteries of Cherokee land. The parcel lay in the heart of huge and valuable marble deposits that became even more valuable with the arrival of the Marietta & North Georgia Railroad in the 1880s. The line built in the obsessive drive to reach Ducktown ran right past the marble quarries, allowing the product of the family's quarries to be easily shipped to distant states. Rail access allowed Stephen Tate, Farish's father, and Colonel Sam Tate, Farish's brother, to build their local operation into a nationally prominent concern. The statue in the Lincoln Memorial at the nation's capitol was carved from Tate marble. Only Vermont produced more American marble. Colonel Sam celebrated the family's fortunes by building an Italianate mansion out of pink marble quarried from the Tate mines.[54]

Yet, the immediate concerns of F. C. Tate's mountain constituents outweighed the fact that the Tate family owed its fortune to the railroad built to serve the copper industry. Ducktown smoke was a political matter, not just a legal one. North Georgia voters had pressured their representatives in the General Assembly into action against smelter smoke. Congressman Tate was a professional politician and knew the prevailing mood in his district toward the copper companies on the other side of the border. It was that knowledge that led him to join Attorney General Hart's group as Georgia's battle against invasive smoke moved to the U.S. Supreme Court.[55]

THE DUCKTOWN DESERT AND GEORGIA'S FIRST SMOKE SUIT

5

On January 24, 1904, Attorney General John C. Hart boarded a train for the six-hundred-mile trip from Atlanta to Washington, D.C. The occasion was his first appearance before the U.S. Supreme Court in Georgia's fight for an injunction against the copper companies. His companions included Ligon Johnson, a young Atlanta lawyer he hired to serve as special counsel on the case, and Farish Carter Tate, a Georgia congressman whose district embraced the smoke-damaged counties. The journey across four southern states gave Hart many hours to consider his two-phase strategy, a plan that combined legal and technological approaches to reduce or, better, to eliminate sulfur pollution from the burning roast heaps of Ducktown.[1]

The first phase of his strategy was the Supreme Court lawsuit, filed under the caption *State of Georgia v. State of Tennessee; the Ducktown Sulphur, Copper & Iron Company (Ltd); and the Pittsburgh and Tennessee Copper Co.* (Hart's caption contained a factual bobble that he soon amended: the Tennessee Copper Company purchased the holdings of the Pittsburgh firm before the suit was filed.) Georgia's case rose from a welter of smoke suits filed by scores of Georgia mountaineers in the Tennessee courts. The suits were, at the moment, bogged down in a procedural and substantive morass of motions and appeals because of the skillful and determined defense by James G. Parks for the Ducktown Company and Howard Cornick for Tennessee Copper. Hart intended to circumvent the Tennessee courts by filing Georgia's lawsuit directly in the U.S. Supreme Court as an original jurisdiction action, and he needed the Court's permission to do so. It was to that end that he and his companions traveled to Washington for a hearing on his motion for leave to file bill of complaint.[2]

The Court's approval to file the bill of complaint would indicate that it was willing to consider a smoke injunction. Hart needed the threat of an injunction to accomplish the second, technological, phase of his strategy, the elimination of the traditional practice of open heap roasting of copper ores. As he stated from the outset, "It was not the purpose of the State of Georgia to suppress and drive out of business this enterprise, representing an investment of over a million dollars." His intention was "to suppress this method of roasting the ore" to be replaced by a new method of furnace smelting, known as pyritic smelting. This, he hoped, would make it possible for the companies to continue operations while reducing the volume of sulfurous smoke that roiled through the atmosphere of the Ducktown Basin. If the new method produced less of the toxic sulfur dioxide fumes, then farmers could resume growing crops for the local market, and loggers could put the woods to productive use. The lawsuits and the political pressures they generated might then subside, and maybe the relentless transformation of Southern Appalachian greenery into western badlands would come to an end.[3]

As the train rolled over the miles of tracks, the attorney general and his young colleague were aware of the legal challenge before them in the U.S. Supreme Court, an experience that only fools took lightly. Judge Hart, age fifty, grew up on a piedmont cotton farm in Greene County, Georgia, during the years of the Civil War. As a young man, he committed himself to the well-worn path of Georgia politicians that took him from the farm to the University of Georgia, followed by the practice of law in his home county, service in the legislature, elevation to the bench as a superior court judge, and now election to state office as attorney general. His position made him the chief legal adviser to Governor Joseph M. Terrell. This placed the new attorney general in an awkward position on the Ducktown case because Terrell was a former attorney general with a brilliant record of wins before the Supreme Court. It nevertheless fell to Hart to devise a legal strategy to transform the pleas of smoked-out Georgia mountaineers and the General Assembly's mandate into a winning case of constitutional law.[4]

His thirty-one-year-old assistant in the cause, Ligon Johnson, was part of the postwar generation. Johnson was born into a prominent white Alabama family in Tuskegee. After college at Emory and law school at the University of Virginia, the attractions of a larger city led him to begin private practice in Atlanta. He was young, single, and had only a decade of legal experience, but his intelligence earlier led Hart to associate him in a pending corporate tax case against the Louisville & Nashville Railroad. Johnson's back office work in that matter persuaded the U.S. Supreme Court to review it on a grant of certiorari. A Supreme Court victory on either that case or the Ducktown litigation would secure the young man's reputation as a rising legal star that might propel him out of Atlanta to an even larger professional arena.[5]

The two men represented different models of legal education. Hart entered the profession in the traditional manner of reading for the law, which was essentially an apprenticeship under a practicing lawyer. As a young college graduate, he performed legal chores around the office and shadowed his mentor in conferences and court appearances. He devoted quieter moments to self-study over law books, with an emphasis on the *Georgia Code* and practice-oriented treatises. Johnson followed the newer path of formal education in a law school. He spent his days at a considerable remove from the workaday legal world by attending law classes on the campus of the University of Virginia. Some of his professors probably lectured. Others may have adopted the case study method, devised by Christopher Columbus Langdell at Harvard in the 1870s and gradually adopted by other law schools. The case method combined the detailed study of case opinions issued by appellate judges with a vigorous Socratic dialogue between professor and student on the facts, holding, and rationale of each decision.[6]

Both models had their merits if done well. Reading for the law gave the prospective attorney valuable exposure to the participants in the legal system—lawyers and clients, cops and criminals, judges and juries. It was an education in applied law where an attentive student with a good mentor could learn which trial tactics succeeded and, just as important, which ones failed. It was also superior training for future politicians like Hart because the student gained a working familiarity with the people and politics of elected office. At the University of Virginia, Johnson's law professors sought to instill a conceptual framework of the law and to teach the advocacy skills useful for practice in appellate courts. Though his training left him short of exposure to the rough-and-tumble of daily law practice, he nonetheless acquired an analytical approach to the law that was broader and deeper than what might ordinarily be acquired by trailing behind a country lawyer in the routine matters of a county seat law practice. It was good that the two men combined their respective strengths on the copper case because it was a high-stakes, politicized venture into a poorly charted area of constitutional law. Johnson had the analytical tools for work in the appellate courts. Judge Hart (former judges retained the title as a matter of courtesy and custom) was alert to the political implications of high-profile cases like the Ducktown matter.[7]

It was easy to stir public interest against smelter smoke. The reports and photographs from the North Georgia high country, with descriptions of barren landscape and dead forest blighted by fumes from across the border, all made for dramatic copy. Adding to the drama was the knowledge that, win or lose, the case would change the lives of many. The *Atlanta Constitution* observed that a Georgia victory could force the closing of the copper mines unless "some method different from the present one of abstracting the copper from the crude ore is discovered." If not, it "will mean that two of the largest corporations in Tennessee and Great Britain will lose an investment of something like $2,000,000." If Georgia should lose, "it will mean the loss of about twenty to thirty square miles of territory" by the killing of all vegetation upon it. In Tennessee, the *Chattanooga News* responded that a Georgia victory "would be a deplorable event" because "nearly every family at Ducktown, among the working classes, draws sustenance from this enterprise, and to destroy it would prove to be a great disaster and would mean untold suffering." The *Knoxville Tribune* asked, "What is Georgia wanting to clean up Ducktown for? Let Georgia wash her own children and keep her hands off Tennessee's." It was thus that reporters in 1904 anticipated the jobs-versus-environment dichotomy typical of environmental news coverage that would appear at the end of the century.[8]

The legal novelty of the case added another set of pressures upon Hart and Johnson. Their immediate task was to gain the Supreme Court's permission to file Georgia's smoke suit as an original jurisdiction action. The Court normally functioned as an appellate body by reviewing cases that originated in lower trial courts. Article 3 of the Constitution authorized it to depart from its appellate role and to act as a trial court in limited circumstances, notably matters "in which a state shall be a party," specifically cases between two states or a dispute between a state and citizens of another state. This was called original jurisdiction because such cases began and ended in the Supreme Court. Most original jurisdiction cases were between states, and most of them concerned disputed boundaries. The boundaries of the former British colonies were often poorly defined, and in states where a river defined the boundary, a shift in its course could alter a border by carving land from one state and adding it to another. Mark Twain devoted a chapter of *Life on the Mississippi* to the boundary havoc created by the writhing river.[9]

If Hart and Johnson were handling a boundary case, they could have relaxed and played cards during their trip to Washington in the confidence that original jurisdiction would be granted. Instead, they had the challenge of persuading the Supreme Court to expand the scope of original jurisdiction to embrace, for the first time, a case of interstate air pollution. With no precedent for it, there was no assurance that the Court would permit it. If the Court refused, the case would end before it began. Georgia had few viable judicial alternatives. It would not file suit in the Tennessee courts because that would require Georgia to subjugate its sovereign authority to the courts of another state. This was something that no state would willingly permit. Conversely, Georgia could not subject the opponents to its own judicial system because they were, as Hart and Johnson alleged, "beyond its power and control." The state of Tennessee was no more likely to submit to a Georgia court than Georgia would submit to one in Tennessee. The copper companies were also out of reach. They were chartered in London and New York and conducted their operations in Tennessee. Georgia thus had no basis for asserting jurisdiction over them under the more restrictive jurisdiction doctrines of that time.[10]

The framers of the Constitution devised original jurisdiction to address quandaries of that sort, recognizing that only the U.S. Supreme Court had the independence and institutional dignity to serve as an appropriate forum when a state was a party. Even so, the Court accepted original jurisdiction reluctantly and required litigants to first seek its permission before filing because it considered original jurisdiction to be an awkward deviation from its usual function as an appellate court. The Court's modus operandi and institutional apparatus were heavily oriented toward appellate review of decisions rendered by lower courts. Trial courts were better equipped to serve as finders of fact, thanks to the procedures and architecture that allowed judges and juries to observe the direct and cross examination of witnesses. The Supreme Court operated without juries and found it necessary, when adjudicating original jurisdiction actions, to take testimony in the written form of depositions and affidavits or to defer fact finding to special masters.[11]

Another cause for its reluctance, and for its insistence on a request for permission, was the potential for abuse presented by original jurisdiction actions. The problem had its roots in Georgia history and led to one of the first great Supreme Court cases, *Chisholm v. Georgia* (1792). The legacy of that case directly impacted the present mission of Hart and Johnson. Article 3 of the Constitution made federal jurisdiction available in actions "*between* a state and a citizen of another state," which was at first taken to mean that it applied in both directions: when a state, acting as plaintiff, sued a citizen of another state, and the reverse, when a citizen of one state initiated a lawsuit against the government of another state. In 1792 Alexander Chisholm, a South Carolinian, filed suit in federal court against Georgia to collect a debt for supplies sold to the state during the American Revolution. Georgia objected that it was immune as a sovereign state from suits to which it did not consent and thus refused to submit to jurisdiction in the matter. The Supreme Court ruled for Chisholm on the rationale that article 3 jurisdiction trumped Georgia's claim of state sovereignty.[12]

The decision caused a constitutional uproar. Other states quickly realized that the decision also robbed them of their sovereign immunity against out-of-state suitors. In response, Congress and the states immediately passed the Eleventh Amendment to prevent more claims like Chisholm's. The amendment narrowed article 3 by providing that "the judicial powers of the United States shall *not* be construed to extend to any suit in law or equity commenced or prosecuted against one of the United States by citizens of another state." Georgia lost the battle and won the war. Its role in forcing adoption of the amendment was a matter of pride to its citizens and received detailed consideration a century later in the *Atlanta Constitution*'s coverage of the smoke suit.[13]

The new amendment restored sovereign immunity to the states but frustrated individuals with otherwise plausible claims against another state. Clever lawyers sought to circumvent it by recasting prohibited claims of private citizens against another state into a new form, as permitted claims between states. New Hampshire sought the Court's original jurisdiction in 1882 when it tried to collect on overdue bonds sold to its citizens by Louisiana. In an 1899 case, New Orleans shippers prompted Louisiana to sue Texas to bar enforcement of the latter's quarantine laws. The Supreme Court rejected original jurisdiction in both cases on the ground that the putative actions between states were actually state-level efforts to pursue the private claims of individual citizens. In the second of the two cases, *Louisiana v. Texas*, the Court ruled that in order to maintain original jurisdiction between the two states "it must appear that the controversy to be determined is a controversy arising directly between the State of Louisiana and the State of Texas, and not a controversy in the vindication of grievances of particular persons."[14]

The language in *Louisiana v. Texas* created a significant hurdle for Georgia's smoke suit. It was all too easy for the copper companies to argue that Georgia's proposed Supreme Court action was nothing more than an attempt to vindicate the grievances of individual Georgia citizens. The sequence of recent events established that the state of Georgia was a latecomer to the smoke litigation, and that it acted only in response to complaints of its citizens regarding the troubles they experienced in getting their cases heard in the Tennessee courts. North Georgians had scores of pending individual cases for smoke damages. They had battled the copper companies for the better part of a decade before the state began its own lawsuit in 1904. The timing allowed for a strong argument that Georgia's new case was an abusive attempt by the state, acting on behalf of its private citizens, to circumvent the Tennessee courts by starting a parallel action in the U.S. Supreme Court. Both of the copper companies argued the point throughout the proceedings. The Tennessee Copper Company stated the matter plainly when it asserted that the state "is in fact lending its name to said individuals." Many years later, Ligon Johnson admitted as much in a paper to the American Institute of Mining Engineers, in which he stated, "First, the farmers sued the Tennessee and Ducktown companies, and upon the failure of these suits, the State of Georgia took up the cudgel."[15]

It was a strong argument, and if Hart and Johnson were to overcome it, they needed to show that the state of Georgia had a legally recognized grievance distinct from the claims of its citizens in the mountain counties. To that end, they based their motion on the idea of invasion, the same word used in the press to stir popular support for their cause. Their motion began with a description of the invasive smoke and the damage it caused. "Vast quantities of smoke, sulfur fumes, and noxious and poisonous fumes and vapors" rose from the open roast heaps operated by the mining companies. The sulfur smoke was "discharged" (a military verb carefully selected to evoke cannon fire) "for a radius of thirty miles or more from . . . Ducktown" onto five North Georgia counties. Sulfur fumes ruined crops, pastures, orchards, and trees, creating a "zone of destruction growing each month." The attorneys then discussed the consequent damage to Ducktown's soil and watershed by drawing from the 1903 report of State Entomologist Wilmon Newell and, by extension, from the conservation doctrines articulated by George Perkins Marsh in his 1864 manifesto, *Man and Nature*. Overall, they considered property damage to be their strongest argument and devoted nine paragraphs of the petition to the description of various forms of it. They argued that loss of plant cover destroyed the watershed by creating a "barren waste subject to sudden, severe, and dangerous floods" that carried topsoil from previously fertile bottom lands and washed away roads and highways. They then tied the property damage to the state's direct interest by alleging that smoke-damaged lands lost value, which in turn caused a reduction of property taxes that had already reduced the state's revenue in the district "more than one-half."[16]

The attorney general and his young assistant then transformed the facts of invasion into a constitutional mandate that required the Supreme Court's grant of jurisdiction under the oldest and most fundamental principles of American federalism. First, they alleged that the actions of the respondents "constitute and are a hostile invasion on the part of Tennessee and its citizens upon Georgia." Next, they meekly asserted Georgia could do nothing about it on its own. The state had no jurisdiction over the defendants and "all diplomatic and amicable negotiations have failed." War against Tennessee and the copper companies was not an option because the Constitution forbade Georgia "from any invasion, aggressive operation or other direct action." The lawyers concluded by citing article 4, section 4 to invoke the constitutional guarantee from the United States to the individual states that it "shall protect each of them against invasion." It followed that the Supreme Court should honor the guarantee by exercising original jurisdiction to resolve the dispute in lieu of a show of arms and the loss of blood.[17]

The two Georgia lawyers presented their arguments with eyes cast to the past and the future. An invasion is the hostile entry upon the lands or person of another and is thus the ultimate violation of sovereignty. The present invasion of toxic smelter fumes into its northern counties invoked an almost mythic sense of injured honor reflecting a series of historical challenges to the state's sovereignty. Georgia fought against the British to secure its independence during the Revolution. Its position in *Chisholm v. Georgia* (1793) forced the adoption of the Eleventh Amendment to the Constitution in 1795 to preserve state sovereign immunity. Georgia then pressed to extinguish Indian tribal sovereignty within its borders. In the 1802 Articles of Agreement and Cession, it insisted on the federal government's promise "to extinguish the Indian title to all the other lands within the state of Georgia" in exchange for release of the state's claim to western lands in what are now the states of Alabama and Mississippi. When the federal government failed to eliminate Cherokee tribal sovereignty by treaty, Georgia then tried to accomplish the same end by legislative fiat, an action that forced a confrontation in the Supreme Court in *Cherokee Nation v. Georgia* (1831) and again in *Worcester v. Georgia* (1832).[18]

Georgia's present insistence upon a federal remedy to preserve its sovereign integrity against invasive smoke was done with heavy, if unspoken, historical irony. It was only four decades earlier that the state rejected the concept of federal authority by seceding from the Union, followed by four terrible years of civil war. Several justices had personal memory of the conflict: Justices Oliver Wendell Holmes and John Marshall Harlan were veterans of the Union Army and Justice Edward Douglass White fought for the South. By 1904 the South had long since returned to the national fold, and the old veterans of both sides sat together on the Supreme Court bench, ready to entertain Georgia's prayer to invoke federal sovereignty on its behalf.[19]

The use by Hart and Johnson of the rhetoric of invasion was more than a jurisdictional necessity; it also served as a prescient insight that interstate pollution was a physically intrusive threat to state sovereignty that required federal intervention. The novelty was in the federal aspect of their argument, for the basic idea of government action against pollution was not new. Governments had the power to act against pollution if the people causing the pollution and the ones who suffered from it were within the same jurisdiction. In Great Britain, sovereignty was undivided throughout England and Wales, and so it was legally simple to enact pollution laws at the national level. Acting centuries before the Industrial Revolution, Edward I, Richard III, and Henry V each attempted to reduce the level of smoke in London by enacting laws to limit the burning of high-sulfur sea coal mined at Newcastle. In the Victorian era, Parliament passed legislation to create London's municipal sewage system after sewage in the Thames River reached intolerable levels with the Great Stink of 1858.[20]

Similar national measures were more difficult to enact in America because of the complications arising from dual federal-state sovereignty. The Refuse Act of 1899 was a tentative first step at federal pollution control. The act barred disposal of solid trash and debris into navigable streams and harbors without a federal permit, but it expressly excluded the flow from "streets and sewers and passing there from in a liquid state" from the scope of the law. The distinction turned on the limited construction Congress placed on its constitutional power to regulate commerce. Congress clearly had power to regulate navigable waters because they were a major avenue of interstate commerce. Solid trash posed hazards to navigation, by damaging ship hulls and clogging navigation channels, in ways that liquid waste did not. Within its modest scope, dumping solid trash and debris into navigable waters became a crime that could be prosecuted in federal trial courts, but liquid sewage could continue to enter streams without fear of the law.[21]

At the turn of the nineteenth century, Congress had yet to seriously address other aspects of pollution at the national level. It was not until 1948, with passage of the Federal Water Pollution Control Act, that Congress embraced the Commerce Clause as authority for regulation of the quality and health of water as opposed to its importance to navigation. In the meantime, the states were on their own regarding other forms of interstate pollution. Their efforts to combat it inevitably provoked constitutional battles over state and federal sovereignty.[22]

Hart and Johnson knew that the Supreme Court had addressed interstate pollution on only one earlier occasion, in *Missouri v. Illinois and the Sanitary District of Chicago* (1901). The case arose when Chicago had a great stink of its own. Sewage generated by its growing population, the wastes from the hundreds of thousands of cattle and hogs in its stockyards, and the slaughter room filth from its meatpacking houses all flowed down the lazy Chicago River directly into Lake Michigan, the city's source of drinking water. Municipal engineers sought healthy water by building a system of fresh water intakes far out into the lake beyond the polluted area near the shore, but the expanding area of contamination frustrated their efforts.[23]

The Chicago Sanitary District then devised a new solution on a scale that would have impressed Egypt's pharaohs: they reversed the flow of the Chicago River. Between 1887 and 1900, they dug a channel through a low ridge that separated the Chicago watershed from the Mississippi River system. Water that once flowed east into Lake Michigan now flowed west all the way across Illinois via the Des Plains River, thence into the Illinois River, and finally into the Mississippi River at a point forty-three miles above St. Louis. The Mississippi served as the common border between Illinois and Missouri, with the line running down the center of the main channel. Chicago pollution that entered the Illinois side of the river fouled Missouri's portion. Missouri officials complained that "fifteen hundred tons of poisonous undefecated [i.e., unpurified or untreated] filth of said Sanitary District of Chicago will be daily carried ... into the Mississippi," where it will "pollute and poison said water with the germs of diseases of various and many kinds."[24]

Missouri filed a Supreme Court original jurisdiction action seeking an injunction to stop the flow of sewage from Chicago, to which the defendants responded with a vigorous demurrer alleging that the Supreme Court lacked jurisdiction to hear the matter. The Court ruled for Missouri in a 6-to-3 decision written by Justice George Shiras. The majority conceded that the Court's previous original jurisdiction cases concerned state borders, state property rights, and issues of interstate commerce. It then extended the scope of original jurisdiction by declaring that "it must surely be conceded that, if the health and comfort of the inhabitants of a State are threatened, the State is the proper party to represent and defend them."[25]

The opinion in *Missouri v. Illinois* provided partial support for Georgia's Ducktown smoke case while leaving other issues unresolved. Although both cases concerned pollution, the differences outweighed the similarities. Missouri's case was presented and decided as a public health issue. The Court readily grasped the state's concern that "contagious and typhoidal diseases introduced in the river communities" by the reversed Chicago River "may spread themselves throughout the territory of the state." Missouri's argument and the Court's ruling both reflected the extent to which the link between waterborne pollution and disease had become a matter of accepted scientific fact. John Snow established the empirical link between disease and contaminated water in his famous 1855 study, *On the Mode of Communication of Cholera*, which traced deaths in a London cholera epidemic to specific contaminated wells. He ended the crisis with a dramatic gesture by removing the handle from the worst of the wells. Louis Pasteur's work in microbiology, Karl Eberth's identification of the typhoid bacillus, and Robert Koch's 1883 discovery of the *Vibrio cholerae* germ were all widely known at the turn of the century. This allowed Missouri's attorney general to draw upon well-established scientific knowledge when presenting his concerns about waterborne disease. Chicago's lawyers could and did argue whether the volume of diverted sewage posed a health threat hundreds of miles away in St. Louis but did not controvert the paradigm of water pollution as a health hazard.[26]

Georgia's lawyers also gave a nod to public health with a line or two in their petition about smoke and respiratory problems, but could not press the point because the scientific link between sulfur dioxide pollution and public health concerns had barely developed at the time. All they had to offer was anecdotal evidence from Ducktown smoke suitors and the contradictory evidence of local doctors. J. H. Verner testified that smelter smoke "caused me and my family to cough, sneeze, vomit, and was very disagreeable.... I was advised by the doctor to take one boy that had had the measles out of the smoke." He did as advised without being able to save the child. When pressed by a company lawyer, he admitted that he could not swear that the smoke caused the death. Testifying on behalf of Ducktown Sulphur, Copper & Iron Company (DSC&I), Dr. H. A. Rogers stated that "it is true that at times the smoke settles down strong enough to cause considerable coughing" but "without any detrimental effects." Noting that sulfur is a germicide, he vaunted the curative benefits of exposure to smelter smoke: "I have frequently recommended patients who were suffering from hay fever and asthmas to make daily visits . . . and place themselves where the smoke was thick, and remain there for some time." Dr. L. E. Kimsey retorted that the smoke was "extremely irritating to the organs of respiration" and increased the impact of lung diseases. His brother and fellow physician, Dr. Fred M. Kimsey, advised consumptive patients to leave Ducktown, though without blaming tuberculosis on the smoke.[27]

An understanding concerning the health effect of sulfur dioxide would have to await developments in the new discipline of industrial hygiene. In the early twentieth century, Alice Hamilton and other hygienists would begin systematic empirical and experimental studies to establish the health hazards of industrial toxins, and to define safe levels of exposure. Their work is now carried on by a host of federal agencies collected under the umbrellas of the Department of Health and Human Services and the Department of Labor. It was not until 1998 that the Agency for Toxic Substances and Disease

Registry released its "Toxicological Profile for Sulfur Dioxide." The authors of the 233-page report evaluated hundreds of scientific studies to confirm that sulfur dioxide is hazardous to health. Short-term exposure to high levels "can be life-threatening" and may cause "burning of the nose and throat, breathing difficulties, and severe airway obstruction." Long-term exposure to sulfur dioxide was observed to cause significant loss of pulmonary function among workers at copper smelters and pulp mills. Though the report came ninety years too late to impact the Ducktown litigation, the findings did prove the Kimsey brothers to be in the right, and Dr. Rogers in the wrong.[28]

Hart and Johnson knew that their public health argument was weak given the evidence available to them. They also knew that the argument about the state's damaged property interests was also weak because the state lacked a titled property interest to lands in the Ducktown Basin. Their best hope was that their characterization of pollution as an invasive threat to state sovereignty would carry them to victory. They made their arguments before the justices, and the Court took the matter under advisement. Because they could do nothing more at the moment, they boarded the train for the long trip back to Atlanta.

One week later, on February 1, 1904, the Supreme Court granted Georgia's request for permission to file its smoke suit as an original jurisdiction action. The ruling was not the Court's final word on the issue. It was merely a preliminary approval that could be modified or revoked in later stages of the case in response to additional arguments from the copper companies. Even so, Georgia's attorney general and his young colleague had reason for confidence that the initial ruling would be upheld in future proceedings because the Court was more inclined to affirm its initial ruling than to reverse it at a later hearing.[29]

Johnson and Hart had other reasons to be pleased. They won a major point in their first appearance in the Supreme Court on a difficult issue of first impression. This placed Hart in good stead with the governor and the voting public. Johnson added to his profile as a rising young attorney. Strategically, the victory secured the legal phase of their two-phase plan. The grant of jurisdiction meant that Georgia could now request an injunction against the copper companies in terms that might force the end of mining operations. That threat gave the state the leverage to pursue the second, technological, phase of its strategy.

Hart wanted an engineering fix for Ducktown smoke. There were only three ways to abate a nuisance: move it, end it, or control it with some form of technology. In the earliest days of the Ducktown copper industry, mine operators removed the problem by shipping raw ore for processing to the Revere Smelting Works in Boston, or across the Atlantic to Swansea, Wales, the largest copper smelting complex in the British Isles. Processing ore at distant smelters was rarely an economical solution because the cost of shipping a ton of raw ore was always greater than the cost of shipping the small amount of metal extracted from it. Ducktown miners understood the problem and abandoned the practice in the late 1850s as soon as they gained the ability to smelt ore on site. If local smelting was cost beneficial during Ducktown's first era, when using rich 25 percent ore, it made even more sense in the industry's second era with the use of low-grade ore containing only 2 to 3 percent copper. Apart from the costs, shipping ore for processing to distant smelters simply transferred the smoke problem from one locale to another. Swansea had enough smoke problems of its own. British authorities were pressuring local firms to reduce smelter smoke at the same time that Hart was litigating against the Ducktown Company and Tennessee Copper.[30]

Attorney General Hart had already rejected the second option of terminating the industry and instead sought a technical fix that would reduce the smoke while allowing the industry to prosper. He and Johnson had done their part in court. Now it was the turn of the engineers. There was no shortage of suggestions. One idea was dispersal. The members of the 1903 legislative commission proposed the use of a tall chimney of two-hundred-fifty feet "so that the poisonous smoke would be carried far into the heavens and dissipated by the winds." Dispersion had a powerful logic to it. Many substances become less potent and obnoxious when mixed into a large volume of air or water. Cigar smoke and heavy perfume are easier to tolerate out-of-doors than in a small, stuffy parlor. The first urban sanitation systems worked on the same principle by discharging untreated filth into large bodies of moving water in the expectation that it would be diluted to tolerably safe levels. Tall smokestacks accomplished the same effect by sending smoke up away from ground level into the winds high above cities.[31]

Dispersion worked to a degree, but often failed when the volume of sewage and smoke exceeded the capacity of wind and water to disperse it. The slow-moving currents at the foot of Lake Michigan frustrated Chicago's effort to dilute sewage into that huge body of water. Whenever winds died in periods of calm, killing smog occurred in cities with heavy concentrations of coal burning, as happened in the steel town of Donora, Pennsylvania, in 1948 and in London in 1952. Advocates of dispersion also failed to account for the way that dispersion regimes led to the gradual accumulation of toxins in air and water. At best, dispersion merely removed pollutants from one locale to another, less favored, locale downstream or downwind. Historian Joel Tarr summarized dispersion efforts as the "search for the ultimate sink."[32]

Another technological approach was the removal of sulfur from the smoke. An armchair scientist from Nashville wrote the editor of the *Ducktown Gazette* with his plan for smelting ores inside one hundred six-by-ten-foot ovens, each connected by a common flue to collect the smoke. The smoke would then be sprayed by steam and forced through a vat of cold water to convert sulfur dioxide gas into useable sulfuric acid. Though he was right about the goal, he fell woefully short on the complex techniques of chemical engineering needed to bring it about. Mining experts in Ducktown, and in the industry as a whole, could only wish for a solution that easy. They knew that open heap roasting hurt industry profits by allowing a valuable commodity to literally go up in smoke and by exposing mining companies to lawsuits for all of the damage it caused. They longed for a technique of sulfur extraction that was both technically and commercially viable.[33]

Ducktown's copper companies knew they could sell all of the sulfur they expelled from their ores, if only there was a way to capture it during smelting. The chemical had thousands of industrial uses in everything from fertilizer, petroleum refining, steel processing, cloth dyes, and dynamite. When August Raht (brother of Captain Julius Eckhardt Raht) contemplated Ducktown's prerailroad decline in the 1870s, he saw the harvesting of sulfur as one of the keys to the industry's salvation. "Unless a plant could be built at Ducktown for the production of sulfuric acid . . . Ducktown is gone without redemption." Unfortunately, an economic process for extracting sulfur from smelter smoke did not exist then, nor did it exist two decades later when Carl Henrich, a noted mining engineer, considered the waste and damage from open roasting. "If the sulfur were only present in more concentrated form and capable of economical utilization," then "its value would greatly exceed that of the copper in the ore." The trouble was that "no practical method of saving it is now available," and until there was, he suggested dispersion by the use of tall smokestacks to at least mitigate the damage.[34]

Miners could not send smoke up a chimney until they first discovered a successful method for pyritic smelting within a furnace. W. H. Freeland, the general manager at DSC&I, and his engineers had a well-designed furnace and knew the correct ingredients—ore, coke, and quartz—to feed into it, yet each time they fired it up, the mixture caused the fire to die and the contents to congeal into an unprocessed mass. They were like a chef whose every soufflé collapsed despite the use of a proper oven and ingredients. Serendipity, the trump card in the inventor's hand, provided the answer when engineering experiments failed. As told by Ducktown historian R. E. Barclay, a workman stumbled upon the solution when he decided to toss more quartz onto a waning fire at the end of his shift before he went home to bed. He learned next morning that the furnace continued to burn properly through the night. When he repeated the step on the next nights, the furnace continued to run until the ore was thoroughly processed.[35]

Whether discovered by calculated experiment or by luck, pyritic smelting was a major achievement that brought significant advantages to the copper industry. The new method ended the production bottleneck caused by open heap roasting. The old method was the slowest stage in the mining process, taking three or four months to roast the amount of ore that secondary smelters processed in a few days. Pyritic smelting combined primary roasting and secondary smelting into one process and accomplished both stages in a matter of days. Production costs for labor and fuel declined as the pace of smelting increased. With these benefits in mind, DSC&I abandoned open heap roasting on August 16, 1902, and completed the transition to pyritic smelting a year later on October 5, 1903. The Tennessee Copper Company adopted the process at its Copperhill works for the same reasons. It announced to the *Wall Street Journal* on October 29, 1903, that in light of DSC&I's success, it too would make the switch to pyritic smelting. The younger company explained that the new method would lower costs of production to eight and a half cents per pound of copper. It would also save four pounds of copper per ton of ore lost during roasting. This was a significant boost in the yield when processing ore that, at best, contained thirty-two pounds of copper to the ton.[36]

Hope stirred that use of pyritic smelting in lieu of open heap roasting would end Ducktown's smoke problem. A Knoxville newspaper reported that the new process "will free the community of the deadly sulfur smoke" that had killed "all vegetation for miles around Ducktown." A. B. Dickey, a prominent farmer from the Hot House Creek community wrote DSC&I counsel James G. Parks to say, "I am now satisfied that the company is going to abandon the roasting of ore," and he suggested that the time was now right to settle the smoke damage suits awaiting trial in the Polk County District Court. "Now if the smoke is taken off, and a reasonable damage [is paid] to people who are really damaged, it looks to me that this would be much the best way to settle it." The Tennessee Copper Company (TCC) told the financial community that "the new system will save ... heavy damages to farmers by reason of the smoke nuisance causing damage to farming lands."[37]

Technological progress, economic motivation, and the grant of original jurisdiction led to a rapid settlement of the Supreme Court case. On February 10, 1904, Attorney General Hart, Ligon Johnson, and Governor Terrell met in Atlanta with W. H. Freeland, general manager of DSC&I, and J. Parke Channing, president of TCC, along with their respective counsel. Freeland stipulated for DSC&I that the company had completed its transition to pyritic smelting and that "the roasting of ore will not again be resumed." Channing acknowledged that, although it was still roasting ore on open heaps, TCC would not light any new fires in the roast yards after completion of the new furnaces in April. He also cautioned, "It takes from three to four months to roast ore after it has once been lighted and therefore it is a physical impossibility to stop roasting when it is once under way." When the present fires burned out at summer's end, the era of open heap roasting in the Ducktown Basin would be over. For Georgia's part, Hart, Johnson, and Terrell agreed that the state would not ask for a temporary restraining order. They also dismissed the state of Tennessee, realizing that it was not a necessary party to the action. The Supreme Court approved an agreement upon these terms and then dismissed the case without prejudice in April.[38]

The little two-word phrase "without prejudice" gave the state great power by leaving open the threat of future injunction proceedings. A dismissal *with* prejudice would permanently end the case. A dismissal *without* prejudice allowed the state to renew its lawsuit if the present arrangements failed to solve the smoke problem. Before the settlement, the lawsuit was like a bomb with a burning fuse that threatened to blow up the local copper industry. Settlement without prejudice snuffed out the burning sparks while leaving the fuse in place. The state could relight the fuse when it chose. The bomb would not be defused until 1937, long after many of the present actors had died.

The settlement was bizarre in one respect. It appeared to be a typical agreement by one party to suspend litigation in exchange for the other party's promise to stop an objectionable practice. The surrounding facts showed it to be otherwise. Hart filed Georgia's lawsuit on January 24, 1904, after the Ducktown Company had already closed its roast yards, and after TCC publicly announced its intention to do the same. The question thus arises why Judge Hart filed the suit when he did. The answer can be explained only in political terms. The case began as a citizen protest amplified by the General Assembly resolution. Terrell and Hart were on the spot and had to respond with demonstrable action in the courts. Invasion had to be answered with victory. James Parks, attorney for the Isabella company, later commented that Hart "was anxious to get rid of his suit ... so as to get as much political capital out of it as possible for himself and Governor Terrell, both of whom were at that time candidates for re-election."[39]

Georgia's Ducktown case occurred in a regulatory vacuum that increased the political pressure on the two state leaders. The political environment was much different from that in 1970 when Congress established the Environmental Protection Agency (EPA). The EPA and its state-level counterparts function as bureaucracies that provide a significant layer of political insulation to elected officials on environmental matters. They make their own regulations in a rule-making process that is conducted at a remove from normal legislative politics. Disputes arising from their regulatory activities are channeled through several rounds of administrative hearings before they ever reach a court of general jurisdiction. There were no environmental agencies in 1903. The political and legal vectors at that time went in a straight line directly from angry voters, through a sympathetic General Assembly, and then to Terrell and Hart, all without a lengthy bureaucratic detour.[40]

Intransigence by the copper companies also provoked the suit. The Ducktown firm squandered the credit earned by its discovery when its lawyer suppressed news of the switch to pyritic smelting. W. H. Freeland, the general manager, considered publicizing the company's commitment to pyritic smelting by submitting an article to a professional mining journal. James G. Parks responded with an April 21, 1903, letter expressing his concern that news of the change would undercut the defense argument made in scores of pending smoke cases that roasting was the best available technology. He wrote that "our position heretofore has been that we have been using the only known method of successful treatment; and we have put it to the courts to say whether we should be permitted to follow accepted standards, or whether we should be driven out." An article about pyritic smelting could be used by the plaintiffs to argue that DSC&I and its competitor willfully continued open roasting despite knowledge of a new and presumably less destructive process. He suggested that "it would be good policy to let the impression remain for a while that pyritic work in Ducktown is still in the experimental stage." Freeland followed the advice by delaying the article. It appeared a year later, after settlement of the Georgia case, in the May 26, 1904, issue of *Engineering and Mining Journal*.[41]

Parks also warned his counterpart at Tennessee Copper about the need to sound a consistent note about the new method. In a letter to Howard Cornick in May 1903, he advised, "There is a wide-spread impression among people that the companies could use other methods, if they would; but that they use this [roasting] because it is the most profitable." Cornick responded the following day with similar concerns and speculated that "Mr. Freeland might be able to testify that the new process is really an experiment ... with no absolutely fixed understanding ... as to its ultimate success." The efforts of both attorneys to manipulate the state-of-the-art defense provided little reason for Hart to forgo his planned Supreme Court case.[42]

The Tennessee Copper Company gave an even worse signal to Hart by adding 150 more roast heaps to its Burra Burra yard in 1903, the year after the DSC&I success with pyritic smelting. TCC knew about the discovery. Its works at Copperhill were only three miles from the DSC&I works at Isabella, making it easy for each firm to monitor the activities of the other. Nonetheless, TCC did not announce its intentions to stop roasting until the fall of 1903, when the last roast fires at Isabella were already dead or dying. Ducktown historian R. E. Barclay considered the delay from his perspective as TCC's longtime treasurer and wrote that "apparently the growing agitation over the smoke nuisance was not being taken seriously by Randolph Adams," the TCC general manager at the time. The manager's stance only confirmed the suspicions of smoke suitors that TCC found roasting too profitable to abandon until forced to stop. Although competitive pressures forced the company to make its announcement in the *Wall Street Journal* concerning the transition to the new method, political pressures did not allow Hart to delay suit to see the plans become a reality.[43]

In the end, Hart's 1904 Supreme Court actions made little difference to either company. The case settled quickly on terms that required them to do nothing beyond what they were already committed to doing for solid business reasons. Hart may have advanced the Tennessee firm's timetable for the conversion but otherwise left the company unscathed. The Supreme Court case was, in its current posture, a sideshow to the much more threatening legal battles in the Tennessee appellate courts. Parks and Cornick were more concerned about the status of the three injunction cases filed by a trio of farmers: William Madison, Avery McGhee, and Isaac Farner. The farmers demonstrated a far greater commitment to winning a final decree of injunction than Hart and Johnson had yet demonstrated in the Georgia case. Each plaintiff was a smoked-out farmer with long-standing grievances against the copper companies. They were veteran fighters in the smoke wars with multiple pending actions filed under their names. The battle had long since turned personal.[44]

The copper companies posted initial victories by defeating each of the injunction cases in the Polk County Chancery Court. The smoke suitors then met with success before the Tennessee Court of Chancery Appeals in October 1903, when it reversed the lower decrees in each case by votes of 2 to 1, thereby allowing the injunctions. The decisions stunned management and counsel at the two firms and sent them scrambling to marshal support in Nashville, and among corporate friends, for the next and final appeal to the Tennessee Supreme Court. Georgia's case had been settled for ten months when the Tennessee's highest court finally ruled for the copper companies on November 26, 1904, in terms that effectively ended anticopper injunction actions in that state. Writing in the afterglow of victory, James G. Parks said to W. H. Freeland, "I believe this decision will have a most wholesome effect upon our Georgia friends." And if Hart decided to renew his case in Washington, then, "I believe the decision of our Court, knowing as it does all the facts and surrounding circumstances, would have great weight with the United States Supreme Court." His words were sensible but would prove wrong, as hot sulfur fumes from the new pyritic furnaces reheated political pressures on Georgia's attorney general.[45]

Fifty years of open heap roasting in the Ducktown Basin were at an end. The fires at Ducktown Sulphur, Copper & Iron roast yards died on August 16, 1902. They continued another two years at the Tennessee Copper Company, finally ending in August 1904. Both companies increased production thanks to the speed of pyritic smelting and to the capital expansion of their smelting works. The Tennessee Copper Company more than doubled its plant by adding four new blast furnaces next to the older three. The new furnaces were the largest copper furnaces in the nation. A massive complex of smelters now spread atop a bluff above the Ocoee River at the company's Copperhill works within site of the Georgia border. DSC&I made similar improvements though on a smaller scale. Together, the annual production of copper by the two companies more than doubled from 8,103,534 pounds in 1902 to 19,475,119 pounds in 1907.[46]

Although pyritic smelting boosted production and increased profits, it did nothing to abate sulfur dioxide emissions. Whether roasting on open fires or processing inside blast furnaces, the goal remained the same: the expulsion of sulfur from raw ore into the atmosphere. More sulfur smoke, not less, rose above the works because of the speed of pyritic smelting coupled with plant expansion. Dr. John T. McGill, a Vanderbilt University chemist, later considered the problem in a 1916 study performed at the request of the Supreme Court. He determined that the new process eliminated the wood smoke from the roast heaps, but the amount of sulfur escaping from the smokestacks "was very slightly, if any, less than in treating the same amount of ore by roasting and charcoal smelting." This was an understatement. Using local production figures, he estimated what the two companies released before and after the end of roasting. In 1902, before either company ceased roasting, they processed a combined 335,864 tons of ore and expelled 184 tons of sulfur into Ducktown skies *per day*. Five years later in 1907, when both companies fully employed pyritic smelting, they processed 557,950 tons of ore and released 286.6 tons of sulfur per day, or more than 100,000 tons per annum. Sulfur dioxide emissions thus increased by 55.7 percent between 1902 and 1907.[47]

Later studies provided special insight about the volume of Ducktown smelter smoke during the period. One researcher determined that acid from Ducktown smelter smoke caused tombstones to deteriorate up to fifty times faster in the Ducktown Basin than in other locations in the Southeast. Sulfuric acid caused gravestones to become "so structurally weakened by granular disintegration and so many have been removed (only their pedestals remain) that even the high measured rates are underestimates." Other scientists drew samples from trees in the Cades Cove section of the Great Smoky Mountain National Park, about fifty-five miles from the TCC smelters at Copperhill. Testing revealed that tree growth declined at a rate corresponding to the worst years of Ducktown smelter emissions. The samples also showed a marked increase in iron in the same pattern because of the high iron content in Ducktown sulfide ores.[48]

The impact of smelter gases upon the forests of the distant Great Smoky Mountains was due in part to the new smokestacks installed by each copper company. Pyritic smelting allowed the companies to channel smelter smoke via flues into new central chimneys built to eject the hot, high-pressure gases further into the atmosphere. The Tennessee Copper Company built a 125-foot stack at Copperhill and replaced it in 1905 with a huge 325-foot smokestack. At Isabella, the DSC&I stack rose only 70 feet but was situated at a higher elevation than the TCC works along the river. The new stacks allowed some of the fumes to exit the Ducktown Basin to reach the Great Smokies and elsewhere, but overall they failed to adequately disperse the fumes within the basin for the simple reason that its mountain rim was higher than even the tallest stack. Formerly, smelter smoke tended to fall upon the areas closest to the smelters or to crawl up and down the valleys of the area's major streams, where, unfortunately, the best farmlands in the district were to be found. Now, because of the new smokestacks "the concentrated volume of smoke reached out to hitherto untouched regions" to trouble the more distant farms and timber stands.[49]

Instead of reducing the number of smoke claims, the effect of increased production and the new smokestacks created more claims from an expanding group of property owners. Pyritic smelting failed as the fix for Ducktown smoke and, in fact, made the problem worse. No significant reduction in sulfur dioxide emissions could occur until the copper companies conquered the next engineering hurdle by devising and building acid plants to convert the fumes into marketable sulfuric acid. The new process was a significant step toward that goal, but pyritic smelting on its own did nothing to abate the smoke—or the attendant lawsuits and political controversies.

In the meantime, smelter smoke from the expanded industry combined with other factors to complete the formation of the Ducktown Desert that would characterize the basin for most of the twentieth century. Estimates of its size varied among observers depending upon how they defined it. In 1906, when the damage was at its worst, the U.S. Forest Service mapped more than five hundred square miles of smoke damage in three zones: an "Area Destitute of Forest Growth," an "Area of Serious Injury," and an outer ring of "Observed Leaf Damage." The combined areas covered the entire tristate basin and extended beyond the mountain rim north almost to the Hiwassee River and south down the Ellijay Valley deep into Gilmer County. The Tennessee Valley Authority (TVA) conducted extensive surveys forty years later during World War II to determine the extent of the heavily eroded areas. A 1945 TVA report with supporting maps determined the area to be 23,000 acres (almost 36 square miles), divided into zones of varying severity. The 7,300 acres of the inner zone were closest to the smelters. The lands within that zone "are practically denuded and intensely gullied." The middle zone of 6,000 acres was also heavily gullied but had a partial cover of sedge grass on the surfaces between the gullies. The outer zone of 9,700 acres "supports a fair cover of grass and scrubby trees, but occasional bare spots and active gullies."[50]

The badlands were so extensive that they produced microclimatic effects. Researchers determined that "wind velocity was thirteen times greater" in the desert portions than in the nearby forests, and "the average soil surface temperature was twenty-two degrees hotter." C. R. Hursh, a U.S. Department of Agriculture researcher, determined that mean temperatures were up to twenty-two degrees higher in the desert than in the nearby woods. Loss of vegetation also hastened evaporation. The Forest Service reported that evaporation was five times greater in the denuded areas than in the forests beyond the zone of smoke damage.[51]

In his 1916 report to the Supreme Court, Dr. McGill identified sulfur dioxide smoke and wide-scale logging as the chief causes of the Ducktown Desert. The two factors reinforced each other. Loggers leveled fifty square miles of timber for use in construction and as fuel for the roast heaps during Ducktown's first era. Though the temporary end of logging and smelting during the 1880s allowed a modest second growth to occur, the new growth quickly fell to the axe when mining resumed in the 1890s and the killing effect of sulfur fumes prevented its return. When open heap roasting ended, residents continued to rely on wood for use in cook stoves and fireplaces, but the workers who lived in the barren zone near the mines found it hard to come by. An old miner named Wallace recalled chopping up stumps to get firewood during a spell of subzero weather in 1905. Others no doubt did the same, thus causing erosion to advance at an even faster pace. When James Smallshaw heard the story thirty years later, he noticed an absence of stumps in the barren area. Close examination revealed that "only fragments of roots remain in the center of the barren area and occasional rotted stumps in the outlying sections."[52]

In a 1943 article in *American Forests*, Kenneth Seigworth extended the list of factors causing the Ducktown Desert. He claimed that "the axe, deadly sulfur fumes from open smelting kilns, forest fires, cloudburst rains upon unprotected slopes, alternate freezing and thawing, and grazing have all contributed to its devastation." Richard Wood, an erosion specialist with the TVA, pointed to another unique factor of Ducktown. He argued that "frequent fires and unrestricted grazing have always been detrimental to natural re-vegetation of the Copper Basin. The two go hand-in-hand." Livestock owners were attracted to Ducktown's open, cutover landscape and the sedge grass growing in the outer zones. It was, until 1946, the only pasturage in the area still governed by the law of open range, whereby anyone could graze livestock on unfenced land owned by others. Area livestock owners trucked in cattle to graze during the warmer months. In winter, herders encouraged new spring growth by setting fire to the sedge after it became tough and course at the end of the growing season. Burning encouraged the growth of tender new grass but at the cost of declining soil fertility and the suppression of new tree growth. Overall, grazing and burning combined to expand the zones of heavy erosion for decades after the copper companies further reduced sulfur dioxide emissions by installing acid condensation units.[53]

M.-L. Quinn pointed to even more factors in a 1991 article regarding the environmental susceptibility of the basin. The idea was that local circumstances can render an activity more harmful in one place than it might be if performed in another. At Ducktown, the list of factors began with reliance on sulfide ores and the topography of the mountain bowl that trapped sulfur smoke. The area's extreme isolation left it totally dependent on local forests for charcoal and construction timber until the arrival of the railroad in 1890. Weather was significant. Frequent light winds and dead calms created temperature inversions that held smoke close to the ground. Abundant rain and persistent mountain fogs hastened precipitation of acid onto vegetation. Summer cloudbursts loosened top soil and swept it away from naked slopes. Local soil was shallow and flaky owing to the large amounts of pyrites within it. When exposed to moisture, it oxidized to create "ferric sulfate and sulfuric acid, forming acid sulfate soils" that "can impede or even prevent vegetation growth." The natural chemistry of pyritic soils was hastened by the massive load of sulfuric acid precipitated by smelter smoke. As a result, bare soil within the basin was especially prone to erosion. The effect was compounded by the steep slopes of most of the area landscape.[54]

The impact of these factors can be appreciated by comparison to the Lake Superior copper district on Michigan's Keweenaw Peninsula. Miners there extracted eleven billion pounds of copper from the 1840s to the 1960s. They left their mark on the landscape by digging up hillsides, clear-cutting forests, and filling in harbors with silt, but without creating another Ducktown Desert. Lake Superior ore contained little sulfur, and so the amount of sulfur dioxide injected into the air from local smelting was much less than in Ducktown. Frequent weather fronts and constant lake winds blowing over lower terrain kept the air much fresher. Together, local factors limited the amount of barren wastelands and allowed forests to eventually recover with little help from humankind.[55]

Dora Galloway, the daughter of a Ducktown miner, added a personal perspective on the desert in a charming memoir she wrote in 1964 for her granddaughter, Charlotte. Most of the Ducktown population came from the mountains of Tennessee, North Carolina, and Georgia, so her recollections spoke for many. She was born on a Christmas Day, "unannounced, unregistered, and unattended by a physician" in a one-room log cabin in the mountains of western North Carolina. She, her parents, and two brothers then moved fifty miles from their heavily wooded homestead to Ducktown in 1906, when it was already a place of smoke and desolation. "After leaving the mountains with the pure pine-scented air, it must have been a severe trial of adjustment for my parents to settle in this place . . . filled with the foul smell of sulfur dioxide." If they missed the woods, they also considered the move a release from sharecropping on a mountain farm to a place where "a man can work for a wage."[56]

The family rented a company house in Cole Town, one of the many villages for mining families located within easy walking distance to various mining facilities. Their house, home to her parents and nine children, was made of unpainted rough lumber and wood shingles that provided imperfect protection against cold, wind, and rain. Construction was so loose that snow blew through the cracks onto bedcovers and "the cat could come and go at will under the kitchen door." Some of the outside wall was missing because "planks were sometimes yanked off to be used as firewood in the cook stove." The lack of firewood was quite a change for a family that once had limitless supplies of timber outside its old mountain cabin.[57]

Though softened by time and her gentle spirit, Dora's account was one of a hard life, lived by a proud people in a strange land, a place different from anywhere else in the Appalachians. She could see trees on the rim of the surrounding mountains, but nearby, on a barren hill across the creek from her home, was a more arresting sight:

> On the opposite hill ... stood a living monument of the past ... the oak tree. It was tall and straight. The trunk was too large for a boy to climb, and the lower branches too high for them to reach. Despite its badly exposed roots, in due season year after year, it put forth those beautiful green leaves. To my knowledge, this is the only tree within the entire Ducktown Basin to survive the onslaught of the copper industry, and its acorns never took root where they fell.

The memory of that tree remained with her always, a part of the mental landscape of her thirty years in the basin. At the far end of her life, while living in the much different environs of Lexington, Kentucky, she sketched a map to fix the memories for her granddaughter. The oak tree appears prominently in the center, in a larger scale than any other feature, as was fitting for the only tree in the only desert east of the Mississippi.[58]

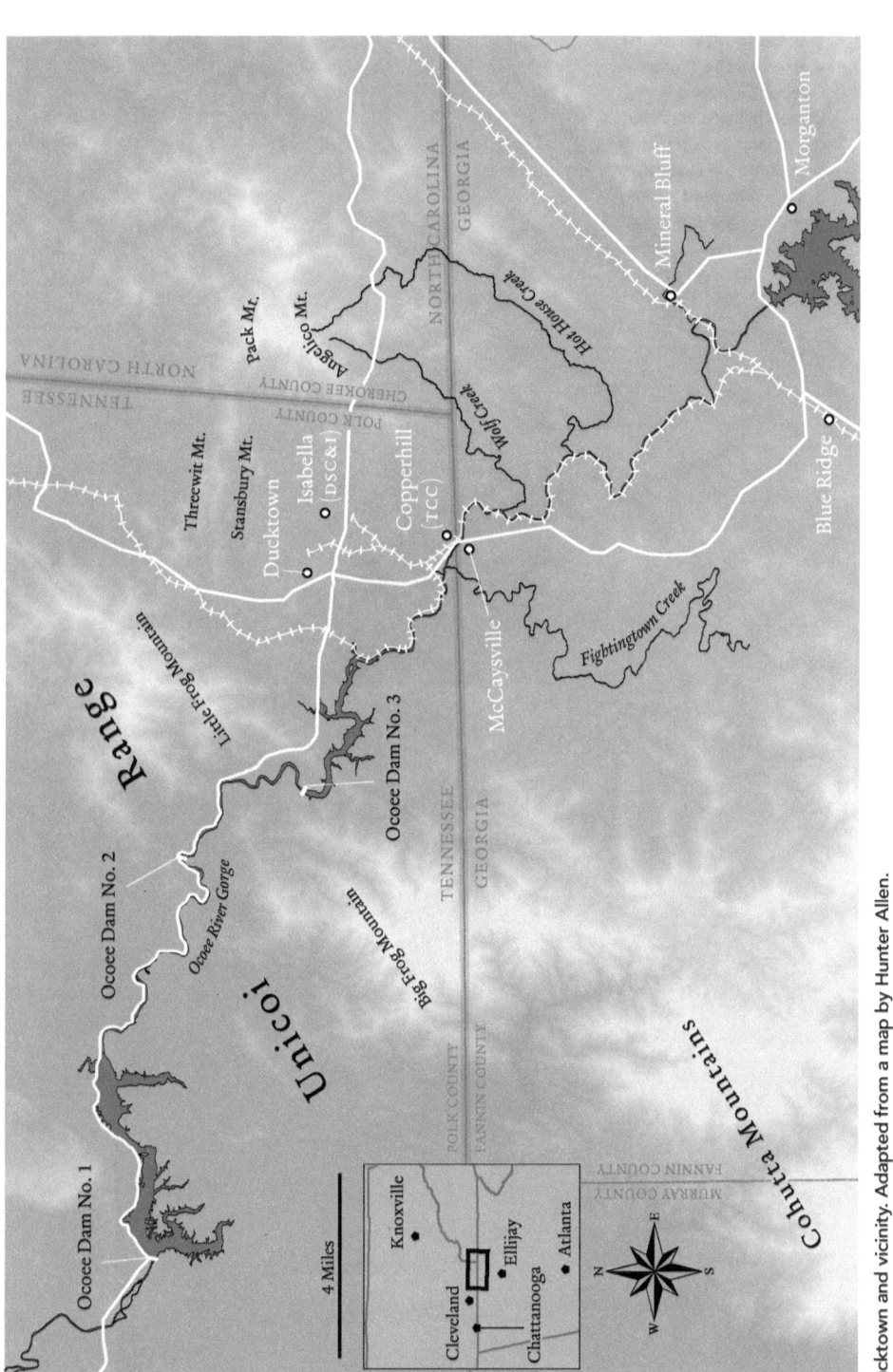

Ducktown and vicinity. Adapted from a map by Hunter Allen.

The 325-foot-tall chimney erected by the Tennessee Copper Company at Copperhill was intended to disperse smelter fumes outside of the Ducktown Basin but instead spread them over a wider area within the basin. Courtesy of the Ducktown Basin Museum.

VIEW OF THE TENNESSEE COPPER MINE.

This pre–Civil War lithograph of an early mine in the Ducktown Mining District depicts a small operation in a forested setting. It was likely romanticized from the beginning, but in any event, neither condition lasted long in an industry noted for its rapid growth and for its total reliance upon wood for fuel. Courtesy of the Tennessee State Library and Archives.

The Ducktown Sulphur, Copper & Iron Company plant at Isabella, Tennessee, circa 1920. Courtesy of the Ducktown Basin Museum.

(above and below)
The erosive power of sixty inches of annual rain and frequent thunderstorms carved extensive gullies into the denuded landscape of the Ducktown Basin. Courtesy of the Ducktown Basin Museum.

Miners roasted copper ore over log fires to expel the sulfur. The process lasted two or three months and allowed the clouds of sulfur dioxide smoke to enter the atmosphere. The shelters protected the burning ore heaps from the rain. Courtesy of the Ducktown Basin Museum.

An aerial view showing the status of reforestation work in the Copper Basin in 1950. The project was challenging because of the scale and severity of the denudation. Photograph by the Tennessee Department of Conservation. Courtesy of the Tennessee State Library and Archives.

Windy Ridge, a row of abandoned housing for miners in the Ducktown Basin. The barren landscape made it difficult for mining families to gather wood and water for their needs. Courtesy of the Ducktown Basin Museum.

The Tennessee Copper Company smelting and acid production complex at Copperhill, Tennessee. The facility, located on the Ocoee River, was only a few hundred yards from the Georgia border. Courtesy of the Tennessee State Library and Archives.

An aerial view showing the area of the worst damage at the core of the Ducktown Desert, 1950. Soil and rock washed from the hillsides and filled in the beds of tributary streams before entering the Ocoee River. Photograph by the Tennessee Department of Conservation. Courtesy of the Tennessee State Library and Archives.

Civilian Conservation Corps tree-planting crew, with the Tennessee Copper Company's Copperhill complex in the background. Photograph by Billy Glenn. Courtesy of the Tennessee Valley Authority.

WILL SHIPPEN, FORESTRY, AND GEORGIA'S SECOND SMOKE SUIT, 1905–1907

Georgia's attorney general, John C. Hart, negotiated the 1904 settlement agreement with the Tennessee Copper Company (TCC) and the Ducktown Sulphur, Copper & Iron Company (DSC&I) in the belief that adoption of the new pyritic method of smelting sulfide copper ore would end the damage to vegetation caused by smoke from the old method of open heap roasting. Events soon proved otherwise: smelter smoke generated by the new method was just as thick and toxic to budding fruit and early ears of corn as before. Because the new method was much faster than the old, the copper companies could now process much larger quantities of ore, and more sulfur fumes entered the atmosphere. Conditions deteriorated still further in 1906 after completion of a giant 325-foot smokestack for the Tennessee Copper Company at its Copperhill complex. Farmer A. B. Dickey complained that "the smoke we have now is a great deal worse than the smoke was from the roast heaps." At the height of the growing season, "there come up a smoke over on the 18th of July ... it just bit everything we have growing on the farm and from all appearances the timber is just about finished."[1]

A sheaf of similar complaints reached Hart's desk in Atlanta, and he knew that legislators from the mountain counties were receiving more of the same. The recent settlement was beginning to shrivel and curl like the smoke-burnt leaves of Ducktown apple orchards. Hart wrote TCC's lawyers on August 4, 1904, to express his mounting fears: "I feel much concerned about this on account of the sufferers in Georgia as well as the disappointment which your own people must feel if the new process has proved a failure." It went deeper than that. He had pushed for the settlement, and as an elected official, he was politically exposed to angry voters. It was only a matter of time before they goaded the General Assembly to force renewal of the state's United States Supreme Court case. With that in mind, he continued, "This letter is personal ... I am writing confidentially to ask of you to investigate and see if the injury is the result of failure in the new process or of negligence in its use."[2]

Hart needed allies and found them in an unexpected letter sent from the

Swannanoa Hotel in Asheville, North Carolina, on the letterhead of the U.S. Department of Agriculture, Bureau of Forestry. The author, a bureau forester named Harold Day Foster, wrote of his meeting with W. H. (Will) Shippen, a prominent timber man from Gilmer County, Georgia. Shippen spoke about smoke damage in the Ducktown area and urged Foster to make an investigation and to contact the attorney general. Foster did both. He made what he called "a hasty survey" and now wrote Hart to volunteer his services. The attorney general grasped at the offer, explaining that "I should be very much pleased to have a report by a competent and disinterested person," and to emphasize the point, he added, "It is part of your duty to do that."[3]

The initiative by Foster and Shippen to Hart signaled the remarkable convergence of copper mining, industrial logging, and forest conservation in turn-of-the-century Ducktown. Will Shippen and his brother, Frank, were in the vanguard of the highly capitalized timber operations that penetrated the Southern Appalachians in the 1890s. Foster was a member of a federal team sent to assess the health of southern mountain forests and to evaluate the related problems of deforestation and erosion so dramatically demonstrated by the creation of the Ducktown Desert. As they worked in the mountains surrounding the basin, loggers and foresters found common cause in their belief that sulfur smoke threatened an important natural resource. The Shippens and their competitors in the logging industry filed a number of major lawsuits against the two great copper companies regarding damage to their timber holdings. Their entry into the smoke litigation added economic and political power to a struggle that had been borne for a decade by mountain farmers at unequal odds. For their part, researchers from the Bureau of Forestry and other government agencies provided the badly needed scientific footing to a cause that had rested primarily upon the anecdotal evidence of smoked-out highlanders. Together, loggers and foresters shaped the course of Georgia's anticipated return to the Supreme Court.

Will and Frank Shippen were born and bred to the rough-and-tumble timber industry, but their ancestors moved in the highest circles of colonial Philadelphia society. Their forbearers included prominent merchants, physicians, judges, and politicians. During the Revolution, one served in the Continental Congress; another was chief physician to Washington's army. There was also a notorious black sheep, Peggy Shippen, who married Benedict Arnold and followed him to England when he turned traitor. Will and Frank's branch of the family eventually moved to western Pennsylvania to pursue logging in the Alleghany Mountains. There, the brothers cut four hundred million feet of timber, and when that was gone, moved to Kentucky to operate sawmills and a retail lumber business. The retail trade did not suit them, so they surveyed timber across the continent from the Deep South to the Pacific Northwest and from Canada to Mexico to determine where to resume commercial logging.[4]

They found, as did other loggers, that the Southern Appalachians presented attractive opportunities. The mountains contained extensive stands of mixed species suitable for a variety of commercial uses. Carl Alwin Schenck, a German forester who would play a major role in Ducktown timber litigation, described them in his 1912 technical manual, *Logging and Lumbering or Forest Utilization: A Textbook for Forest Schools*. Magnificent yellow poplars (also known as tulip poplars) grew straight as columns that could reach two hundred feet high and grow to a dozen feet in diameter. The poplars' smooth-grained wood took paint well and was easily shaped by woodworking machines, making it useful as side panels for railroad cars and wagons, as moldings for doors and windows, and as siding for houses. The strength of red and white oak was ideal for barrels, hardwood flooring, mine timbers, and the wheels and undercarriages of wagons. The chestnut tree provided tannin for curing leather, and its durable wood was excellent for outdoor use as fencing and railroad ties. Another forest giant, the eastern white pine, was once used for masts during the era of wooden ships. Commercial loggers put it to less romantic uses as construction lumber, box parts, and excelsior shavings used to cushion freight in the days before Styrofoam. Highly figured woods such as walnut, cherry, maple, and red oak supplied the needs of the South's furniture and paneling industries.[5]

The letterhead of the Shippen Bros. Lumber Co. listed several species as featured products. The firm dealt in poplar, white pine, and oak, and listed specialties in "oak flooring, yellow poplar bevel siding and moldings." There was also the curious term "box shooks" on the letterhead. Shooks were precut parts for wooden boxes and crates, sold in disassembled sets to shippers. The corrugated cardboard box, invented by Robert Gair in 1890, did not come into widespread use until well into the twentieth century when developments in the pulpwood industry made it economically attractive. Before then, the nation's goods moved in wooden boxes, crates, and barrels. Demand for wooden containers was so great that the box and container industry was the second greatest user of American lumber until World War II. As long as that demand remained, lumbermen hewed the giant white pines and tulip poplars of the southern mountains into box slats to move everything from canned goods to pianos.[6]

The abundant and varied timber was one draw to the Southern Appalachians. The low price of Georgia mountain forestland was another. The brothers arrived in 1895 with cash in their pockets at a time when land prices were low because of the Panic of 1893. They bought their first tract of 40,000 acres in fee simple, plus timber rights for another 20,000 acres, with a sawmill included, for a total of only $65,000, or about a dollar per acre. Land prices increased with demand and the return of better times but even then remained a bargain. When the brothers paid $4.00 per acre for a 30,000-acre tract in 1905, it quadrupled in value in less than a decade. Over time, they acquired 128,000 acres of timber in North Georgia.[7]

Logging was a part of mountain life all the way back to Cherokee times, as Indian farmers and then white settlers cleared the bottomlands and lower slopes for their farms and used the timber for their cabins, fences, implements, and firewood. Iron and copper miners consumed vast areas of timber in the form of charcoal to fuel smelters. Fifty square miles of Ducktown forest ended up as charcoal for the copper industry before the new railroad began to haul cheap coal and coke to the mines in 1890. Yet despite the demands of farms and local industry, the formidable upper slopes resisted wide-scale logging until expensive steam technology made it possible toward the end of the nineteenth century.

Any strong person with a good ax and a crosscut saw could cut down a tree. The challenge was getting the timber out of the mountains to the mill, sawing it into marketable products, and then transporting the results to distant markets. Schenck's manual showed how it was done, with great attention to engineering principles and unit costs. Yokes of oxen dragged logs in some places. On steeper slopes, gravity sent timber down chutes and flumes. A small tumbling mountain stream could be made to carry big logs with the use of a temporary splash dam. The dam held back the water until workers blasted it apart with dynamite to allow the resulting flood to carry logs pell-mell down the mountainside. Down in the hollows, specially designed steam locomotives, such as the Shay and Climax engines, used geared drive wheels and tracks to climb slopes too steep and curving for ordinary engines. The larger timber operators laid arrays of narrow-gauge track up every cove and hollow to bring logs down to the mills. Steam-powered derricks lifted huge logs onto the rail cars. Steam-powered sawmills and wood-shaping machinery hastened timber processing. Overall, steam power greatly increased the economies of scale in every aspect of logging except for the initial felling of the trees—that would await the invention of a practical gasoline chain saw after World War I.[8]

Steam also carried processed lumber to market. The Shippens enjoyed the advantage of two major railroads on either side of the mountains of Fannin, Gilmer, and Polk counties. On the west side, the Louisville & Nashville ran up the Tennessee Valley from Chattanooga to Knoxville and points north. On the east, the Atlanta, Knoxville & Northern Railroad (formerly the Marietta & North Georgia Railroad) ascended up and over the Blue Ridge via the Ellijay Valley. Though built primarily to serve the Ducktown copper mines, the railroad's promoters pointed to timber as one of the many resources it would carry; "as for hardwood timber," they emphasized, "there is no end to it all along the line, including almost every variety" and "the best white oak timber in the world." The Shippens had a practical measure for the importance of rail transport: timberland increased in value in relation to its proximity to the track.[9]

The high costs of mechanized commercial logging made the logging companies quick to view smoke damage from the Ducktown copper smelters as a major threat to their businesses. Their motivations differed from the first wave of smoke suitors, the small-scale mountain farmers who saw themselves as fighting to preserve their livelihoods, their quality of life, and their ability to remain on the farm. The timber barons sued to protect capital investment and profits. They filed their first smoke suits around 1902, about seven years after the first of the claims from local farmers. The delay reflected the way that zones of smoke damage advanced yearly from the roast yards and blast furnaces on the floor of the Ducktown Basin. The initial smoke suitors held farms near the mining operations where severe damage first occurred. The timber companies had no reason to sue until the advancing impact of sulfur dioxide fumes finally reached their timber lots in the encircling mountains.[10]

Some of the large timber concerns belonged to absentee owners who delegated operations and litigation to their managers and attorneys. George Peabody Wetmore was a New England blueblood educated at Yale and Columbia. He served two terms as Rhode Island's governor and was in the U.S. Senate during the smoke suits. Rosine Parmentier was a society figure in Brooklyn, New York. Before the Civil War, she bought fifty thousand acres along Sylco Creek in the Ocoee Gorge to create an experimental colony for European Catholic immigrants. She called it Vineland in the hope that her wine-drinking colonists would establish profitable viticulture in a region better known for corn liquor. The venture failed, but Ms. Parmentier retained most of the land. At the end of the century, she divided her interests between the logging business in the Tennessee forests and an array of New York Catholic charities.[11]

The Shippen brothers differed from Wetmore, Parmentier, and their class by living and working amid their timber holdings in the mountains of Gilmer County, Georgia. Their homes and their largest lumber mill were in the county seat of Ellijay, then a small town of less than a thousand souls. Though they were outsiders from Pennsylvania, they came to know most of the local citizens by virtue of being the largest employer in the county. As hands-on owners of a large timber company, Will and Frank demonstrated the adaptability needed to function in radically different settings, from a mountaineer's cabin to the governor's office.[12]

Andrew Gennett, a contemporary and rival of the Shippens, explained how this was done in his lively memoir, *Sound Wormy* (the title refers to a grade of chestnut timber). Gennett grew up in Nashville, Tennessee, and by turns eventually developed a major timber company in western North Carolina. He often stayed in mountain cabins and ate meals of corn bread and pork when cruising timber. He lived and worked among rough men in the logging camps. Yet the scale of his business required frequent travel to New York and other commercial cities, where he enjoyed fine dining and elegant hotels while negotiating with bankers and industrialists. He was a risk taker who invested heavily in a cutthroat commodity business closely tied to the rise and fall of the nation's business cycles. He litigated constantly over land titles and broken contracts. (Gennett recounted with amusement about how he bested the Shippens in two Georgia land title suits.) He was accustomed to wielding influence and had no hesitation about meeting with the secretary of agriculture in Washington if it served his business interests.[13]

Will Shippen employed the same wide-ranging social and political skills of the hands-on timber baron through his constant activity in the Ducktown smoke litigation. A man of influence in his own right, he pressed political levers to gain appointment to the 1905 Georgia legislative commission charged to investigate Ducktown smoke damage. He then maintained a steady correspondence with the many governors and attorneys general holding office during the long history of the case. He used his familiarity with mountaineers, strengthened by his standing as their frequent employer, to organize their opposition to the copper companies. Whenever the copper companies gathered miners and local merchants to petition against a smoke injunction, Shippen rallied the farmers and loggers in Ellijay and Blue Ridge to sign pro-injunction petitions drafted by his own hand. James G. Parks wrote his fellow copper lawyer, Howard Cornick, about one of those petitions: "It reads to me very like one of Mr. Shippen's lurid literary productions. If it is, it must be plain that he is trying to force us to a settlement by bringing legislative and official pressure to bear." Parks continued by describing Shippen as "a wiry, untiring, and resourceful chap."[14]

The timber man made no secret why he was so untiring: he had a huge financial stake in Ducktown smoke litigation. The state sought injunctive relief, not damages, so a Georgia victory would not directly result in payment to the Shippens and other claimants, but the ongoing threat of a Supreme Court injunction forced TCC and DSC&I to take the claims of the timber barons more seriously. The converse was also true; a defeat in the Georgia case would weaken all the suits for damages. In a blunt 1913 letter to Attorney General Warren Grice, one of Hart's successors, Shippen pointed out that "my corporation is vitally interested in the outcome of the state's suit, since our own case against the D.S.C.&I Company either succeeds or fails with yours."[15]

An even more cynical reason for the brothers' ardent support came from the Ducktown Company's general manager, W. H. Freeland, who claimed that "one of the Shippens is said to have remarked that if we are shut down, they would get labor for fifty cents per day" from among the throng of unemployed miners. Freeland sought an end to Will's multifaceted role as litigant, ringleader, rural agitator, and political gadfly by instructing Parks to offer a $10,000 settlement on terms that "will make it impossible for him at any future time to annoy us in any way." Shippen could be enticed to settle, but not at that price. His firm eventually settled with the rival Tennessee Copper Company in 1908 for $50,000 plus a large additional sum paid directly to the brothers, as would be revealed later. The refusal of DSC&I to make a commensurate offer meant that Will Shippen would remain a "wiry, untiring, and resourceful" opponent of the company—and an ally of Georgia's attorneys general for the remainder of the litigation.[16]

The January 28, 1905, issue of the *Atlanta Constitution*, like other newspapers of the day, often assembled short unrelated articles in a grab-bag column, such as what appeared on page six under the headline "Condemned to Gallows Hilburn Seeks Liberty—Interesting Appeal Made to Prison Commission—State Chemist Makes Report—Judge Hart's Narrow Escape—State Capitol Gossip." The first of the two paragraphs touching on Attorney General Hart described his narrow escape from death in the wreck of the No. 39 train outside of Danville, Virginia. The judge was on the return trip to Atlanta after arguing a business tax case before the Supreme Court in Washington. It was early in the morning, and insomnia kept him awake in his sleeping berth. He tossed and turned, ending up with his feet pointing toward the engine.

That is what saved him when his train collided head-on with another: his feet and legs absorbed the blow against the forward wall of his berth instead of his skull and neck.[17]

Railroad travelers at the time endured a level of risk that would be intolerable for the modern commercial air passenger. Danville gained a special notoriety because of a wreck that happened there a year and a half earlier when the No. 97 train jumped the tracks while trying to make up for lost time on a mail run. The incident inspired a famous train ballad, "The Wreck of the Old 97," with verses such as,

> It's a mighty rough road from Lynchburg to Danville,
> And the lie was a three-mile grade,
> It was on that grade that he lost his air brakes,
> And you see what a jump that she made.

The song, the first American record to sell a million copies, became a country and bluegrass standard recorded by Woody Guthrie, Johnny Cash, the duo Flatt and Scruggs, and many others. Nobody wrote a ballad about Hart's incident, but the wreck of his No. 39 train and that of the Old 97 gave him reason to dread the many future train trips to Washington he would have to make when the Ducktown case made its inevitable return to the Supreme Court.[18]

It was news of that case that led to the fourth item in the grab-bag column: "To Investigate Ducktown." The segment noted that Attorney General Hart had received conflicting reports about smoke damage and that he wrote the Bureau of Forestry "asking that an expert be sent ... to investigate and report just what damage, if any, is now being done." It was a humble announcement to what would prove a major, even controlling, theme of the case, and to the few who knew the full story, it reflected the initiative of Will Shippen and Harold Day Foster.[19]

The Bureau of Forestry honored Hart's request by sending Alfred Chittenden, an assistant forest inspector. His report, prepared after the transition to pyritic smelting, noted that areas that "have hitherto remained untouched are now being damaged by the fumes" spread over a much wider area by tall chimneys. "The sulfur fumes can be plainly smelt at a distance of over fifteen miles if the wind is in the right direction." Abnormally high tree death was observed twenty miles from the smelters, and "timberland owners within thirty miles are deeply concerned."[20]

Chittenden observed that the extent of tree death from sulfur fumes varied among the species. Conifers such as the white pine, yellow pine, and hemlock suffered because their permanent foliage absorbed sulfur fumes throughout the year. The great height of the white pine made it vulnerable because, "extending above the tops of other trees, it is more exposed." Yellow poplar was the most resistant, though whether this was because of the tree's inherent qualities or its preferred environment within sheltering coves he could not determine. Among the hardwoods, hickory and chestnut oak fared worst because they grew on the upper slopes and ridges most exposed to smoke-bearing winds.[21]

Variations of species and habitat sometimes, though, had little significance; "from the absolute lack of vegetation in the immediate vicinity of Ducktown," Chittenden judged "that no species of forest tree can resist the smoke if long or constantly exposed to the full fumes." Even at a distance of twenty miles, "destruction of forest growth seems certain" with long exposure to the fumes. Chittenden then addressed Attorney General Hart's most important question by declaring: "The old method of open roast heaps has already destroyed all vegetation in the immediate neighborhood; the new method will extend the damage." Chittenden's report included statements from several local lumbermen, including Will Shippen. The scientist included a long quotation from a December 22, 1904, letter by the logger: "The situation . . . is appalling in the extreme, and certainly merits the action of both Federal and State aid in its suppression." Shippen added that "the present process is far more destructive than the old one ever was," so much so, that "I have known whole districts here to be blighted in a single night."[22]

Though somewhat flawed because of overt influence of the timber barons, the Chittenden report was only a part of a much larger body of conservationist research by the federal government on threats to the forests of the Southern Appalachian Mountains. The American forest conservation movement preceded Georgia's smoke suit, reached its formative peak during the litigation, and ultimately provided the winning rationale for the case. The sources of the movement extended back to the nation's colonial past but coalesced only after the Civil War, as shrinking forests led to fears of an impending timber famine. Farmers cleared more than 150 million acres from the eastern forests for cropland and pasture before 1860. Loggers fed the country's insatiable demand for wood products; by 1909 the industry produced 44.5 billion board feet of sawn lumber and processed vast numbers of other trees for use as utility poles and railroad ties. Carl Schurz, a former secretary of the interior, warned in 1889, "If the present destruction of forests goes on for twenty-five years longer, the United States will be as completely stripped of their forests as Asia Minor is today" (thus referring to George Perkins Marsh and his book *Man and Nature* concerning the effects of deforestation in the

Mediterranean countries). At the 1908 White House Conference of Governors, Theodore Roosevelt urged, "We are over the verge of a timber famine in this country" and insisted that "it is unpardonable for the Nation or the States to permit any further cutting of our timber save in accordance with a system which will provide that the next generation shall see the timber increased instead of diminished."[23]

The American Association for the Advancement of Science (AAAS), the newly formed American Forestry Association, lumbermen, bureaucrats, sportsmen, and nature lovers all urged federal protection of woodlands. Their efforts culminated in the creation of a system of national forests administered by what became the U.S. Forest Service. (The U.S. Forest Service went by several earlier names, including Division of Forestry and Bureau of Forestry.) It is the U.S. Forest Service that bears upon the story of Ducktown smoke.[24]

Gifford Pinchot became head of the Forest Service in 1898. He was a New Englander born to wealth, went to college at Yale, and then trained at the French school of forestry in Nancy. He began his career by serving as the forester on George Vanderbilt's Biltmore Estate, outside of Asheville, North Carolina, where he conducted the first systematic program of forest management in America. The experience gave him firsthand knowledge of the much-abused woodlands of the Southern Appalachian Mountains. The knowledge served him well as he took up the cause to create federal forest reserves in the East that would match the new national forests in the West. Creation of federal forests reserves in the West was easy from a legal perspective. The federal government, as owner of the vast public domain in that half of the country, simply set aside certain forest tracts to prevent their transfer to private owners. In the East, most forestland belonged to private owners, so creation of eastern forest reserves required compensation to the owners, in keeping with the Fifth Amendment prohibition against the taking of private property for public use without just compensation. This, in turn, required enabling legislation and appropriations from Congress.[25]

The idea of a forest reserve in the Southern Appalachians has been traced back to 1885 when a Boston physician, Henry O. Marcy, presented a paper before the American Academy of Medicine lauding the healthful benefits of the North Carolina mountains. After gaining the backing of Pinchot and the great arborist Charles S. Sargent, author of the fourteen-volume *Silva of North America* (1891–1902), the movement developed into an organized campaign at an 1899 meeting in western North Carolina hosted by the Asheville Board of Trade. The meeting led to a congressional campaign involving an array of national groups including the Appalachian Mountain Club of New England, the Appalachian National Park Association of the South Atlantic States, the American Forestry Association, the AAAS, and numerous commercial groups from Atlanta, Knoxville, and elsewhere. Georgia's legislature gave support by passing six acts and resolutions between 1900 and 1918 to that end.[26]

Congress responded in 1901 with a $5,000 appropriation for a study by Secretary of Agriculture James Wilson "to investigate the forest conditions in the Southern Appalachian Mountain Region of western North Carolina and adjacent states." Ducktown's location at the common border of western North Carolina, Tennessee, and Georgia was well within the scope of the study area. It is this effort and subsequent studies that brought Harold Day Foster, Alfred Chittenden, and their fellow government experts to the basin. The completed study, with a cover letter from President Roosevelt, was issued in 1901 and published in 1902 under the clunky title, *Message from the President of the United States Transmitting a Report of the Secretary of Agriculture in Relation to the Forests, Rivers, and Mountains of the Southern Appalachian Region*. Wilson's report contained almost two hundred heavily illustrated pages, based primarily on the fieldwork of H. B. Ayres and W. W. Ashe, showing the terrible effects of erosion from denuded mountain slopes caused by industrial logging and destructive farming practices.[27]

The report made the case that destruction of the South's mountain forests caused by poor logging and farming practices created an erosion problem that was especially destructive in the southern mountains because of high rainfall. Secretary Wilson wrote that "upon these mountains descends the heaviest rainfall of the United States, except for the North Pacific Coast," and the rain often fell with "extreme violence, as much as 8 inches . . . in eleven hours" and "31 inches in a month." Falling on denuded slopes, it caused soil to wash "in enormous volume into the streams, to bury such of the fertile lowlands as are not eroded by the floods, to obstruct the river, and to fill up the harbors." Photographs showed gullied slopes upstream and once fertile bottomlands covered with boulders and dunes of deposited silt, even on lands two hundred miles downstream from the mountains. The U.S. Geological Survey (USGS) provided the statistical foundation to support the visual impact of the photographs. A significant portion of the Wilson Report described the hydrologic study conducted by the USGS on every major river flowing from the Southern Appalachians and on more than a thousand tributaries. The report concluded that, "if the forests are wantonly cut, . . . all of the soil and vegetation will be washed from the mountain sides," and "nothing will remain but the bare rock."[28]

Secretary Wilson and President Roosevelt insisted that only healthy forests could prevent the evils of erosion. Given the market realities of logging, and the population pressure that led to more new farms on steep slopes, only a government forest reserve could limit future harm and work to repair past damage. Proper forest management was essential to control stream flow and to provide for sustainable timber harvests. Without prompt action, they warned, "within less than a decade every mountain cove will have been invaded and robbed of its finest timber" and the last "remnants of these grand primeval woods will have been destroyed." Their cause eventually prevailed in 1911 with the passage of the Weeks Act, authorizing the federal government to purchase private land to create a forest reserve in the Southern Appalachians. Some of the earliest purchases would occur in 1920 from ruined forestland around the Ducktown Basin in what is now the Cherokee National Forest.[29]

Secretary Wilson's 1901 report did not mention Ducktown or sulfur dioxide pollution, but the Forest Service was committed to a thorough study of the area because it intended to create a forest preserve in the area. The service built upon Alfred Chittenden's 1905 study by sending Forester J. S. Holmes and Assistant Forest Inspector A. B. Patterson to make a detailed survey and map the extent of smoke damage in 1905 and 1906. They determined that it extended through the gaps of the Blue Ridge Mountains on the southern rim of the basin into the Ellijay and Coosawatee river valleys. To the north, the damage spread beyond Stansbury Mountain almost to the Hiwassee River. Another researcher, Charles Keffer, formerly assistant chief under Pinchot and then professor of forestry at the University of Tennessee, made yet another field study and found damage on the Cartecay River, thirty miles south of Ducktown.[30]

The Forest Service also enlisted the support of J. K. Haywood, a chemist with the U.S. Department of Agriculture's Bureau of Chemistry, to analyze the impact of sulfur dioxide upon local vegetation. Haywood was already an expert in the field, having studied the effects of sulfur dioxide fumes in the mountains of northern California before considering the problem in Ducktown. His work in the West involved smelter fumes from the Mountain Copper Company located near the newly created Shasta National Forest. His fieldwork there, coupled with laboratory studies back in the East, led to the publication of his report, *Injury to Vegetation by Smelter Fumes* (1905). Haywood's report and his later field studies at Ducktown became the scientific foundation of Georgia's injunction case against smelter fumes.[31]

Haywood's report first addressed causation. He observed that "for each pound of sulfur burned" in furnace smelting, "two pounds of sulfur dioxide are formed and given off into the atmosphere." Sulfur dioxide then combined with oxygen to form a transitory substance, sulfur trioxide. That substance then combined with moisture in the air or on leaves "to form the highly corrosive compound sulfuric acid, which in its turn acts upon the delicate foliage." He next considered the concentrations and periods of exposure that caused damage to vegetation. German scientists began that line of research in the 1860s and 1870s as they studied damage to forests downwind from the huge mining and steelmaking operations in the Ruhr district. Their experiments established that significant injury occurred from short exposure at a level of one part SO_2 to 100,000 parts of air, and from longer exposure at a reduced level of 1:1,000,000. Haywood conducted his own experiments using a sealed glass cabinet, roughly the shape of a telephone booth, where he exposed plants to SO_2 gas for measured periods. The tests resulted in visible damage to foliage. Chemical analysis showed a marked increase in sulfur trioxide within the plant tissues. He then confirmed those results in the field by testing plant samples gathered near the Mountain Copper smelters.[32]

Haywood supplemented his earlier work with two visits to Ducktown in 1905 and again in 1906. There, he performed calculations to determine that in a single day the two companies generated so much sulfur dioxide that "the atmosphere for a thickness of 100 feet would be contaminated for 520 square miles," equivalent to the area within a circle with a radius of 13 miles in every direction from the smelters. He expected the zone of damage to expand with increased production and the erection of the taller chimneys. Moreover, "with each zone of timber killed ... the carrying power of said fumes is increased, inasmuch as the destruction of timber and vegetation permits the fumes to travel further before being absorbed."[33]

Attorney General John C. Hart and Governor Joseph M. Terrell recognized the importance of the Forest Service's work to their suit against the copper companies. Hart publicly thanked Secretary Wilson, Gifford Pinchot, and the Bureau of Forestry in his 1906 annual report "for aid given the State." Governor Terrell attended the 1906 National Forestry Congress in Charlotte, North Carolina, "to consult with some of the experts on forestry ... regarding the situation in the counties of North Georgia." Hart sent his special counsel, Ligon Johnson, to Washington with instructions to comb the Bureau of Forestry for additional supportive studies, reports, and government bulletins. This was more than a matter of case strategy for the young lawyer; he became a true believer and organizer in forest conservation. Soon after the current round of litigation, Johnson helped to organize the Appalachian National Forest Association and served as its first president. He also served as the southern regional director for the American Forestry Association. He used his positions to lobby for legislation to create the Appalachian Forest Reserve and received the personal thanks of Theodore Roosevelt for his efforts. Johnson summarized his conservation philosophy in an impassioned half-page article published in 1907 in the *Atlanta Constitution* under the headline, "Vast Importance to South of Forest Reserves."[34]

Hart and Johnson recognized that Secretary Wilson's regional report and the many supplemental studies specific to the Ducktown Basin went far toward repairing the greatest weakness in the state's lawsuit before the U.S. Supreme Court: the need to establish a state interest independent of the private claims of smoked-out Georgians so it could maintain its right to invoke the Supreme Court's original jurisdiction. The copper companies recognized this also and had already argued that the state had no interest of its own in the controversy and that it was "simply lending its name to prosecute the suit for and in behalf of a few of its citizens." Attorney General Hart and Ligon Johnson now had a credible response in the form of Secretary Wilson's report on the importance of southern Appalachian forests and in the many subsequent reports specific to conditions in Ducktown as prepared by Alfred Chittenden, J. K. Haywood, J. S. Holmes, A. B. Patterson, and other federal foresters and scientists. Their work, and that of their counterparts within the Georgia government, would enable the lawyers to now argue that the state was entitled to the grant of Supreme Court original jurisdiction to protect its forests and watersheds from the damaging effects of sulfur dioxide fumes in furtherance of the best principles of scientific conservation.[35]

Georgia's suit against the copper companies began its return to the U.S. Supreme Court when the General Assembly introduced another resolution on Ducktown in July 1905. Will Shippen wrote a column in the *Atlanta Constitution* to outline key points in support of passage. Speaking of acid, he noted that, "as every chemist knows, just as soon as dioxide of sulfur comes in contact with moisture sulfuric acid is the result." The case was no longer just about smoke. It was about a dangerous corrosive agent, "this sulfuric acid that does the terrible damage." Next was the phrase "forty miles," to capture the idea that the zone of vegetative death extended that far from the copper works, citing the U.S. Bureau of Forestry for authority. It was no longer just a case of damage in the immediate vicinity of the smelters.[36]

Talking points, favorable press coverage, and a midsession legislative junket to Ducktown accomplished the desired end. The resolution easily passed, and the legislature specifically named Shippen as a member of the new commission, giving him an official platform to continue his personal and public campaign against the copper companies. The 1905 commission confirmed, as expected, that conditions had worsened since the visit of the previous commission two years earlier and noted that the smoke would become even worse when the four new blast furnaces at the Tennessee Copper Company came on line. Local farmers "expressed themselves as being determined to abandon their further attempts to grow crops... until the nuisance is abated." The commissioners then fleshed out their report with frequent references to Secretary Wilson's report and J. K. Haywood's sulfur dioxide studies.[37]

The report appeared in the *Atlanta Constitution* the day after completion. Governor Terrell acted on it by ordering renewal of the suit in September. The Tennessee Copper Company responded by sending its president, J. Parke Channing, and counsel, Howard Cornick, to meet with Hart in Atlanta in the hope of forestalling renewal of the suit, but by then the smoke damage was too great and political pressures too powerful to permit delay. Hart then made another train journey to the Supreme Court in Washington, where he filed Georgia's second injunction suit, *Georgia v. Tennessee Copper Company*, on October 23, 1905. His earlier 1904 case against the copper companies lasted just a few weeks before reaching a tenuous settlement. Thirteen years would pass before the new 1905 case reached its effective end with a settlement between the state and DSC&I.[38]

Like prizefighters facing off for a repeat bout, each party in the second Supreme Court case knew the strengths, weaknesses, and strategy of the opponent. All of the arguments asserted in 1903 reappeared in 1905; beyond that, both parties had reference to a decade of smoke litigation in the Tennessee courts. A competent advocate must be able to wield both law and facts in court, but like a boxer with a stronger punch in one arm than the other, the lawyer often has a stronger position on the facts than on the law in a given case, or might experience the reverse situation on another case. Thus, the old courtroom adage: "When you have the facts on your side, argue the facts. When you have the law on your side, argue the law. When you have neither, holler."[39]

The state's lawyers elected to argue the facts. Hart and Johnson anchored their case on evidence from federal experts: the Wilson report, J. K. Haywood's toxicology study, U.S. Department of Agriculture bulletins on erosion, and the several field reports from federal and state foresters, entomologists, and other technical specialists. Government mapmakers determined the zones of smoke damage. Together, expert testimony by federal and state scientists allowed Hart and Johnson to move their factual case beyond the realm of anecdote and conjecture to the much stronger scientific rationale based on the findings of government experts employing the latest methods of conservation science.[40]

Hart and Johnson supplemented the government evidence with reports from Will Shippen, Frank Shippen, L. D. Rogers, and four other industrial loggers. The brothers took care to frame their affidavits in conservationist terms. Will stated, "We are practicing forestry, cutting out only the mature timber and letting the young trees grow." Frank added that "the forests are not destroyed by cutting of such timber but, on the contrary . . . grow much more rapidly after the old timber is cut away." Will asserted that the copper operations were a threat to good conservation because "our timber would reproduce itself faster than we could grow it" if left unharmed by the smoke. Frank added that "the sulfur fumes . . . not only destroy the mature timber, but kill the young as well, and make re-forestation impossible."[41]

The copper companies responded with testimony from their own foresters, notably Bernhard Eduard Fernow and Carl Alwin Schenck. Fernow was justly honored in American silviculture for his service as the longtime chief of the U.S. Division of Forestry and as founder of the school of forestry at Cornell University. He testified that loss of forest cover did not lead to soil runoff and impaired water flow "except in the steepest places." This was true in Europe where healthy grassy slopes covered the lower slopes of the Alps. Fernow then qualified his testimony by adding, "provided that the lower soil cover and underbrush is not also destroyed by fire or otherwise, or reproduction to replace the old growth is not prevented." Unfortunately, he failed to support his general observations with fieldwork at Ducktown to determine whether the present smoke conditions made healthy ground cover impossible there.[42]

Carl Schenck, another German forester, earned his doctorate in forestry at the University of Giessen and then served under England's leading forester, Sir Dietrich Brandis, before coming to America to succeed Gifford Pinchot as forester to the Biltmore Estate. There, he established the Biltmore Forestry School in 1898, now considered by many to be the "Cradle of American Forestry." The opinionated Schenck eventually fell out with the equally opinionated and vastly more wealthy George Vanderbilt, leading to his dismissal in 1909. He later returned to Germany, fought for the kaiser in World War I, and practiced forestry in his homeland for the remainder of his long life.[43]

Schenck kept abreast of the literature on the toxic effects of sulfur dioxide and set forth his thoughts on the subject in his 1909 work, *Forest Protection*, a compilation of the lectures he first delivered to his students at the Biltmore Forestry School. He agreed that sulfur dioxide was toxic to plants, especially in humid conditions, but cautioned that "a number of injurious influences (frost, heat, desiccation of soil, insects, [and] fungi) bring about, within the leaves and needles, identical or similar alterations." Death from sulfur fumes could be assumed only when there was "death visible to the naked eye," when there was "no other plausible cause of death," and when chemical analysis demonstrated its marked presence within the leaves. Unless those three tests were met, a forester was to assume that a tree died from some other cause.[44]

Dr. Schenck's 1909 lectures mirrored his 1906 testimony in the Ducktown case. He dismissed testimony about smoke damage by decrying the tendency among North Georgians to attribute all timber death to sulfur fumes. Instead, he and his three assistants determined that "in a large majority of cases the deterioration of the timber lands is due to carelessness on the side of the owner . . . to the axe and the saw; to gradual desiccation of the soil; to forest pasture; to forest fires; to insects; and to pathological diseases." These findings deserved some respect as the thinking of a gifted scientific forester, but they were contradicted by government experts.[45]

Schenck undercut the value of his evidence when he minimized the value of local timber. He thought little of it, writing that "the timber in the region under observation, in primeval times, must have been of a very poor description, perhaps of the poorest description anywhere in the Appalachian region." His views reflected the experience and expectations of a forester trained in the highly managed forests of Germany. American lumbermen and the U.S. Forest Service saw things differently. For them, the unmanaged forests of the Southern Appalachian Mountains, for all of their ills, remained one of the nation's most valuable remaining stands of timber.[46]

Georgia's next line of factual argument turned to the proof of the existence of a commercially viable method for condensing sulfur fumes into marketable sulfuric acid. If Hart and Johnson failed to prove the point, the copper companies could defeat an injunction with the argument that their methods represented the best state of the mining and smelting arts. J. Parke Channing, the president of the Tennessee Copper Company, perhaps unwittingly made the state's argument in an article he wrote for the June 12, 1905, edition of the *Engineering and Mining Journal*. Channing lauded the cost savings and increased yield in copper realized from pyritic smelting and then added that "we found that the gases from the furnace are so rich in SO_2

that we can make acid from them by the chamber process." It would "give us a by-product, which will be of great value in the South for the purpose of making fertilizers," while at the same time "reducing the volume of gases to the atmosphere." Best of all, the sale of acid to the fertilizer industry "will prove a source of considerable income." Channing later tried to back away from his bold statements by submitting an affidavit, along with another from the company treasurer, John H. Sussman, to the effect that sulfuric acid production from smelter smoke was still experimental. Sussman added that of all the copper smelting plants in North America, "not one of such works is producing commercial sulfuric acid."[47]

Channing and Sussman also warned there might not be a market for all the acid they could produce. Hart and Johnson responded with a stack of affidavits and reports from experts in the fertilizer industry, the Georgia state chemist, the commissioner of agriculture, and the United States Census. Together, they established the pressing need for sulfuric acid to convert the South's abundant supply of phosphate rock into useable super phosphate fertilizer for the region's worn-out cotton fields. Two Atlanta fertilizer manufacturers, D. B. Osborne of the Armour Fertilizer Works and H. A. Rogers with Swift Fertilizer, testified that they currently relied on sulfur pyrite rock imported at great cost from Spain to serve as the source of the needed acid. Both firms were eager to acquire low-cost sulfuric acid in carload lots from nearby Ducktown.[48]

The state's evidence on acid conversion, coupled with TCC's all-out effort to construct an acid plant at Copperhill, placed the Ducktown Company in an awkward legal position. The Tennessee Copper Company had effectively abandoned the state-of-the-art defense while the Ducktown firm persisted with an increasingly dubious claim to the Supreme Court that acid conversion was impossible. James G. Parks, the DSC&I lawyer, argued that the district's monosulfide ores "have never been successfully used on a commercial scale in the manufacture of sulfuric acid." While true, it did not explain why Tennessee Copper was busily constructing a plant to do just that with the same kind of ore. This was only one of many signs to come that the Tennessee Copper Company no longer marched in lockstep with the Ducktown firm in their defense against Georgia's lawsuit.[49]

The third prong of the state's factual case consisted of citizen statements. Hart and Johnson offered fifteen hundred affidavits from local farmers and residents setting forth their suffering under the clouds of sulfur smoke, followed by another set of more than two thousand statements about the worsening conditions in 1906. The testimony included endless repetitions of how the smoke burned and shriveled the leaves on fruit trees, vegetables, and field crops as if bitten by a killing frost. O. F. Chastine, a dealer in farm implements, spoke of how his wife's sweet peas and other flowers "growing upon fragile stalks are particularly susceptible to the fumes." His real concern was probably the fear that smoked-out farmers would no longer buy his equipment.[50]

The thirty-five hundred affidavits represented a considerable portion of the vox populi in the Ducktown Basin. The copper companies responded with more than six hundred witnesses representing another large segment of the population, the workers and merchants who wanted to keep the mines open to preserve their jobs and their trade. Company operatives conducted a mass meeting of supposed Georgia citizens at Harper's store in Fannin County. The resolution adopted there warned that an injunction, if granted, "will work a calamity" on the region and "there will be thousands of people thrown on starvation, without food and without shelter." Other affiants denied the damaging effects of smoke. F. M. Jones, a merchant at McCays (later renamed Copperhill), J. B. Witt, a local banker, and James Akin, a teamster, all said that vegetation was green and thriving close to the Tennessee Copper smelting complex thanks to the newly installed tall smokestack.[51]

The many statements from citizen witnesses for and against the injunction marked Georgia's smoke suit as a regional conflict that impacted nearly every one of the ten thousand people in the Ducktown Basin. The trouble for the Supreme Court was that the opposing voices tended to cancel each other out. Each side challenged the credibility of the affidavits and resolutions tendered by the other side. The state sent observers to the rally at Harper's store. The observers then matched the signatures on the resolution against the local tax digest and determined that "the greater number of said employees . . . who attended said mass-meetings are persona who have no permanent residence or citizenship anywhere, for the reason that they reside at no one place long enough to obtain either."[52]

The copper companies responded in kind. They offered affidavits from Georgians who recanted their pro-injunction affidavits on the basis that they did not know what they were signing. W. R. Early, presumably illiterate because he signed with an "x," said he did so "because if he did not sign it he would never be able to get damages" against the copper companies. Other affidavits pointed to the role of the Shippen brothers as instigators. G. G. Hyatt, an attorney for the Ducktown Company, described his attendance at a "smoke meeting" in Ellijay, the seat of Gilmer County, where "the only persons who were taking an active interest in the prosecution of the suit against the mining companies at Ducktown were the Shippen Brothers."[53]

The most embarrassing challenge to the state's citizen affidavits came from Martin H. Vogel, general counsel for the Tennessee Copper Company at its New York City headquarters. He left the company offices on Broadway to journey through North Georgia, which he described as a land where "the territory is wild and mountainous, the roads are rough and narrow, the soil is thin and poor," and "the mountaineers live in a crude fashion in the poorest sort of farmhouses miles from each other." In scenes that must have presented a combination of high legal purpose and low social comedy, the Manhattan corporate lawyer traveled the dirt roads of Fannin and Gilmer counties to interview farmers about their affidavits. Vogel discovered that one of the agitators obtaining affidavits was Ligon Johnson, the state's special counsel. Johnson had an unusual flyer posted in towns, hamlets, and country stores throughout North Georgia. The poster urged citizens to sign affidavits about their smoke damage by appearing before any one of the named local politicians, judges, postmasters, and justices of the peace who stood ready to help them. In bold uppercase letters, the flyer stated, "IF THESE CITIZENS WILL FURNISH THE TESTIMONY AS TO SAME . . . AN EARLY RELIEF MAY BE CONFIDENTLY HOPED FOR." Then in lowercase, he explained, "the state is acting for the benefit of her citizens." (This was a very dangerous assertion in light of the anticipated challenge to the state's claim for original jurisdiction.) He ended his cry for justice with more uppercase text: "HAVE YOU GIVEN YOUR TESTIMONY?" With Johnson's flyer in his satchel, Martin Vogel returned to Manhattan full of tales from the southern highlands to share with members of his Manhattan social set.[54]

With the war of affidavits being at best a tie, the overall case remained a contest between the state's strong position on the facts of smoke damage and its significance in terms of forest conservation versus the equally strong legal defenses of the copper companies. Howard Cornick, trial counsel for Tennessee Copper, and his counterpart, James G. Parks, for the Ducktown Company asserted four key legal defenses, strengthened by several recent decisions in other courts. The first defense was the attack on the state's claim to Supreme Court original jurisdiction. This was the gateway issue that, if granted, would end the case at the outset. If it failed, they had three other defenses aimed at preventing the grant of injunctive relief.[55]

The copper lawyers leaned upon the jurisdictional test from *Louisiana v. Texas* (1899), which required a showing "that the controversy to be determined is a controversy arising directly" from state interests "and not a controversy in the vindication of grievances of particular persons." Georgia answered with dubious claims that state tax revenues declined with the growing number of farms ruined and abandoned because of smoke damage, that erosion from denuded hillsides damaged state roads, and that smoke threatened the health of its citizens. Copper lawyers effectively countered each allegation. Tax records showed that revenues actually rose substantially with the return of the copper industry. That made sense because land values tended to increase with population, and the local population had increased from a few hundred during the industry's collapse to almost ten thousand with the return of prosperity. The road issue failed with the observation that there were no state-owned highways in Fannin and Gilmer counties; instead there were only local roads maintained at the local level. They were all unpaved dirt lanes subject to constant washouts from the locally heavy rain—with or without the added consequences of smoke-related erosion. The health issue failed for lack of a creditable scientific basis. Yet for all that, the original jurisdiction defense appeared increasingly shopworn as the case progressed. The Supreme Court effectively rejected the arguments when it granted the state leave to file its first and second lawsuits and then again when it overruled the motions for dismissal filed by the copper companies. The Court had thus upheld the state's claim to original jurisdiction on three separate occasions before the case entered the final hearing.[56]

If the original jurisdiction defense failed, it was then likely that the Supreme Court would find sulfur dioxide pollution from the copper smelters to be an actionable nuisance, as the Tennessee Supreme Court had done in *Ducktown Sulphur, Copper & Iron Co. v. Barnes* (1900). Such a finding would lead to either an injunction or an award of damages. Though the copper companies wanted neither, an injunction was their least favorite alternative. They raised three equitable defenses to avoid that result.[57]

Two of the defenses rested on ancient maxims of the law of equity. The defense of laches, or inexcusable delay, was an equitable doctrine analogous to a statute of limitations. The applicable maxim stated "equity aids the vigilant, not those who slumber upon their rights," because to do otherwise might work an unfairness upon the defendant. On behalf of Tennessee Copper, Howard Cornick claimed such negligence by arguing that "the State of Georgia, by sleeping upon her rights...has allowed a community to become built up" and "has allowed thousands of people to acquire their homes and make their investments and establish their family ties." The Ducktown firm and the Pittsburgh and Tennessee Copper Company (predecessor to TCC) both began smelting in the basin in 1891, thirteen years before the state filed its first suit. A Georgia defeat for its laches was a real possibility.[58]

Another maxim held that an injunction will not be granted where there is an adequate remedy at law, meaning the right to recover monetary damages. By 1907 individual smoke suitors had broken through the procedural barriers that delayed their claims for damages in the Tennessee courts. Now, nearly every term of the Polk County Circuit Court resulted in another batch of monetary verdicts to Georgia citizens. It followed that if the citizens managed to win compensation for their losses, the state should also adjust its property claims by the payment of damages instead of an injunction.[59]

The best hope of both companies lay with the application of the balancing-of-interests test that required a court to weigh the economic harm of the injunction against the harm caused by the sulfur dioxide pollution. This test defeated the injunction suits filed by William Madison, Isaac Farner, Avery McGhee, and their fellow farmers in *Madison v. Ducktown Sulphur Copper & Iron Co.* (1904). There, the Tennessee Supreme Court refused to grant an injunction for the farmers when that would destroy a huge industrial complex employing thousands of workers in order to protect "several small tracts of land, aggregating in value less than $1,000."[60]

The U.S. Ninth Circuit Court of Appeals expressly followed the *Madison* balancing test to reverse the grant of an injunction in *Mountain Copper Co. v. United States* (1906), involving the federal government's attempt to enjoin sulfur dioxide pollution in northern California. The court of appeals acknowledged the toxic effects of fumes on four thousand acres of trees in the nearby Shasta National Forest but characterized the damage as being to land that was "mountainous in character, with little or no soil" and "practically worthless for agriculture." The loss of timber did not warrant closure of a mining complex that produced ten thousand tons of copper per annum and employed more than a thousand workers. The court also rejected the government's contention that Mountain Copper should extract sulfur fumes to make sulfuric acid. "That it is possible to convert the sulfur fumes into acid is ... conceded," but it was economically impractical because of the "enormous cost of such production" and the "impossibility of disposing of such product" on the market.[61]

It was a dramatic win for the copper industry in the West. If the Supreme Court applied it to conditions in the Ducktown Basin, the state of Georgia would lose its case. TCC counsel Howard Cornick gleefully wrote DSC&I attorney James G. Parks to say that the *Mountain Copper* opinion "is a daisy, and we are pleased to note follows pretty closely the *Madison* case." Hart and Johnson knew that their star scientific witness, J. K. Haywood, wrote his monograph "Injury to Vegetation by Smelter Fumes" for the failed California case. The failure of his evidence to sway the Ninth Circuit in *Mountain*

Copper gave the Georgia lawyers reason to fear that Haywood's visits to Ducktown and the mass of other forestry evidence assembled for the case might be a wasted effort in a similar lost cause.[62]

Lawyers for the copper companies gained additional legal encouragement from recent developments in *Missouri v. Illinois*, the suit over the diversion of Chicago sewage into the Mississippi River via the Chicago Sanitary and Ship Canal. The Supreme Court issued two major opinions in the case, the first in 1901 and the second in 1906. In the 1901 opinion, the Court granted original jurisdiction to the state of Missouri in its action for an injunction to stop the diversion. Georgia's lawyers relied heavily on the first opinion because it established a vitally important precedent for the application of original jurisdiction to problems of transborder pollution, but that is all it did because the justices had yet to hear the evidence on the merits of the case. Missouri lost when the case returned to the Supreme Court for a full hearing in 1906.[63]

Missouri's petition suffered, as recently argued by law professor Robert Percival, from two weaknesses. First, its proof of actual disease consequences from Chicago's diversion of polluted waters was weak. Everyone agreed that the diversion occurred and also agreed that contamination of drinking water by untreated sewage *could* cause disease. The question was whether it actually caused a significant outbreak of disease downstream in Missouri. Writing for the majority, Justice Oliver Wendell Holmes found "the now prevailing scientific explanation of typhoid fever to be correct." However, acceptance of germ theory came at a subtle price: Holmes noted that a case built on microscopic evidence "depends upon an inference of the unseen." In effect, Missouri had no case without a statistically significant increase of death and disease along the river to complete "the inference of the unseen" germs. The Court then found the numbers wanting. Second, Missouri placed itself in an awkward equitable position by allowing local cities to dump sewage into the state's own rivers. A party suing in equity for an injunction is held to a standard of fairness expressed as the doctrine of "unclean hands." The doctrine acts to deny an injunction to a party if it comes before the court with hands dirtied by the same sort of conduct that it accuses the defendant of doing. The Court would not enjoin Illinois and Chicago from the same sewage practices that Missouri tolerated by its own citizens.[64]

The 1906 opinion in *Missouri II* was legally fresh news as lawyers for the state of Georgia and the copper companies returned to the Supreme Court in 1907 in their smelter pollution case. Would Georgia be granted original jurisdiction as in *Missouri I*? If so, would the Court refuse to issue an injunction as it did in *Missouri II*? And what role would Justice Holmes play in the copper case? The respective strengths of the parties on the facts and the law made for a closely matched bout. Appreciation of the close odds led Hart to offer additional delays to test other technological cures. The state agreed to waive the scheduled December 1905 hearing for a preliminary injunction to allow for evaluation of the impact of Tennessee Copper's new giant smokestack. Hart's lawyerly prudence became a point of mockery among the defendants. Writing to J. G. Gordon, the English general counsel for DSC&I, James G. Parks compared Attorney General Hart to the grand old Duke of York of the nursery rhyme: he climbed the hill with an offer to do battle and then "marched down the hill again" without landing a blow. When settlement talks and smokestack technology proved unavailing, he made another trip to Washington and girded his loins for battle in the Supreme Court.[65]

The two-day final hearing took place in February 1907. Ligon Johnson opened for the state, and Attorney General Hart gave the closing argument. The justices peppered both with questions on jurisdiction, with attention to the difference between the state's case and the individual smoke suits. Hart argued that, even though Georgians had the right to press their individual claims in the Tennessee courts, the state had the right to original jurisdiction in the Supreme Court to protect its own domain. DSC&I attorney James G. Parks concluded afterward that the copper companies "had the best of the argument," though not without misgivings. He wrote a candid letter to W. H. Freeland expressing his hope for a win on jurisdiction, but warned that "if jurisdiction should be taken, then I think the result would be doubtful" because there was "no case on record where the injury has been so wide-spread." This was a grudging compliment to the quality and extent of Georgia's expert testimony from the government scientists.[66]

The Supreme Court rendered its judgment on May 13, 1907, in a unanimous opinion written by Justice Oliver Wendell Holmes, with a concurrence by Justice John Marshall Harlan. It was a typically brief opinion for Holmes. Thousands of pages of evidence resulted in an opinion that was only four pages long, apart from the head notes, syllabus, and recitation of counsels' arguments. Brief as it was, it fulfilled the worst fears of James G. Parks: the Court affirmed Georgia's claim to original jurisdiction and then granted it the right to an injunction to abate the sulfur smoke.[67]

From the opening sentence, Holmes carefully framed the case as a state-level action concerning cross-border air pollution. It was filed, he wrote, "in pursuance of a resolution of the legislature and by direction of the Governor of the State," for the purpose of seeking an order to "enjoin the defendant Copper Companies from discharging noxious gas from their works in Tennessee over the plaintiff's territory." The second sentence described the scope of the harm, that "a wholesale destruction of forests, orchards and crops is going on, and other injuries are done and threatened in five counties of the State."[68]

It followed that the dispute was not an ordinary suit in nuisance. Holmes observed that "the case has been argued largely as if it were one between two private parties, but it is not." Georgia's claim of damage to its property interests was "merely a makeweight" because "the State owns very little of the territory alleged to be affected" and the monetary damage directly suffered by the state "is small." This was exactly what the copper companies had argued all along, but that was not what the case was about. Instead, Holmes declared,

> This is a suit by a State for an injury to it in its capacity as a quasi-sovereign. In that capacity the State has an interest independent of and behind the titles of its citizens, in all the earth and air within its domain. It has the last word as to whether its mountains shall be stripped of their forests and its inhabitants shall breathe pure air.

The case was not about highways and tax revenues, nor was it a mere subterfuge to advance the private smoke damage suits of its citizens. It was about Georgia's sovereign right to protect the natural resources within its dominion. Georgia's claim of injured sovereignty received resounding vindication.[69]

The third and fourth paragraphs addressed the jurisdictional issues. Again, the distinction between an action by a state and one by a private party was crucial. Justice Holmes began the section by naming *Missouri v. Illinois* (1901) (*Missouri I*), as the relevant standard for original jurisdiction. It was the only case he cited in the opinion. He did not mention *Louisiana v. Texas*, the strongest jurisdiction case for the copper companies. The citation to *Missouri I* established that Georgia's concerns over polluted air from copper smelters were analogous to Missouri's complaint about polluted water for purposes of original jurisdiction. It was the first time that the Supreme Court declared interstate air pollution to be a constitutional issue.[70]

Holmes worked through the jurisdiction analysis along the lines considered by Attorney General Hart and Special Counsel Johnson in 1904. The matter turned on fundamental concepts of dual federalism whereby each state surrendered its right to make war upon a neighboring state, or that state's citizens, in exchange for access to the Supreme Court under article 3 of the Constitution. The justice wrote, "When the States by their union made the forcible abatement of outside nuisances impossible to each, they did not thereby agree to submit to whatever might be done." They retained the right to make "reasonable demands on the ground of the still remaining quasi-sovereign interests." The only constitutionally permissible way to enforce their interests was in federal court, because "the alternative to force is a suit in this court," citing *Missouri I* for the second time.[71]

State sovereignty precluded the wooden application of nuisance remedies devised for private litigants for, as Holmes commented, "some peculiarities necessarily mark a suit of this kind." For one thing, "if the state has a case at all, it is somewhat more entitled to specific relief [i.e., an injunction] than a private party." The state "is not lightly to be required to give up quasi-sovereign rights for pay." Again, the remedies framed by the common law of nuisance had to be tailored to the requirements of federalism, because "the States, by entering the Union did not sink to the position of private owners subject to one system of private law." It followed that the Court "has not quite the same freedom" to apply the balance-of-interests test that would weigh the harm of an injunction against the copper companies against the harm to the state caused by smelter emissions. And for the same reason, "we cannot give the weight that was given to them in argument to other common factors of equity jurisprudence," such as "the commercial possibility or impossibility of reducing fumes to sulfuric acid," or to the "special adaptation of the business" to its locale. All of the lawyers recognized the careful phrasings for what they were: a veiled criticism of the Ninth Circuit's application of the balancing test and commercial viability considerations to bar the federal government's claim for injunctive relief in the *Mountain Copper v. United States* smelter case. Holmes did not mention the case by name, nor did he need to in order to make his point.[72]

The justice concluded his constitutional analysis with a powerful affirmation of a state's right to protect its natural resources from interstate pollution:

> It is a fair and reasonable demand on the part of a sovereign that the air over its territory should not be polluted on a great scale by sulfurous acid gas, that the forest on its mountains, be they better or worse ... should not be further destroyed or threatened by the act of persons beyond its control, that the crops and orchards on its hills should not be endangered from the same source.

As a matter of federal constitutional law, a state always had the right to protect natural resources from activities originating within its boundaries. Georgia now had the constitutional right to protect its resources from pollution emanating from beyond its borders.[73]

Having addressed the constitutional issues, Holmes turned to the factual proof of alleged harm from smelter fumes. This he accomplished in a mere five sentences, the first being the simple declaration, "The proof requires but a few words." The mass of reports, affidavits, maps, and experimental data by chemists, entomologists, geologists, and Gifford Pinchot's foresters convinced the Court that "the pollution of the air and the magnitude of that pollution are not open to dispute." Moreover, the damage from the fumes was "on so considerable a scale to the forest and vegetable life, if not to the health, within the plaintiff state as to make out a case within the requirements of *Missouri v. Illinois*." This was Holmes's third citation to the 1901 decision in his terse opinion.[74]

In short order, Holmes had rejected most of the legal and factual defenses raised by TCC and DSC&I. The last legal defense, laches, fared no better. Georgia had not slept on its rights. It filed suit in 1904 when smoke damage became intolerable and then, in a settlement with the copper companies, entered a dismissal without prejudice to assess pyritic smelting and tall-stack technologies. It then refiled the case only when damage from new methods proved to be even worse.[75]

Holmes made no reference to the opinion he wrote the previous year in *Missouri II*. That left the copper companies wondering why Georgia won the right to an injunction where Missouri failed. Georgia succeeded by avoiding the two weaknesses of the Missouri case. It lacked a significant smelting industry of its own, so it was not subject to the doctrine of unclean hands. It also enjoyed the advantage of being able to present the tremendous visual impact of sulfide dioxide pollution upon local vegetation. Where Holmes struggled with "the inference of the unseen" in *Missouri v. Illinois*, he seized upon the photographs and other depictions of the miles of naked gullies in the Ducktown Basin to declare that "the pollution of the air and the magnitude of that pollution are not open to dispute." The disturbing visual impact of the Ducktown Desert secured Georgia's victory.[76]

Upon reading the opinion, Howard Cornick's partner, John H. Frantz, groused to James G. Parks that "the Court dodged every question in the case and has decided it upon a ground which never occurred to anybody on either side of the case." There was a certain amount of truth to his comment in the way that the Court brushed aside all of the tried and true legal defenses so often applied in private nuisance litigation. The decision was nonetheless consistent with the judicial philosophy articulated twenty-six years earlier by Holmes in his book, *The Common Law* (1881). There, he wrote that "the life of the law has not been logic: it has been experience." And he added that "the felt necessities of the time, the prevalent moral and political theories, intuitions of public policy ... even the prejudices which judges share with their fellow-men, have a good deal more to do than the syllogism in determining the rules by which men should be governed."[77]

The copper lawyers understandably embraced the powerful logic of nuisance law as developed over centuries of private disputes. They had the *Mountain Copper* opinion and many other favorable cases on their side. Hart and Johnson had the burden of arguing that the traditional principles of nuisance law should be expanded to encompass cross-border pollution on the regional scale manifested in Ducktown. The genius of their advocacy lay in their wholehearted embrace of the national movement for natural resource conservation. This was the ultimate fruit of the brief exchange between Will Shippen, the logger, Harold Day Foster, one of Gifford Pinchot's government foresters, and Attorney General Hart at the outset of Georgia's second Supreme Court case.

Georgia's brief on final hearing revealed the real heart of the case. Hart and Johnson cited only two cases in support of their case. Instead of the usual copious legal references, they devoted almost every sentence of their sixty-eight-page brief to expert testimony presenting Ducktown smoke as a conservation nightmare that threatened timber supplies in the Southern Appalachian forests and imperiled one of the nation's great watersheds with the manifold evils of denudation and erosion. In so doing, they aligned Georgia's claim with the forest conservation movement, which was surely one of "the felt necessities of the time." The obligation to preserve forests and watersheds for future generations had become, by 1907, one of the "prevalent moral theories" shaping the mission of the Forest Service. By welcoming government technical expertise of every sort, Georgia's lawyers imbued the Ducktown story with the "intuitions of public policy" that animated Theodore Roosevelt's administration. Justice Holmes was a Roosevelt appointee and could not be expected to be wholly free of "the prejudices which judges share with their fellow-men."[78]

There remained the question of a suitable remedy. The Court's ruling that Georgia was entitled to an injunction came with a heavy note of reluctance. Holmes wrote, "Whether Georgia, by insisting upon this claim is doing more harm than good to her citizens is for her to determine," adding that "the possible disaster to those outside the State must be accepted as a consequence of her standing upon her extreme rights." The Court then declined to issue an immediate injunction and instead ordered Georgia to return five months later in October with a proposed decree of injunction "after allowing a reasonable time to the defendants" to complete the acid plants "to stop the fumes."[79]

Georgia won the legal battle by gaining the formidable power to shut down the copper companies. With it came the power to destroy thousands of jobs and a major portion of the North Georgia economy. That power would embroil Hart and his successors in a politically charged war for another decade.

7 ATTORNEY GENERAL HART, THE NATIONAL FARMERS UNION, AND THE SEARCH FOR A REMEDY, 1907–1910

Some legal victories bear the sweet sense of finality. The criminal defense attorney who wins an acquittal for a murder suspect can celebrate with the added pleasure that comes from closing a file. Other lawyers, especially those who are paid on an hourly basis, are happier when a file remains open for additional lucrative work. The chancery case at the center of Charles Dickens's *Bleak House* lasted for the duration of his nine-hundred-page novel as lawyers on both sides milked the decedent's estate for their fees. The case ended only when the drain of fees exhausted the assets of the estate, rendering the litigation moot. Attorney General John C. Hart won a great legal victory for the state in *Georgia v. Tennessee Copper Co.* but enjoyed neither the sense of finality nor the financial rewards of an ongoing billable case.[1]

There was a time when Georgia's attorneys general received fees for court appearances. That ended when the state constitution of 1877 placed the office on a salaried basis. Hart earned $2,000 per annum from the state, a handsome salary compared to the earnings of laborers at the time, but a pittance compared to the large fees received by the corporate lawyers he regularly bested. His income bore no relationship to his caseload. Georgia's constitution mandated that his understaffed office handle every appeal of capital murder cases and every civil claim to which the state was a party. As his caseload increased, his income remained flat.[2]

The situation would have been a little more tolerable if his victory in *Tennessee Copper* allowed him to close the voluminous file. It did not. The Supreme Court's decision of May 13, 1907, granted the state the right to an injunction to abate sulfurous smelter smoke released at the Tennessee Copper Company (TCC) facilities in Copperhill and at the Ducktown Sulphur, Copper & Iron (DSC&I) works in Isabella. At the same time, the Court reset the case for the following October to determine the form of the injunction, doing so in language that questioned the wisdom of issuing an injunction at all. In his 1907 opinion, Justice Oliver Wendell Holmes wrote, "Whether Georgia by insisting upon this claim is doing more harm than good to her own citizens is for her to determine." He then warned, "The possible disaster to those outside the state must be accepted as a consequence of her standing upon her extreme rights."³

The momentous decision to be made about the injunction case left Hart and Georgia's entire political establishment subject to pressure from all sides concerning the course to be taken. Citizens in North Georgia were at odds with each other. Smoked-out mountaineer farmers massed their voices to insist upon an injunction, often with the encouragement of the timber baron Will Shippen. Other Georgians insisted upon forbearance because they owed their livelihoods to the copper industry as mine workers or as merchants to the mining community. The National Farmers Union (NFU), a prominent agrarian populist organization, would soon add its political weight to the controversy. And underneath the clamor lay the issue of whether advancing technology might offer a practical means to achieve an accommodation among the parties.

News of a recent smelter case in Utah gave all parties in the Georgia case added concern about the effect of an injunction against the copper industry in Ducktown. In *Godfrey v. American Smelting & Refining Co.* (1906), a group of more than four hundred farmers joined suit in federal court against American Smelting (ASARCO) and four other smelting firms that operated in the Salt Lake Valley. The court noted the impact of pervasive sulfur dioxide and arsenic pollution on local crops and livestock. It then issued an injunction against all of the smelting firms, prohibiting "the further roasting or smelting of sulfide ore carrying over 10 percent sulfur" and banning "the further discharging into the atmosphere of arsenic." The five smelting firms were given leave to apply for modification of the decree if they could show that new technology or changing conditions no longer warranted the injunction. In the meantime, the decree effectively shut down the smelting industry in the Salt Lake Valley.⁴

The *Godfrey* decision had serious implications for both sides of the Georgia case. The Ducktown copper industry relied on copper ore containing 25 percent or more sulfur, far in excess of the 10 percent limit set for Utah smelters in *Godfrey*. It followed that if the United States Supreme Court followed a similar standard in the Georgia case, then the Tennessee Copper Company and the Ducktown Sulphur, Copper & Iron Company would have to close their operations. For their own part, Attorney General Hart and fellow Georgia officials had reason to be concerned about a backlash similar to the one that the *Godfrey* decision generated against the Utah farmers. The injunction against the five Utah smelting companies eliminated thousands of industrial jobs in the Salt Lake City area. Some of the smelters closed permanently; others relocated, forcing their workers to uproot or find other work. Only one, ASARCO, sought and achieved modification of the decree to continue operations at its existing location. The *Salt Lake Tribune* observed that "it began to dawn on many of the affected communities . . . that the smoke nuisance had its advantages as well as disadvantages, that the permanent removal of the plants would stop their source of revenue."[5]

As Georgians struggled with similar issues, partisans from both camps came to Atlanta to lobby the General Assembly when it convened its 1907 session in July. The prohibition of alcoholic beverages was the dominant issue for the Baptist-dominated legislature, but the Ducktown matter also required attention. The legislature initiated Georgia's suit by passing resolutions in 1903 and 1905 requiring the investigation of smoke damage in the Ducktown Basin and directing the governor and attorney general to file suit against the copper companies if necessary. Everyone now wondered what instructions the General Assembly would issue as to how the state should proceed when the case returned to the Supreme Court in October.[6]

The copper companies sought to influence the legislature by preparing and circulating petitions against the injunction. J. G. Gordon, managing director of DSC&I at its London headquarters, instructed the company's general manager in Ducktown, W. H. Freeland, and local counsel, James G. Parks, "to take some pains (and expense if necessary) in getting petitions signed both in Georgia and Tennessee, setting forth the great loss and suffering which would be caused by the discharge of the employees of the Companies." It is doubtful that the General Assembly knew that the instructions for the latest petition came from the pen of an Englishman writing aboard the ocean liner *Lucania* as it crossed the Atlantic. That information might have added jingoist fervor to the prevailing anticorporate attitudes of the pro-injunction forces.[7]

W. A. Daves, a probate judge in Fannin County, wrote Attorney General Hart on behalf of the anticopper crowd to warn about the company petition and to request that it not be considered until local citizens affected by the fumes "can be fully heard on the matter." Hart declined the request and sought to position himself above the political fray: "I consider that I am proceeding under instructions heretofore given by the General Assembly," adding, "I shall take no part in the political side of this law suit. That is to say that I shall not advocate, nor oppose any action which the Legislature may see fit to take."[8]

Hart's response may be read as either the principled statement of a legal advocate awaiting directions from his client or an expression of his wish that the legislature take the political heat on the injunction issue. If the latter, he was soon disappointed. Legislators heard from both camps and then washed their hands of the matter. They passed Resolution No. 1, "insisting upon the state's right to a final decree of injunction" (a nod to Georgia's mountain farmers and timber interests), but also "recognizing and realizing the vast interests involved to the copper companies and to the people dependent thereon" (another nod, this time to Georgia's mine workers and merchants in the Ducktown District). The competing interests required that the state "act in a spirit of wisdom, justice, and moderation," the virtues depicted by the three-columned arch embossed on the great seal of the state (and again by an identical arch serving as the entry to the University of Georgia campus in Athens). Therefore, and "having every confidence in her Attorney-General," the lawmakers resolved "that the matter of final procedure is left entirely to his discretion." They further recommended "that he proceed liberally in the matter, to the end that no unnecessary hardship shall be imposed upon the copper companies." Yet "no unnecessary time shall be allowed them to complete the structures they are now building to stop the fumes, to the damage and injury of the citizens of Georgia and to her public domain." Whether or not he wanted it, Hart now had full power and responsibility to decide the fate of thousands of citizens in North Georgia. He and his successors in office lost their political cover and were left exposed to pressure from all sides for the duration.[9]

The copper companies had reason to be pleased with the resolution because they knew that Hart had a pattern of delaying legal action whenever a new form of technology promised a cure for Ducktown's sulfurous smoke. He did so in 1904, when he withdrew the state's first smoke suit just a few weeks after it was filed, to assess whether the new pyritic smelting process would yield any improvement. When the new process failed to abate the smoke, he filed the state's second smoke suit and then waived a hearing for a preliminary injunction to see whether Tennessee Copper's huge 325-foot smokestack would disperse sulfur fumes out of the Ducktown Basin. That also failed. The latest hope was for a method to extract sulfur dioxide gases from smelter fumes and to convert them into marketable sulfuric acid.[10]

The idea was not new. John Roebuck, an Englishman, invented the lead chamber process around 1740, and Peregrine Phillips, another Englishman, devised the contact process ninety years later. Though the chemistry and sequencing of the processes differed, they had similar functions: the cleansing of furnace gases to remove soot and other contaminants, the chemical transformation of sulfur dioxide into usable sulfuric acid, and the collection and processing of the resulting acid at desired concentrations. Both processes worked well in small-scale experiments under controlled conditions. The question remained whether they could be duplicated on an industrial scale with smelter gases of various compositions, pressures, temperatures, and so forth. One major problem was devising a suitable container to collect and hold highly corrosive sulfuric acid. The first experiments used glass vessels, but the smelting industry required larger, less fragile containers. Roebuck decided to make his acid collection chambers out of lead. With enough lead, he could make a durable, noncorroding chamber of any size to receive and condense the transformed gases. Many other technological problems remained before acid conversion would ever move from the laboratory to the great smelting operations in Ducktown, but if solved, there was money to be made from a by-product that presently consumed company resources in endless smoke litigation.[11]

Sulfuric acid was, and is, the nation's most important industrial chemical. It is used in petroleum refining, metal processing, automobile batteries, explosives, dyes, and as a feedstock for the production of a wide array of chemical products. Sulfuric acid was necessary to transform the South's abundant deposits of phosphate rock into superphosphate, a potent fertilizer readily absorbed by cotton and most other crops. It was the South's great need for superphosphate fertilizer that made sulfuric acid a potential bonanza product for the Tennessee Copper Company, and eventually for the Ducktown Sulphur, Copper & Iron Company.[12]

The South made cotton its king, but the king imposed a heavy tax in the form of soil depletion. Antebellum growers constantly cleared new fields to replace worn-out older fields. When all the tillable land on a given farm was exhausted, owners took their slaves with them and moved west to repeat the cycle on virgin lands. Soil infertility from cotton monoculture impelled westward expansion across the southern tier of America, all the way from the Atlantic Ocean to Texas and, in turn, proved to be a major factor in the national debate over the expansion of slavery. As James Henry Hammond of South Carolina lamented, "Our most fateful loss, which exemplifies the decline of our agriculture, and the decay of our slave system, has been owing to emigration" from the older states along the Atlantic to the newer states along the Gulf Coast.[13]

The Civil War ended slavery, and the complete settlement of the South's cotton lands eliminated westward expansion as a viable response to soil infertility. Farmers in the region were now more receptive to the use of fertilizer to restore their existing lands. Phosphate had been valued as a fertilizer long before agriculturalists understood its chemistry and the science of plant nutrition. It is one of the three primary nutrients listed on every bag of fertilizer, the others being nitrogen and potassium. On a bag of 10-10-10 fertilizer, phosphate is the middle number. The substance is an essential component of all forms of plant and animal life and is found in trace amounts within every living cell. Phosphate can thus be obtained from biological remains. A German, Hennig Brandt, discovered elemental phosphorous in 1669 by boiling down urine.[14]

At the end of the eighteenth century, Europeans obtained phosphate fertilizer by grinding the bones of animals—and humans. Justus von Liebig, a pioneering agricultural chemist, noted England's aggressive effort to import bones for its fertilizer mills: "Already in her eagerness for bones, she has turned up the battlefields of Leipzig, and Waterloo, and of the Crimea; already from the catacombs of Sicily she has carried away the skeletons of many successive generations." England annually imported a supply of bones "equivalent of three million and a half of men.... Like a vampire she hangs from the neck of Europe." American fertilizer mills acquired much of their supply, for a time, from the bone pickers who roamed the Great Plains to gather millions of buffalo skeletons left in the wake of hide hunters. The gigantic piles of bones stacked along railroad tracks for shipment back to the East stood as mute testament to the size of the former herds and to the rapidity of their destruction.[15]

The supply of bones was insufficient to meet the South's pressing need for fertilizer. Farmers turned to other sources such as guano, the deposited droppings of seabirds. It proved to be an even better fertilizer than bone meal because it contained significant amounts of nitrogen in addition to phosphorous. Though it was excellent fertilizer, it was also expensive because it was mined in the Pacific Ocean on desert islands off the shores of Chile and Peru and then shipped around Cape Horn to American ports. Another fertilizer, cottonseed oil, was a homegrown product. Hundreds of Georgia fertilizer dealers dealt in both substances by the end of the nineteenth century, as reflected in company names such as the Davisboro Cotton Oil and Fertilizer Co. and the Elberton Guano Co. Nonetheless, limited supplies and high prices left tenant farmers and landowners with a shared longing for a cheaper source of fertilizer.[16]

The discovery of enormous deposits of phosphate rock in South Carolina, Florida, Tennessee, and North Carolina gave promise of a local fertilizer that was easy to mine and cheap to ship. The rock was full of fossil bones—

an 1870 visitor to a South Carolina mine said, "very often well defined bones are found, heads of giant mastodons, teeth of sharks, ribs, etc."—leading to the accurate belief that it contained the same nutrient as bone meal. Though abundant, it was also difficult to use because the insoluble nature of phosphate rock kept nutrients from reaching plants. Sir John Bennett Lawes, an Englishman, and Sir James Murray, an Irish physician, discovered that mixing crushed bones with sulfuric acid produced superphosphate fertilizer that was readily absorbed by plants. The process also worked with phosphate rock. (Each inventor received an English patent for the process on the same day, May 23, 1842, a coincidence that gave rise to years of litigation to determine primacy of invention.)[17]

The South had phosphate rock, and its cotton industry promised a huge market for superphosphate fertilizer. The *Atlanta Constitution* gave regular coverage as early as 1869 to superphosphate and numerous schemes to manufacture it. The missing component for commercial success was a cheap and abundant supply of sulfuric acid. Fertilizer manufacturers first turned to sulfur mined in a relatively pure state from the bowels of Sicily's volcanoes. They also imported pyrite ores from Spain, Norway, and Portugal for conversion into acid in America. It would have been cheaper to make acid in Europe and ship it in liquid form, rather than shipping raw ore to America, but Dr. N. P. Pratt, a chemist from Roswell, Georgia, noted that "this acid is the troublesome component. It is dangerous to handle, expensive to transport, and must be made on the spot where it is to be used."[18]

Dr. Pratt, like many others, cast wishful eyes to the abundant supplies of iron pyrites and copper pyrites to be found at Ducktown. In 1882 a railroad promoter pointed to the great deposits of pyrites found there: "The sulfur for the making of sulfuric acid which is now so enormously used in the treatment and preparation of fertilizers made from phosphates found in the southern states, and which was formerly wasted, is worth more than the copper." The sulfur "is alone sufficient to pay for the mining of both." It was a prescient comment that became a reality in the twentieth century. The Tennessee Division of Geology reported in 1966 that "sulfuric acid is the most important product of the copper mining industry in Tennessee, exceeding copper in dollar value."[19]

Decades passed before the dream of an acid industry in the Ducktown mining district could be realized. In 1902 DSC&I solved one important technical problem when it developed the pyritic method for smelting ores. The new process made it possible to conduct initial smelting within an enclosed furnace, which in turn made it possible to collect smelter gases through flues and chimneys for later processing into acid. The financial concerns were also great. Though promising, the creation of an acid industry required a huge financial investment in untried technology. Both of the copper companies made satisfactory returns from copper without it and were loath to invest in acid condensation plants until forced to do so.

Judge Hart's success in the *Georgia v. Tennessee Copper Company* litigation proved to be the necessary goad to compel the companies to begin constructing acid conversion plants. R. E. Barclay, Ducktown's leading historian and the longtime treasurer at the Tennessee Copper Company, acknowledged that "production of copper and the attendant waste of sulfur would have gone on so long as copper showed a profit." The rising cost of defending and paying claims for smoke damage and the risks of a loss in Georgia's Supreme Court injunction case then caused TCC to reconsider its position. The company started construction of its first acid plant in 1906, in the middle of Georgia's second smoke suit "because the citizens of Georgia had the industry here all but shut down because of alleged smoke damages." Officials admitted that "when the company began construction of the plant, it had primarily in mind the relieving of the injury to vegetation," and that it "was skeptical about the profits to be earned." Still, the enormous size of the new project suggested that they were firmly committed to the creation of what they hoped to be a profitable new industry of acid production.[20]

When the Tennessee Copper Company began construction of its first acid plant, it did so on a colossal scale. The new chamber acid plant erected atop the knoll at Copperhill was the world's largest at the time, with a capacity of 300 to 400 tons of acid production per day, or about 100,000 tons per annum. It contained 26 large chambers, each measuring 50 by 50 feet at the base and 75 feet high, and another 72 smaller chambers measuring 10 by 10 feet at the base and 50 feet high, with a total volume of 2.1 million cubic feet. The interior surfaces of the chambers were lined with 8 million pounds of half-inch thick lead sheets. P. J. Falding, a young chemical engineer who had yet to establish his reputation, solved most of the technical problems, though at the cost of his health. One account states that "for a week he stayed night and day at the plant, watching the machinery with eagle eye, testing every valve, pipe, screw and compartment of the great venture until he broke down, a nervous wreck." The company sent him on vacation and brought in another engineer, Utley Wedge, to solve the remaining issues.[21]

In nearby Isabella, the Ducktown Sulphur, Copper & Iron Company made a slower and more tentative entry into the acid business. Its English directors refused to authorize an acid plant until the Supreme Court's May 1907 decision compelled them to change their minds. They realized that their failure to build an acid plant at Isabella, when TCC had one under construction at Copperhill, would be construed by Georgia and the Court as an act of corporate defiance that warranted injunctive remedy. The board instructed Freeland to proceed with the construction of a modest ten-ton acid plant using the contact method.[22]

The DSC&I board made its decision without any confidence that acid condensation from sulfur smoke was practicable or profitable. A larger plant with a capacity of a hundred or more tons would make better economic sense, but the board did "not feel justified under any circumstances, in taking the risk of erecting a large sized contact plant ... before they have had an experimental plant at work for some time." When General Manager Freeland wrote the London headquarters from his office in Tennessee to urge construction of a larger plant, he received a stern rebuke. The directors replied that "we wish to say that we are disappointed that you take such a strong line in your letter ... as to the possibilities of making even a moderate profit out of this product, and to tell you that we do not in this matter agree with you at all." The reprimand, harsh on its face, must have been all the worse, coming as it did from directors who were three thousand miles and an ocean away from the great acid plant that the Tennessee Copper was building only three miles from Freeland's office.[23]

If the lawsuit was the immediate prod for the construction of the acid plants, the great scale of the TCC project suggested that its management had confidence in the economic potential of the venture. Its New York directors sensed a business opportunity that the more remote London directors of DSC&I were too conservative to embrace. Tennessee Copper gambled big with new technology at great expense and stood ready to reap the profits that soon followed. As reported in one article, "scarcely had it decided upon building the plant, when offers to purchase the acid began to arrive from all parts of the surrounding country."[24]

Where the Tennessee Copper Company saw a valuable new product line, Attorney General Hart saw a happy technological end to the state's smoke litigation. Acid conversion had a powerful logic to recommend it. All agreed that sulfur dioxide was the prime toxic agent in smelter smoke. If the companies could remove it from the fumes and then convert it to acid, the remaining smoke would be cleaner and less damaging, to the benefit of the farmers and timber owners in the Ducktown Basin. Better still was the prospect that an abundant local supply of sulfuric acid would bring down the price of superphosphate fertilizer to the benefit of Georgia's cotton farmers.

TCC officials and their counsel did all they could to encourage Hart's favorable view of the new plants. As the October 1907 Supreme Court hearing on the terms of the injunction approached, they encouraged his visit to the TCC works at Copperhill in September and coordinated schedules so that he would be escorted by P. J. Falding, the brilliant young chemical engineer in charge of the operation. Upon his arrival, they also presented him a detailed letter setting forth the economic importance of TCC to the citizens and economy of Georgia. A total of 2,125 men worked for the company during the first eight months of 1907, including many who work full time and others "who lay off on their own accord" to tend to farming and other interests. Three-fifths of the company's employees came from Georgia. As for purchasing, "this entire region is very closely related to the State of Georgia in a commercial way," with "by far the larger part of the supplies consumed here by the population coming from the State of Georgia."[25]

The great size of the acid plant under construction at Copperhill made a strong impression on Hart. The smaller test facility being constructed at the DSC&I works in Isabella was not so convincing but was at least under way. TCC promised to have its plant in operation by December 1, and DSC&I would soon follow. The attorney general returned to Atlanta the following day, September 28, 1907, and announced to the press, "I was more than pleased with the progress that is being made in this direction, and I am satisfied that these companies are consciously and earnestly endeavoring to equip their plants so as to prevent any further damage." He then declared his intention to leave the question of the injunction open when the Supreme Court convened the following month to allow for completion of the plants and time to assess their impact upon the smoke problem.[26]

Hart's decision to delay the injunction was the product of "the spirit of wisdom, justice, and moderation" that the General Assembly had urged upon him. He sought to balance the needs of Georgia smoke victims with the needs of the Georgia mine workers and merchants who depended upon the copper companies. It was good politics and good economics if it worked. It was also good legal thinking under the principles of equity. As would any competent lawyer, he had to anticipate what the Supreme Court might do at the injunction hearing. The Court was sitting as a court in equity and, as such, had broad discretion to frame an injunction ranging from permissive to punitive, and more likely something in between. Hart's decision to postpone an injunctive remedy represented his best guess about the Court's next move.[27]

The opinion by Justice Holmes strongly hinted that he and the Court wanted to fashion a remedy that served both sides. The Court supported

Georgia's right as a sovereign state to have "the last word as to whether its mountains shall be stripped of their forests and its inhabitants shall breathe pure air." It also delayed its final order to allow the companies "to complete the structures they are now building." Though it was too soon to tell if the acid plants would work, the pace of construction evidenced a cooperative spirit by the copper companies. The Court would be loath to punish their present good behavior with a harsh injunction. Hart's decision to postpone the state's request for a remedy did no more and no less than the Court would have done on its own. The Court wanted a technical solution to the smoke problem as much as he did. His decision preserved the good opinion of the Court that would have been lost had he requested severe action before the plants were finished.[28]

With these thoughts in mind, Hart filed his motion for leave to postpone entry of final decree. He declared that Georgia continued to "stand upon her rights" and to adhere "to her determination that the fume injury within her territory shall be remedied," but the state would not "unduly press her rights for an injunction" in light of recent work on the acid plants. He then wisely recited the 1907 General Assembly resolution for political support before making his formal request for an order "preserving complainants' full rights herein but postponing the entry of decree for injunction" for an indefinite time, "until further application by complainant." It was so asked and so ordered, bringing the case to another temporary pause.[29]

The threat of a federal injunction from the U.S. Supreme Court was dormant for the moment. Nor was there any more threat of an injunction from the Tennessee courts. The Tennessee Supreme Court put an end to smoke injunctions against the copper companies when it ruled against the farmers in *Madison v. Ducktown Sulphur, Copper & Iron Co.* Lawyers for the Tennessee Copper Company and the Ducktown Sulphur, Copper & Iron Company could now devote their full attention to the scores of smoke cases for monetary damages then pending in the Tennessee Courts.[30]

Cases for smoke damage during the first decade of litigation, roughly 1895–1905, were stymied by clever defense work by the copper lawyers until lawyers for the smoke suitors learned how to frame and litigate cases that would stand on appeal. They learned in the *Padgett* case (1896) to be alert for defective summons and return of process. The *Barnes* decision in 1900 confirmed that they could establish a substantive case in nuisance for smoke damage, thanks to the court's ruling that smelter smoke in the Ducktown Basin met the definition of nuisance per se. In *Fain* (1902), farmers defeated the DSC&I tactic of enjoining suits at law on the grounds of vexatious litigation and collusive multiplicity of actions. The *Jones* case, another 1902 decision, taught them to be careful to join spouses as co-plaintiffs when the spouses held joint title to the damaged tracts. This was followed by *Swain* in 1903, where plaintiffs learned the hard way to file suits separately against each copper company rather than suing them as co-defendants in a single action.[31]

As plaintiffs learned to dodge the pitfalls, defense attorneys lost their ability to delay trials. James G. Parks acknowledged in 1903 that "the delays hitherto have not been the result of continuances from term to term, but by injunction. It is plain that we have about exhausted this means." Both companies grudgingly accepted that each term of the Polk County District Court would result in the award of damages to several more farmers. Parks wrote in 1909 that "we have tried so many of these smoke suits that the trials have become largely a matter of routine. It is seldom that any question of law comes up that amounts to anything." This state of affairs allowed for quick trials of cases involving the typically small mountain farms. They averaged about one per day, and "of the smaller cases we sometimes try more than one a day—generally, about three in two days." Jury verdicts were usually small, from $50 to $300, rarely more. Many suitors chose to settle out of court for small amounts once the community became familiar with the likely range of recoveries. James Parker, a farmer in the community of Postelle, wrote an anguished letter to TCC's Howard Cornick on September 16, 1908, asking compensation for his farm that "no one wants to buy as all or about all the timber is killed." He offered to settle with both companies less than a week later for a mere $112 in exchange for a permanent smoke easement that barred future suits for damage to the property.[32]

One Georgia smoke suitor sought to escape the pattern of delayed and paltry judgments in the Tennessee courts with a novel strategy to bring his case before a Georgia judge. All agreed that Georgia courts could not obtain personal jurisdiction over the two copper companies because they did their business in Tennessee and were incorporated in distant jurisdictions; New Jersey was the charter state for TCC, and DSC&I was chartered in England. A plaintiff named Williams sought to establish jurisdiction on a different basis. Personal jurisdiction depended upon the location of a person (whether an actual human or the accepted legal construction of a corporation as a person). Williams realized that *in rem* jurisdiction turned upon the location of the defendant's property. He studied the flow of TCC wagons making deliveries back and forth across the Georgia border and then had a Fannin County sheriff garnish one of them with its contents on the Georgia side of the line. Copper lawyers for TCC and DSC&I probably admired a clever legal turn when they saw one but considered it "a bad precedent, in that it will educate Georgia claimants to make a special study of the ways and means by which they can obtain jurisdiction for their courts."[33]

Unwanted jurisdiction in Georgia was one problem. Another was that the tactic might hinder the movement of goods by company wagons and, worse, the movement of company product, whether sulfuric acid or copper, by rail. The copper companies faced a Hobson's choice: if they shipped sulfuric acid into Georgia, they subjected themselves to Georgia jurisdiction on the smoke suits, and, conversely, if they avoided Georgia jurisdiction by refusing to ship acid into the state, they would lose a major market for their new product. Parks devised a solution even more clever than the jurisdiction by garnishment tactic used by Williams. Parks borrowed a principle from the law of sales and counseled the copper companies to revise their sales contracts to read "FOB Copperhill" or "FOB Isabella." The term FOB—free on board—determined the point at which title for the goods transferred from the seller (the copper company) to the buyer (the purchaser of acid). If the sales contract read FOB Copperhill or FOB Isabella, then title to the acid transferred to the buyer at the acid plants on the Tennessee side of the border. The upshot was that the acid no longer belonged to the copper companies when it crossed over the border into Georgia and could no longer be garnished as copper company property. That solved the shipping problem for TCC and DSC&I and compelled Georgians to continue litigating their smoke suits in the Tennessee courts.[34]

The typically small judgments awarded in the Polk County Circuit Court could be absorbed by the copper companies as an item of overhead, given that copper mining was booming and acid production promised a new source of wealth. Profits at the Tennessee Copper Company tripled from 1905 to 1906, and that was before its new acid plant was finished. Suits from the more prosperous farmers and from the timber companies were a matter of a much greater concern. The best farms in the Ducktown Basin were along Hot House Creek as it ran through Georgia and North Carolina. Parks advised DSC&I's General Manager Freeland that "these farmers have been pretty badly injured." Their claims would yield higher jury verdicts because of the extent of damage, the quality of the farms, and the communal standing of the owners. "We have thus far obtained very low verdicts in nearly all of our suits. We may reasonably expect heavier verdicts in this lot of cases." Parks and his TCC counterpart, Howard Cornick, nonetheless wavered about settlement.

An agreed resolution lowered the risk of excessive awards by runaway juries but would "serve as a precedent in other suits thereafter to be tried"—or, in plainer speech, "there will be trouble in undertaking to settle with people who are really injured—there will be a swarm of others who think they might just as well have a little something while it is going."[35]

Parks engaged in lengthy settlement correspondence with A. B. Dickey on behalf of his Hot House neighbors, T. J. Willson, John J. Withrow, J. W. Anderson, F. M. Harper, J. M. Withrow, J. T. Fry, and W. L. Harper. The plaintiffs demanded settlements as high as $10,000. The companies balked and instead prepared for trial by sending inspectors to the farms and by comparing the demands to listed valuations of the properties on local tax registers. When the cases reached trial in 1908, jurors awarded Willson only $1,090 upon a final demand of $4,200. J. M. Withrow received $1,200 on his demand for $10,000, and their leader, A. B. Dickey, won only $1,065 after demanding $9,000. Cornick concluded that "we were fortunate in the outcome of the Hot House cases," and "the plaintiffs were certainly very badly disappointed."[36]

Ducktown's attorney considered the Hot House cases as "a kind of skirmish line beyond which it is not likely that we shall have any heavy suits, except from owners of large bodies of timber lands, such as Shippen Bros., Vestal Lumber Company and such people." Suits by the large timber companies posed risks of a greater magnitude. The great timber barons and their companies, George Peabody Wetmore, Paul E. Stevenson, the Shippen brothers, Rosine Parmentier, J. P. Vestal Lumber Co., Ocoee Timber, Alaculsy Lumber, Conasauga Lumber, and others held enormous tracts of valuable standing timber. Many holdings exceeded 50,000 acres, and a few, such as that of the Shippens, were more than 100,000 acres. It followed that smoke damages, if proved, could be vastly greater than the amounts awarded to mountaineer farmers, even the substantial farmers of Hot House Creek. Jury awards in the first wave of timber suits proved this to be true. Stevenson won $7,000 and Parmentier won $6,000. Stevenson's award was roughly equal to the total awarded in all of the Hot House cases. The next wave of suits promised to be much larger.[37]

With recoveries such as this, the timber cases could not be handled in the same routine manner as ordinary farmer suits, even though suits by both types of plaintiffs involved the same legal principles. The copper companies expected to lose on the issue of liability if the cases reached the juries, so they turned their focus on limiting the amount of damages. They needed solid expert testimony to know the number of trees that died from sulfur smoke compared to other causes such as fire, insects, and fungal diseases. They also needed to know when the trees died for the purpose of the statute of limitations. Last, they needed to determine the value of trees lost to smoke damage, a question that required knowledge of the value of various species at various ages. Mature poplar timber, for example, was worth far more than immature cordwood. To answer the questions, they once again engaged the services of the German forester, Dr. Carl Alwin Schenck, at the Biltmore School of Forestry. His services did not come cheaply. He demanded and received $9,065.74 for his expertise.[38]

It all came down to the problem of how to count trees on huge tracts of mountainous land with a combined area of more than 400,000 acres. Many homeowners cannot accurately recall the number of trees on their suburban lots. Schenck had to make a defensible estimate for a tract exceeding 625 square miles. The plaintiffs' experts, W. D. Hale and John Williams, neither of whom had any formal forestry training, advocated the sample acre method by which they would count the number of living and dead trees on small plots representing the different composition of forests on the north slopes of the mountains, the south slopes, the valleys and coves, and the ridgeline tracts. Schenck objected that "no such thing as a 'sample acre' exists." Moreover, the copper companies shared the reasonable suspicion that the plaintiffs would select only the most damaged sample plots to calculate their losses. Instead, he argued for the strip method, used by Gifford Pinchot's Division of Forestry, by which each member of a trained crew measured trees within a strip one chain (sixty-six feet) in width and followed the strip over whatever terrain it crossed. The results of a coordinate pattern of strip surveys were then averaged. The method came with the recommendation of Henry Solon Graves, an American forester trained in Germany and Pinchot's chief assistant. Graves described it in his 1906 treatise, *Forest Mensuration*, (mensuration being derived from the Latin for measurement).[39]

Schenck acted according to the state of the forestry art, yet Park and Cornick both requested what they called "practical saw-mill men" to accompany the survey and to assist in courtroom testimony. Schenck no doubt bristled at the phrase because he prided himself on establishing the first American school for industrial foresters, but he was wise enough to appreciate that what the lawyers really wanted was someone who could identify with jurors from East Tennessee a little more readily than a German forest scientist with a doctoral degree. He wrote to Cornick that "cooperation of practical saw mill men with our cruisers seems advisable," because their views "are more forcible [i.e., convincing to a jury] than those of trained cruisers, scientists, and timber surveyors."[40]

Parks and Cornick pursued other defenses against the timber barons as Schenck's men walked their measured strips through the Southern Appalachians. The lawyers never lost sight of fundamental issues such as jurisdiction and ownership. Cornick defeated J. Harvey Ladew and George Peabody Wetmore in the U.S. Supreme Court for their failure to establish diversity jurisdiction as required for suit in federal court in Chattanooga. (Neither the plaintiffs nor Tennessee Copper had standing as a citizen of Tennessee; the company was chartered in New Jersey and the plaintiffs resided in Georgia or the North.) Parks and DSC&I then attacked the validity of timber options claimed by the Shippens, because they could not claim damages for the loss of lumber they did not own.[41]

In another approach, Cornick attempted to suppress evidence from federal foresters, a tactic stirred by bitter memories of the powerful impact of government experts in Georgia's Supreme Court injunction case. He wrote Parks that "we are endeavoring to prevent the foresters from giving evidence and hope to succeed." J. K. Haywood was Cornick's special target because he was the author of a government report, *Injury to Vegetation from Smelter Fumes*, that Attorney General Hart cited to great effect in the Georgia case. Cornick told Parks that Haywood intended to publish another bulletin for the Bureau of Chemistry "explaining the situation, with even greater details." If published, "it will be quasi evidence," so "we are attempting to prevent its publication." The effort failed. Haywood published another U.S. Department of Agriculture bulletin, *Injury to Vegetation and Animal Life by Smelter Fumes*, in 1908 (revised in 1910) and shorter articles on the same topic in 1907.[42]

Both lawyers remained alert for tree diseases that could explain tree loss in the local forests. Parks discussed the white pine blight with Schenck after reading about it in a Department of Agriculture bulletin. Cornick noticed a learned article by William Alphonso Murrill about the chestnut blight, a fungal disease discovered three years earlier in the New York Botanical Gardens. Writing to Parks, Cornick promised to send Murrill's article and expressed his intention to take Murrill's deposition. This was good defense work. It was also a sad portent of the biological havoc the chestnut blight would soon bring to the eastern hardwood forests. Its impact would become more apparent in the later stages of the smoke litigation as it advanced from New York City to the forests of Ducktown.[43]

Clever defense work and the best of forestry science combined to defeat some timber suits and to limit the monetary awards in others. Copper lawyers also suffered stinging defeats. The executors of the Paul E. Stevenson estate won $32,371.11 against TCC in 1911. The old timber baron's last smoke suit was his best one. His living rivals in the industry were eager to win even greater amounts.[44]

As a remedy, actions for monetary damages rarely accomplished much for smoke victims. Some, though not all, of the timber barons recovered a great deal of money. Most of the mountain farmers received very little. The award of a couple of hundred dollars was small compensation for those whose farms lay in the most heavily damaged areas; the money could not replace a way of life made impossible by sulfur smoke. Suits for damages were also inefficient. A court of law could award damages only for past smoke damage, not for prospective harm. The statute of limitations made it necessary to periodically file new lawsuits for subsequent damages. The *Swain* decision required separate actions against each company rather than joining them as co-defendants. As a result, scores of local farmers and timber owners filed two, three, and sometimes even four smoke suits.[45]

Most of all, actions for monetary damages did little to stop the destructive impact of smelter smoke in the Ducktown Basin. The two copper companies learned to accept verdicts in ordinary smoke suits as a cost of doing business. Copper lawyers succeeded in keeping verdicts low in the Hot House cases, dampening the litigation ardor of the more prosperous farmers. Only the timber suits truly hurt the firms, but huge awards to the logging interests were too few in number to force production changes to reduce the amount of sulfurous smoke.

Georgia's Supreme Court injunction case had proved to be the only effective means of compelling the Tennessee Copper Company and Ducktown Sulphur, Copper & Iron Co. to change their operations. So it was to the state, and specifically to Attorney General Hart, that frustrated smoke suitors turned in their search for a different remedy. Hart had the discretion and power to renew the application for a final decree of injunction. He was the only one with the ability to obtain an injunction to shut the plants down. Some smoke victims viewed his power as a source of relief from the fumes. Others, particularly timber owners and plaintiffs' lawyers, considered his power more for the potential boost it gave to the value of their claims for damages. Either way, a large segment of the basin's population, particularly Georgia's farmers and timber owners, wanted action from Hart to improve their situation in the smoke wars. And there were the many Georgians with ties to the copper industry that wanted matters to remain as they were. Hart heard from both sides in a daily stream of letters and petitions that covered his desk in the state capitol in Atlanta.

Smoke suitors wanted action. Hart counseled patience. He insisted that time be given to complete the acid plants and to assess their effectiveness in reducing sulfur fumes. The wait-and-see approach was difficult for many because until the plants became operational, both companies continued to emit untreated sulfur smoke at historically high volumes into the Ducktown skies. Property owners continued to sustain damage throughout the 1907 growing season while the plants were under construction at both companies. Construction delays extended the emission of untreated sulfur fumes into the 1908 growing season. It was the second growing season after Hart's May 1907 Supreme Court victory, so patience was wearing thin.

Hart withheld publication of his 1907 annual report until the following summer in the hope of being able to share positive news about the acid plants. To that end, he sent Dr. John M. McCandless, the state chemist, to investigate matters in Ducktown. The report was only mildly encouraging. The Ducktown plant remained unfinished. The TCC plant was operational, though at just three-fifths' capacity. That meant that TCC condensed only 30.5 percent of sulfur dioxide gases it generated by smelting. The rate would improve to 50.1 percent when the plant was fully operational. In the meantime, "there will continue to be discharged into the atmosphere 347.2 tons of sulfur dioxide from the Tennessee Copper Company, and . . . 166 tons from the works of the Ducktown Sulphur, Copper & Iron Co., who as yet, are doing nothing to actually relieve the situation." McCandless concluded that "we still have a total of five hundred thirteen tons of sulphur dioxide being daily discharged into the atmosphere from the stacks of the two copper reduction companies." Ducktown counsel James G. Parks wrote Hart an apologetic letter about delays on the company's new acid plant, promising the first test run in 1908. He added that they would soon commence construction of a second plant with a daily capacity of two hundred tons. Neither development provided any immediate encouragement. The first DSC&I plant was too small to do much good, and the proposed larger plant was a distant dream.[46]

Slow progress at the plants increased the volume of angry mail to Hart from smoke victims, legislators, and members of Congress. Though he usually exercised courtly manners toward others, his patience was getting frayed. When a Fannin County judge complained of smoke damage to his garden, the attorney general responded with heavy sarcasm: "I note with much sorrow the destruction of your sage bush. . . . This of itself is very sad, but the blight of smoke has not even spared your honey suckle vine. This is distressing... and in this severe loss you have my profound sympathy." And then, "I do not however exactly feel... the fact that your sage bush and honey suckle vine had been killed could be plead as complete justification" for throwing "five thousand people out of employment."⁴⁷

The attorney general's correspondence included mail from a new petitioner that required far more careful handling: the National Farmers Union. Mountaineer farmers who formerly acted alone or in ad hoc coalitions now began to speak as organized political groups through their county chapters of the NFU. The Farmers Union of Fannin County, Georgia, and its sister chapter in Gilmer County both urged Hart to end his policy of patience.⁴⁸

Hart held a statewide elected office and had the politician's appreciation for the rising power of the National Farmers Union in Georgia. The Farmers Union arose in Texas in 1902 out of the ashes of the Farmers' Alliance and the Populist Party after the movements self-destructed in the election of 1896. As a new organization, the NFU inherited the goal of framing an economic and political response for its members caught in the web of market forces governing southern cotton and western wheat. Links between the Union and the Alliance were strong: both organizations began in the cotton country of central Texas, and most of the Union's founding officers were Alliance veterans. The Union naturally borrowed from the old Alliance gospel to preach a message of farm cooperatives, reform of cotton finance and marketing, corporate regulation, and agricultural education. The unusually low dues structure of the NFU encouraged its growth among smaller farmers of middling status.⁴⁹

Farmers Union organizers found a receptive audience among Georgia farmers; by 1908 R. F. Duckworth established units in 135 of the state's 147 counties. The Gilmer County chapter organized on June 20, 1907, a few months after the Supreme Court decision in *Georgia v. Tennessee Copper Co.* NFU influence in the state increased still further when Georgia became the center of the national movement: Duckworth and his successor, Charles S. Barrett, both served as national presidents, with Barrett holding office from 1906 to 1928. Barrett moved the national headquarters to Union City, Georgia, in the heart of the Piedmont cotton belt. Duckworth also moved to Union City, where he edited the *Farmers Union News*.⁵⁰

The NFU's headquarters were just south of Atlanta, close enough to monitor politics in the state's capitol while far enough in the countryside to foster ties with the farmers. Its rapidly growing membership and Barrett's rise to national prominence on farm issues drew the public support of Georgia's leading politicians. Governor Hoke Smith and Rebecca Latimer Felton, a noted suffragist and the first woman to enter the United States Senate, both addressed NFU state conventions. Tom Watson, Georgia's Populist leader, provided the introduction to Barrett's history of the movement, writing that the Farmers Union was essential to provide farmers an organized collective voice because a farmer acting alone was powerless against the agriculture system. As he put it, "a naked swimmer, trying to make shore through a swarm of man-eating sharks would have just about as good a chance for his life as a Southern cotton grower has to prosper under the present conditions." Watson was a frequent plenary speaker at NFU state and national conventions. He devoted a monthly column to the group in his *Watson's Jeffersonian Magazine*. Hart knew that, with support like that, he needed to act carefully in his dealings with the NFU.[51]

Cotton dominated the Farmers Union agenda, with Barrett calling the crop "the absolute commercial despot of civilization." The NFU's membership in the North Georgia mountains had different concerns. Few of them grew any cotton at all. They remained largely untouched by King Cotton and its burdensome system of crop liens, supply furnishing, tenant farming, and falling markets. Even so, NFU leaders knew that the organization's strength lay in its numbers and thus had no hesitation over raising chapters outside the cotton regions. The NFU's membership canvassers were paid on a per capita basis, so they also had a personal incentive for expanding the movement into the mountain counties.[52]

Historians have engaged in an extensive debate about the impetus for farm movements. Some point to shared economic concerns, others argue geographic and class cohesions, and one thought the binding factor was a shared but irrational "complex of fear and suspicion" rooted in nativism and Jeffersonian nostalgia. Robert C. McMath provided the answer that best explains NFU's success in Fannin and Gilmer counties: the organization built on existing interlocking rural social networks of kinship, church, shared work, and face-to-face trade and held the group together with the "expectation of relief for its members." The farmers already shared communal ties, strengthened during the early years of smoke litigation. They embraced the NFU because it promised to give them the legal and political power they needed against the copper companies and to goad the attorney general. The letters received by Hart in 1908 from the Fannin and Gilmer chapters were just the beginning of the NFU's role in the case. Jesse A. Drake, the NFU's state counsel for Georgia, and C. T. Ladson, its general counsel, participated directly in the litigation by serving as plaintiffs' counsel for their constituent farmers. Their direct financial interest as attorneys for the smoke suitors guaranteed that the organization would remain active in the case for years to come.[53]

There remains the question of whether the Farmers Union had a stake in the smoke litigation beyond the desire to serve the needs of its mountain constituents. Existing records from its national conventions and issues of its national newspaper, the *National Field*, fail to reveal significant devotion to conservationist issues, at least until 1914, when it took a strong position in favor of Gifford Pinchot and his campaign to build the Hetch-Hetchy dam and reservoir in Yosemite National Park. Plans to construct the project within one of America's earliest and most spectacular national parks generated a heated national debate between advocates of wilderness preservation, led by John Muir and the Sierra Club, and conservationists, led by Pinchot, who sought to make what they considered the wise use of natural resources.[54]

The Farmers Union aligned with Pinchot and the pro-dam interests through editorial support in the *National Field*. In January 1914 the editors set aside the debate between preservationists and conservationists to recast the issue as a populist anticorporate campaign. The dam, to be constructed on federal parkland, would provide a publicly owned water supply for San Francisco, allowing "over 400,000 citizens out of the grip of private water concerns." The NFU was fundamentally opposed to "giving any private corporation a franchise which would enable it to gouge the consumers of a vital necessary [sic]." Hydroelectric power from the project would enhance the city-owned trolley system "so as to put the old United Railroads gang out of business." With cheap power from Hetch-Hetchy, "the town can give its citizens transportation, light and power service at the actual, nominal cost." Pinchot adopted a similar approach in his letter of thanks to the editors. The Hetch-Hetchy issue was, he said, "just one phase of the great issue in the whole conservation controversy." The issue was "whether the public welfare shall be made subservient to the profits of the magnates, or whether the natural resources shall be used primarily for the benefit of all the people."[55]

The Hetch-Hetchy fight was three thousand miles away from Ducktown and occurred several years after the NFU's entry into Georgia's suit against the copper companies. The two battles were also distinct in terms of ownership issues. Hetch-Hetchy involved the use of public property; the Ducktown litigation arose from damage to private farms and woodlands. There were also similarities. Hetch-Hetchy and Ducktown were both conservation issues insofar as they concerned the sustained use of natural resources for the public good, though the pressures were from different angles: the former contest was a struggle with preservationists to put protected resources to public use; the latter was a fight against corporate pollution that threatened the health of agricultural and natural resources. Yet, for the NFU, the Ducktown litigation, like Hetch-Hetchy, was ultimately a battle against abusive corporate power, one that impacted its mountain constituents in Fannin and Gilmer counties and its larger constituency of southern cotton farmers. The key was fertilizer. The NFU recognized fertilizer costs as one of the two greatest expenses born by cotton farmers and worked to reduce prices by building a cooperative phosphate fertilizer plant at Union City. The two copper companies were the chief source of the sulfuric acid necessary to manufacture the product. The combination of Ducktown acid and NFU fertilizer manufacturing might have promised a mutually beneficial business relationship, one that Hart could endorse as a useful result of his patient management of Georgia's lawsuit. Instead, relations between the NFU and the copper industry proved to be an aspect of the agrarian populist battle against the fertilizer trust. Attorney General Hart would soon be crushed between the combatants.[56]

In 1908 the Farmers Union supported a resolution by a Fannin County legislator, William Butt, requiring Hart to seek a final decree of injunction against the copper companies. The debate turned upon the familiar complaints of crop and tree damage and the newer theme of the promised benefits of the acid industry in the form of reduced fertilizer prices. Both sides argued the technical merits of acid condensation, as each side considered the percentage of sulfur dioxide the plants would remove when operating at designed levels. Rep. Butt's resolution failed on a 12-5 vote in committee and a 97-12 vote in the House.[57]

Hart's opposition to the measure placed him in alignment with the copper companies, exposing him to attack by the anticopper *Blue Ridge Post*. "This censure is undeserved," he wrote. "The only difference between the people of Fannin County and myself is that of procedure." They wanted to close the mines and force people out of work, "while my idea is to preserve the interests of Fannin County without doing irreparable injury to other interests." What Hart saw as a principled stand for a balanced remedy actually distanced him from Georgia's smoke suitors. Even well-meaning Georgians failed to grasp his vision. When one wrote for suggestions about making a smoke claim, Hart responded in dry legalese, "the State's suit does not contemplate recovery for individuals but suppression of the nuisance complained of."[58]

He complained to TCC's Howard Cornick in October 1908, saying that Georgians "have made it very embarrassing for me this summer and have annoyed me greatly." At DSC&I, he received more support from James Parks, a Tennessee attorney, and W. H. Freeland, a naturalized Englishman, than he did from fellow Georgians. He needed the support and cooperation of the copper companies to accomplish what he desired most: the completion of fully operational acid plants that would end the smoke controversy and put an end to political pressures from Georgia mountaineers. As he told Cornick, "I sincerely trust that before another season rolls around that the companies will . . . have the matter in hand." He urged Cornick and TCC to "make an extra effort to take care of the situation" so he could "avoid a repetition of criticism and abuse" that he suffered in the cause.[59]

The next year, 1909, did see real progress. Tennessee Copper brought its first plant to full capacity and began building another just as large. The Ducktown firm completed its full-size acid plant on May 6, 1909, almost two years from the date of the 1907 Supreme Court decree. Still, Hart's correspondence on smoke matters ranged from the mild to the antagonistic, tending mostly to the latter. The attorney general's position as holder of the injunction weapon placed him in the ridiculous position of fielding inquiries as to whether Mr. Dean's cabbage patch or J. D. Northcutt's apples succumbed to smoke or to other agencies. The state's chief attorney was responsible for life and death issues in murder cases. He now spent time reading letters about unhealthy vegetables and then passed them to state experts for their considered opinion. A typical letter from the state's entomologist informed him that "a careful examination of the bean which you submitted me shows that it is affected with a common bean disease known as bean anthacnose, technically known as *Collectrichum lindemuthianum*," a condition unrelated to smoke damage. It was not a role that the framers of Georgia's constitution had in mind for the office of attorney general, but in the absence of an environmental paradigm, much less the scientific bureaucracy to implement it, his was the office where such matters landed.[60]

Apart from vegetables, Hart engaged in an increasingly heated exchange with the National Farmers Union that carried into 1910. J. T. DeWeese, leader of the Gilmer Farmers Union, complained of "the present increasing damages done to our farms and farm products by the continuation of the 'poisonous fumes.'" W. T. Buchanan, leader of the Fannin chapter, insisted that Hart act to "relieve our people from this distressing and dangerous situation and restore to us our rights under the law." NFU attorneys C. T. Ladson and Jesse Drake announced that they were retained by "four score people" in their smoke suits against the copper companies. Further pressure came from the NFU's state and national headquarters in Union City. Drake made formal inquiries to Hart. The attorney general forwarded the same to the copper companies for their response, along with a note saying, "the Farmers Union is a large and respectable organization and I desire to treat it with the utmost courtesy and consideration."[61]

Political deference fell to the wayside when R. F. Duckworth attacked him through the *Farmers Union News* at the approach of the 1910 General Assembly. Hart once again ignored the advice attributed to Mark Twain, to "never pick a fight with a man who buys ink by the barrel." He wrote a four-page letter justifying his stance and pointing to the reduction in fertilizer prices caused by the abundant supply of Ducktown acid. His argument failed to persuade Duckworth. NFU officials knew more about fertilizer and acid pricing than did Hart. Duckworth acknowledged the drop in fertilizer prices but denied it was due to a decline in the price of acid; instead, the price of acid remained flat despite production at TCC and DSC&I. Worse, the NFU knew from published accounts that the Tennessee Copper Company contracted to sell its entire acid output to the Independent Fertilizer Company, a trust created "to absorb independent fertilizer companies in the South."[62]

An even larger fertilizer trust, the International Agricultural Corporation (IAC), soon absorbed Independent Fertilizer and acquired the acid contract in the deal. It was a masterstroke for IAC because it was also the largest miner of phosphate rock in Florida and now had secured all of the acid produced at Ducktown needed to process it. After the merger, TCC granted IAC a ten-year extension on the original contract made with Independent Fertilizer. None of the acid produced in the Ducktown District could be put to use at the NFU's Union Phosphate Plant without first passing through IAC as middleman. The arrangements between TCC and the fertilizer trusts were thus considered an anticompetitive thrust at the cooperative ethic that animated the Farmers Union and the entire agrarian populist movement.[63]

Duckworth thundered, "The Farmers Union has made a fight on two lines of corporation, both using products produced by these plants; one using copper, the other using sulphuric acid; the one being the Rail Roads, the other being the Fertilizer Company." In that light, Attorney General Hart's policy of continued patience was counterproductive because it served the NFU's corporate enemies without alleviating the suffering of southern farmers. Duckworth closed his rebuttal to Hart, saying the Farmers Union had "asked that the Copper Companies pay the people for damages done, or that they be shut down. But they have so far refused to pay all damages, and you have so far refused to shut them down." Therefore, "we are pursuing the only course left for us that of making a fuss about it."[64]

That was Hart's last exchange on the matter. He resigned two weeks later on June 29, 1910, and left the fate of Georgia's Supreme Court smoke litigation to Governor Joseph M. Brown and future attorneys general. Hart's work on the preeminent case of his career ended on a bitter note. His increasingly defensive spirit was that of a legal warrior made weary by a case that seemed to have no end. His fervent desire to accomplish smoke reduction without shutting the plants left him politically isolated and often despised. His friendly supporters among the lawyers and managers at the copper companies could do him no good in Georgia, especially after he antagonized the National Farmers Union and the smoke suitors within its membership. He had become a political liability to Brown's administration.[65]

Yet his accomplishments in *Georgia v. Tennessee Copper Company* surpassed those of any other Georgian affiliated with the case before or after. He won a landmark decision that established the state's sovereign right to protect its people and resources from interstate air pollution. The copper companies were still roasting ores on open heaps when he entered office in 1902. His legal skills, and those of his chosen assistant, Ligon Johnson, forced them to end the practice. The ongoing threat of an injunction compelled the companies to build acid condensation plants that promised to extract at least some of the sulfurous fumes and to convert them into a valuable product for farmers. Every tank car of acid that rolled out of the Ducktown Basin on the Louisville & Nashville rails represented tons of harmful sulfur dioxide removed from the sky and transformed into useful fertilizer for the worn-out cotton fields of Georgia and the South.[66]

THE SMOKE INJUNCTION AND THE GREAT WAR, 1914–1918

8

The resignation of Attorney General John C. Hart in 1910 did nothing to quell the controversy in North Georgia over injunctive relief. The issue dominated the fall elections in Fannin County when pro-copper Republicans turned out all but one of the anticopper Democrats to gain control of the county for the first time in twenty-six years. Fannin was the Georgia county closest to the copper works and suffered the worst of the smoke damage. At the same time, it had the most to lose if an injunction ended the flow of copper dollars into the local economy. All of the successful Republican candidates were well-known businessmen and professionals who saw their livelihoods at risk if the mines closed. Democrats explained the elections results by pointing to rampant vote buying, notably in the Hot House district where several leading smoke suitors lived. One witness told of an approach by a Republican campaign worker: "This is a mighty good, juicy apple. And here's the ticket you ought to vote. After you vote that ticket eat that apple." The voter did as he was told and discovered that "inside that apple was a five-dollar bill." Another lifelong Democrat switched his vote when offered thirty dollars and the cancellation of his promissory note held by a bank in Copperhill. All that money had to come from somewhere, so Democrats quickly pointed fingers to the several senior mining officials currently under indictment for creating a public smoke nuisance.[1]

The local election results demonstrated that Hart's strategy of patient delay on the injunction had run its course. When the Supreme Court ruled for the state in *Georgia v. Tennessee Copper Co.* (1907), it suggested that the copper companies be given six months to complete their acid condensation plants before it issued the final decree of injunction. Hart extended the six months to three years because of his deep reluctance to destroy the industry and its thousands of jobs. The entire thrust of his strategy had been to force adoption of technologies to abate sulfur fumes while leaving the industry viable, and acid condensation was his last and best hope because it converted harmful smelter gases into the sulfuric acid needed for the fertilizer industry.

As he stated in his 1909 annual report, "It means a great deal not only to the people of Georgia in the immediate vicinity of these works, but the entire State, and the South ... especially the agricultural interests, to develop this industry."[2]

The rationale for Hart's strategy of patience ended with the completion of the acid reduction plants. The giant plant constructed by the Tennessee Copper Company (TCC) at its Copperhill complex in 1907 had a daily production capacity of 200 tons of acid. In 1909 the Ducktown Sulphur, Copper & Iron Company (DSC&I) finished the 160-ton plant it built in nearby Isabella to replace its tiny experimental 10-ton plant. The great volume of acid produced each day left no doubt that the process worked, nor was there any doubt that acid production had succeeded beyond expectations as a money-making enterprise for both companies. The question now was whether the new technology was equally successful in reducing the amount of sulfur fumes entering the skies of the Ducktown Basin. The issue plagued Hart's successors as they worked through piles of correspondence and petitions for and against an injunction over a period of eight years following his 1910 resignation. Attorney General Clifford Walker, Hart's fourth successor, spoke for his predecessors and himself by paraphrasing the gospel of Matthew: "This litigation, like the poor, we have with us always."[3]

In 1910 the General Assembly dealt again with the issue when it considered another resolution for action in the Supreme Court case. The National Farmers Union (NFU) lobbied for passage through its attorneys, C. T. Ladson and J. A. Drake. Lawyers and officials for the two copper companies responded with an excursion to Ducktown for the House Judiciary Committee. The legislators toured the new plants, became favorably impressed, and then tabled the resolution on their return to Atlanta. A newspaper headline summarized their conclusion: "No Fumes Now at Ducktown."[4]

Governor Joseph M. Brown and his new attorney general, Hewlett Hall, were not convinced, nor were they ready to dismiss the concerns of the powerful Farmers Union. Hall selected William M. Bowron, a consulting engineer from Chattanooga, Tennessee, to make an investigation. It was an inspired choice. In its earlier investigations, the state usually turned to chemists, geologists, entomologists, and foresters, all of whom examined the sulfur smoke for its impact upon Ducktown vegetation. Bowron was an engineer, not a life scientist. He spent little time examining crops and trees for smoke damage and instead studied sulfur smoke as a matter of production statistics, a mathematical problem involving inputs and outputs in the smelting process. He framed the study by asking four questions. First, how much sulfur entered the smelting system? That was easily answered by knowing the amount of ore smelted and the sulfur content of the ore. Second, where did the sulfur go once it entered the system, that is, how much was captured and condensed to acid, compared to the amount released up company smokestacks? Third, how did the amount of sulfur released in 1910 compare to the amount in 1903 when the state filed its first case? And, fourth, what production and economic factors explained the amount of sulfur released in 1910? His answers to the four questions changed the course of Georgia's injunction suit before the United States Supreme Court.[5]

Bowron devoted most of his report to the Tennessee Copper Company, by then the larger of the two copper producers in the district. During the month of October 1910, the company processed 39,000 tons of ore containing 22.5 percent sulfur, yielding 8,775 tons of sulfur. Of the 8,775 tons of sulfur, 2,574 tons went to acid and 6,201 tons went into the air. In short, 70 percent of the daily production of sulfur entered the atmosphere notwithstanding use of the acid plants. When annualized, the sulfur escaping from the TCC works amounted to 74,000 tons in 1910, when acid production was fully operational, compared to 34,056 tons released in 1904, before the first brick had been laid to build an acid plant. Bowron's surprising numbers proved what many local smoke suitors had suspected: the amount of sulfurous fumes in the skies over the Ducktown Basin had more than doubled *after* the acid reduction plants had become operational.[6]

He explained his findings in two ways. TCC simply mined and smelted a great deal more copper ore in recent years than in the past. Production roughly doubled from 241,855 tons in 1904, at the start of the case, to 439,365 tons in 1909. It followed that even if the acid plants captured a portion of the increased amount of sulfur, a far greater amount nonetheless entered the air. The second explanation related to the first. Acid condensation could capture most of the sulfur, but only if plant capacity corresponded to the amount of sulfur gases generated by smelting. Bowron determined that TCC "deliberately increased their production year by year with the full knowledge that the acid plant that they were erecting could not at best collect the acid fumes even then made." He further asserted that "these gentlemen have not acted in good full faith in attempting to remove the poisonous gases." The imbalance could be corrected either by increasing the capacity of the acid plants or by reducing the amount of ores smelted. TCC did neither. It was content to allow a large amount of untreated fumes to escape into the atmosphere because the copper and acid it produced were highly profitable, notwithstanding the waste of sulfur dioxide gases. Bowron then pointed to the shared interests between TCC and the International Agricultural Corporation (IAC). (Headquarters for both were at the same address, 61 Broadway, New York City.) TCC declined to increase acid production, he suggested, because greater production would cause a fall in the market price for sulfuric acid, and thus for superphosphate fertilizer. And lest that happen, the copper company determined, "it pays to run the risk of possible penalties and adverse verdicts from the courts."[7]

Governor Brown reinforced the Bowron report with a more conservative report by R. E. Stallings, the current state chemist. Stallings determined the percentage of escaped gas to be 55 percent of the sulfur content of smelted ores, rather than the 70 percent calculated by Bowron. Still, that yielded a daily release of "191 tons of sulfur, equivalent to 381 tons of sulfur dioxide." The findings of both reports substantiated the concerns of the National Farmers Union. Acid condensation worked, but only to a point; the rest of the fumes entered the air as always. The claim "No Fumes Now in Ducktown" fell under the weight of the numbers.[8]

Attorney General Hall sent the Bowron and Stallings reports to Tennessee Copper counsel, Howard Cornick, on December 10, 1910. Cornick and the company's officers immediately grasped the implications of the findings and elected to settle the Georgia case. The terms of agreement centered on the issue of overproduction identified by Bowron. TCC agreed to limit smelting during the growing season, from May 20 to August 31, to a level that matched the capacity of the acid plant. During the season, the company "shall not operate more green ore furnaces than are necessary to permit of operating its sulfuric acid plant at its normal full capacity." It also promised to verify compliance by submitting monthly production reports to the state during the season. In exchange, the state agreed to suspend injunction proceedings against the company until the October 1913 term of the Supreme Court.[9]

The terms reflected an understanding that the greatest consequence of sulfur dioxide pollution was to seasonal crops and to deciduous timber. If the gardens, orchards, field crops, and hardwood timber stands were protected in the spring and summer months, nothing would be harmed by allowing excess production in the fall and winter. The parties failed to heed the warning contained in the 1905 report of U.S. forester Alfred K. Chittenden, where he noted that white pines, hemlocks, and other evergreens in the Ducktown Basin were especially susceptible to sulfur dioxide fumes because they retained their foliage throughout the year. Nor did the agreement account for the cumulative acidification of local soil and water.[10]

The settlement, for all of its weaknesses, stabilized the shaky truce that had existed between the state and TCC since 1907; it also suggested that a permanent settlement could be reached on similar terms. The 1911 General Assembly easily confirmed the agreement after enjoying another successful round of junket diplomacy arranged by TCC to discourage any resolutions calling for resumption of the Supreme Court case. The company also placed illustrated articles in the newspapers to win over popular opinion during the session. The full-page headline of one read, "Conservation of Injurious Gases of Ducktown Copper Plants Results in Enormous Benefit to Farmers of the South."[11]

Unlike its neighbor, the Ducktown Sulphur, Copper & Iron Company refused to settle with the state. Company officials misconstrued the Bowron report as an exoneration of their operations in Isabella. Bowron did in fact deal lightly with the company by noting that DSC&I produced only half the volume of smoke generated by TCC; the former ran only two blast furnaces compared to the latter's seven. The English firm also enjoyed a closer ratio between the amount of ore smelted and the volume of acid produced, meaning that it processed a higher percentage of smelter gases. Bowron noted that the ratio would improve beyond current levels if the company ever completed several partially constructed acid chambers. That was unlikely to happen soon because, as local managers told him, "their English directors did not want to assume the expense of finishing them ... with an injunction from the Supreme Court of the United States hanging over their heads."[12]

Despite the findings of the Bowron report, Georgians were in no mood to make allowances for the English firm as tensions over the smoke issue increased in 1912 and into 1913. DSC&I had better numbers than its rival, but it still released a tremendous amount of sulfur fumes into the local skies. Timber baron Will Shippen, the Farmers Union, and North Georgia legislators all pressed for action against DSC&I and found Governor Brown to be responsive. The company rejected or evaded settlement overtures from the governor and from Attorney General Felder. With settlement attempts unavailing, Felder wrote Ligon Johnson to ask if there were any other technological remedies to be considered. Johnson had gained extensive knowledge on the subject from his service as special counsel for Hart in the earlier Georgia litigation and during his present service as special counsel for Theodore Roosevelt's administration in its litigation against smelting companies in the West. Johnson replied, "Up to the present time we have been unable to ascertain anything which now offers any real remedy, other than converting the fumes into acid." The lack of a new technological fix made it all the more important that DSC&I agree to limit its production during the growing season—or that it be punished for its refusal to do so.[13]

In the meantime, the Tennessee Copper Company was eager to extend its settlement with the state when the initial agreement came up for renewal in 1913. The company sweetened the deal by agreeing to participate in regular arbitration to resolve citizen claims for smoke damage, and further agreed to tender $16,500 per annum to fund payment of approved claims. Governor Brown and Attorney General Felder were delighted. So was the General Assembly. It passed a resolution authorizing acceptance of the new TCC agreement and then, in a second resolution, issued an ultimatum to DSC&I. The second measure instructed the governor to offer settlement with the Ducktown Company on the same terms offered to Tennessee Copper Company, "or to take a decree of injunction." The English firm again refused settlement, leaving the governor (now John M. Slaton—the administration changed on the eve of the legislative session) no choice but to direct Attorney General Felder to renew Georgia's case before the U.S. Supreme Court.[14]

The Ducktown Company's refusal to settle was a colossal legal blunder. The company consistently overestimated the strength of its legal position. It wrapped itself in the now three-year-old pages of the Bowron report, having become convinced that the document eliminated its liability exposure for the smoke. It also placed too much reliance on a recent victory in a timber suit brought by J. P. Vestal, another of the industrial loggers in the Ducktown Basin. The company cited the victory in a letter to Governor Slaton as an explanation for its refusal to settle. It was a bad political move for an English-owned corporation to crow to the governor about a victory over a prominent Georgian. It was worse as an exercise in legal analysis. DSC&I lawyers failed to recall the state sovereignty rationale in *Georgia v. Tennessee Copper Co.* (1907). When the Supreme Court granted the state's right to an injunction, Justice Oliver Wendell Holmes took pains to emphasize that Georgia's smoke suit rested on a different footing from that of a private action for nuisance. Georgia sued "for an injury to it in its capacity as a quasi sovereign," which "was more entitled to specific relief than a private party might be." The company's success in a private suit by a timber baron did not speak to the company's exposure in a suit by a sovereign state.[15]

Beyond its legal position, the Ducktown Company's refusal to settle was typical of its conservative business philosophy. The company gained a ten-year head start over the Tennessee Copper company when it began mining in 1891 and then squandered its lead by pursuing a policy of cautious financing and moderate growth. The Tennessee Copper Company began mining operations in 1901 with the intention of dominating the local industry. In the space of just four years, it had seven blast furnaces compared to DSC&I's two. It was the first to build an acid plant and took the risk of building a two-hundred-ton facility with unproven technology and would build another just as big in 1916. TCC's directors gambled big on business but were not inclined to gamble on a possibly fatal lawsuit when it could be resolved at a tolerable cost without interrupting the stream of profits to the shareholders. Reduced production in the summer months and placing an ante of $16,500 on the arbitration table were a small price to pay to avoid an injunction. The DSC&I directors in London, being confident in the law, considered the settlement terms an unwarranted burden on earnings.

Distance also impacted the decision. The London management of the Ducktown Company was deaf to the political voices in Georgia and instead rested upon its legal and factual defenses. Ducktown had logic on its side regarding the smoke: it generated only half the amount of smoke made by TCC; it was three miles from the state line compared to TCC's location just a few hundred yards away; the DSC&I smokestacks were short; and TCC's rose more than three hundred feet to cast smoke throughout the basin. The facts had some bearing in the courtroom but did not alter the political climate in Georgia. Smart lawyers know that major cases occur within a political context that may have as much bearing upon the client's business as the logical merits of legal arguments.

TCC's American officers and directors settled because they understood that Georgia's politicians needed to deliver a victory to their mountain constituents. Voters wanted a tangible remedy. Hart provided a partial one with his 1907 win that led to construction of the acid plants, but that was six years ago, and matters had yet to proceed to an injunction. The smoke problem was getting worse, not better. Elected officials needed to deliver fresh proof of their commitment to constituents. They needed and wanted to hunt big political game. Will Shippen, the Farmers Union, and mountain legislators were happy to serve as the guides. Tennessee Copper saw the hunters coming and settled while it could on terms that amounted to a consensual injunction during the summer months plus a politically popular, well-funded arrangement to promptly settle new claims through arbitration. With that accomplished, the hunters pointed their guns at DSC&I as the only remaining target. And the stubborn English company refused to dodge the bullet.

The Ducktown Sulphur, Copper & Iron Company stood alone when Georgia returned to the U.S. Supreme Court in 1914 for a final order of injunction. It had been seven years since John C. Hart and Ligon Johnson won the right to obtain an injunction. That fight had required difficult arguments involving novel and untested issues of constitutional law. The state's burden in 1914 was much simpler: it had to show only that there was still a need for a smoke injunction and that DSC&I deserved to be sanctioned.

To that end, Attorney General Warren Grice repeated the successful strategy used by Hart and Johnson: he relied on scientific evidence from government experts. Three federal foresters, Dr. George W. Hedgecock, H. L. Johnson, and E. B. Clarke, testified for the state. Most of their testimony consisted of the same sort of surveys about the geographic extent of tree damage as used in earlier phases of the Supreme Court case. This time the foresters spoke from the perspective of a neighboring landowner. By virtue of the 1911 Weeks Act, the federal government now owned more than twenty-five thousand acres of forestland in the Unaka Mountains near Ducktown, the beginnings of what are now the Chattahoochee and Cherokee National Forests. The government was not a party to proceedings against DSC&I, but with its closest forest holdings being only four miles from the company's Isabella complex, it now had a motive to provide testimony on the impact of Ducktown sulfur smoke.[16]

In addition to the new forest reserve on the edge of the basin, the federal government continued to study the area as part of its ongoing research into forest conditions in the Southern Appalachian Mountains. The U.S. Geological Survey described it at length in a regional study of forest loss and erosion that culminated in a book-length report written by Leonidas Chalmers Glenn, *Denudation and Erosion in the Southern Appalachian Region and the Monongahela Basin* (1911). Glenn observed within the Ducktown Basin that the bottomlands so valued by farmers have "been buried beneath the rapid accumulation of waste from the hillsides, so that both hill slope and flood plain have been destroyed, the one by erosion and the other by sedimentation or aggradation." Then, as predicted by George Perkins Marsh, he noted that "abnormal denudation and erosion has also affected the underground water level in the region." Despite the sixty inches of annual rain, "wells have been doing dry," and many springs that "supply water to the miners' families, flow less than formerly."[17]

In Potato Creek, which ran through the heart of the mining district, silt accumulated so rapidly that "telephone poles have been buried almost to their cross-arms, and highway bridges, roadbeds and trestles have either been buried by the debris or have been carried away by the floods." The normal flow of the creek was now "half as large as it used to be." Downstream on the

Ocoee, every large flood clogged the river with silt "as to prevent the running of the two ferries at the smelter until the river has had time to scour its channel again." Glenn concluded, "The Ducktown region is, then, not only an impressive object lesson, but an emphatic warning of the extent and character of the disaster that may result in these southern mountains from the thorough destruction of the forests."[18]

DSC&I attempted to blunt the testimony of the government experts by offering its own expert, Professor S. M. Bain, a University of Tennessee botanist and the author of "The Action of Copper on Leaves, with Special Reference to the Injurious Effects of Fungicides on Peach Foliage." The extent and topical focus of his research was not sufficient to effectively counter the mass and specificity of the evidence from the federal scientists, especially when the Supreme Court had relied on the latter in its 1907 decision. The copper company then attempted to raise a question of causation by arguing that local forest damage was at least in part the result of the dreadful chestnut blight that was just then entering the southern mountains.[19]

First discovered in New York City in 1904, the blight was already present in the Appalachians at the time Georgia renewed its Supreme Court case in 1914. It was a terrible scourge. Environmental historian Ralph H. Lutts states that "in less than fifty years" it "killed an estimated 3.5 billion trees, the equivalent of over 9 million acres of pure chestnut stands." Chestnut trees once constituted a fourth of the Southern Appalachian forests. Their loss was disastrous for the health of the forests and equally so for the many people who relied on them. They gathered the edible nuts for food and for sale. Their hogs grew fat on the nuts, or mast, left scattered on the forest floor. They logged the trees for their valuable timber, a durable wood for outdoor uses, and for the tanbark used by the leather tanning industry.[20]

J. A. Fowler, a DSC&I defense attorney, cross-examined each of the government foresters about the disease. They all knew about it, and some had conducted research on it. They knew of its presence near Knoxville, Tennessee, and in western North Carolina but did not yet detect it in the forests around Ducktown. The answers disappointed Fowler, so he then asked Dr. Hedgecock whether sulfur fumes would help chestnut trees by acting as a fungicide. The scientist replied, "I doubt it very much."[21]

Beyond the scientific testimony, both sides offered hundreds of affidavits and depositions from witnesses regarding the extent of smoke damage, or the lack thereof, to local crops and forests. Attorney General Grice appointed NFU counsel Jesse A. Drake to serve as special counsel for the state to obtain the testimony of farmers and timber owners. The appointment placed Drake at the center of events where he could monitor the state's case for its impact upon the scores of individual smoke damage claims he handled for members of the National Farmers Union. At the same time, the move transformed the role of the Farmers Union from that of an outside critic to a comrade in arms. The NFU and the state would share the praise or share the blame, however the case turned out.[22]

Drake sat through the hundreds of depositions from foresters, engineers, timber owners, miners, merchants, and farmers, a task that combined high-stakes legal practice with repetitious, mind-numbing tedium. Each property owner spoke about his or her experience of the smoke and stated the distance in "air miles" from the farm to the smelters. It was on this occasion that B. H. Sebolt, the elderly farmer of 110 acres in the Blue Ridge Mountains of Fannin County, spoke of watching smelter smoke flow south across the basin from the smelters to his vantage point at Mr. Trammel's mountainside orchard. The lawyers first stipulated that he lived eight miles from the TCC works at Copperhill and ten and a quarter miles from the DSC&I smelters at Isabella. When asked his age, he replied, "Going on seventy-eight years old; if I live until next Christmas, I will be seventy-eight."[23]

Though the statement of his age went without discussion, it bears mention here because his life embraced the entire chronological span of Ducktown's copper industry. Sebolt was born in 1836, at the end of President Andrew Jackson's administration. He was a toddler at the time of the Cherokee Removal in 1838 that opened the way for white settlement and the birth of the copper industry. He survived the Civil War, and then began farming his mountain spread around 1870 when Captain Julius Eckhardt Raht still dominated the first era of Ducktown mining. He lived in the basin through the twelve-year suspension of the copper industry after 1878 and remained when it revived in 1890. He saw the renewed industry grow year by year and watched the smoke extend further and further southward until it reached his farm. And now, as an old man, he was a witness in the Supreme Court's first air pollution suit, where he spoke of smoke damage to his garden of peas, beans, potatoes, cabbage, and corn.[24]

Defense attorneys cross-examined Sebolt and other plaintiff's witnesses about the number of times they had sued the copper companies or had testified on behalf of another claimant. This proved a source of embarrassment to Will Shippen when he made his familiar appearance as a star witness. Ducktown's lawyers knew about his many lawsuits and business dealings and, with the documents in hand, portrayed him as a plaintiff who led public opposition to the copper companies as a means of personal gain. They forced him to acknowledge an affidavit he gave in a shareholder suit where he boasted of his services to his lumber company by serving as a ringleader in the smoke wars. In the document he stated that "after almost numberless appeals to the public authorities, to the newspapers, to scientific magazines, and using every influence in the world," he and his brother, Frank, "finally got the movement into a compact, well-organized condition where it presented a formidable front to the further infliction of the damage." He boasted that his support of Hart's 1907 victory boosted company stock values: "I consider the Shippen Brothers Lumber Company worth 25 percent more today than it was with this thing going on." Other documents established that Tennessee Copper settled with Shippen Brothers Lumber for $50,000 and paid the brothers an additional $175,000 for stock in their lumber company. Ducktown counsel could now argue that Shippen and TCC were in league as fellow shareholders and that Will's personal claim against DSC&I was actually a subterfuge for TCC's attack upon its rival.[25]

DSC&I lawyers also established another line of argument, a blame-thy-neighbor defense by which it blamed the Tennessee Copper Company for the smoke damage. In dry legalese, the argument ran thus: "Whatever injuries are now, or have been for the last five years, done by sulfur fumes to the forests, crops, gardens, and all other forms of vegetation in North Georgia, are being and have been produced by smoke escaping from the plant of the Tennessee Copper Company, and not by smoke emitted by the furnaces of this defendant." The lawyers then made the familiar comparisons about the relative size of the two companies, the different heights of their smokestacks, their respective distance from the Georgia border. Strategically, the argument was a version of the familiar trial tactic known by lawyers as the "empty chair" defense, in which blame is cast upon a party that is not in court. The defense is often successful, which is why plaintiffs' attorneys often seek to name as many potential defendants as they can to prevent its application. In the present context, TCC had been effectively removed as a party by virtue of its settlement with the state, leaving the Ducktown firm free to blame it for all of the damage. The argument was easy to make but hard to evaluate because of the difficulties in distinguishing between the smoke generated by one copper company and that created by the other. Both companies processed similar sulfide copper ores, and both produced basically the same toxic smoke. Their proximity within a mountain bowl allowed the smoke of both firms to frequently blend into an undifferentiated whole.[26]

The Court was disinclined to wade through the competing expert and lay testimony. Justice James Clark McReynolds expressed the Court's frustration, writing that the evidence "does not disclose with accuracy the volume or true character of the fumes which are being given off daily." Still, the weight of government testimony, the proven toxicity of sulfur dioxide to vegetation, and the dramatic reality of the ever-expanding Ducktown Desert prompted the justices to rule in favor of the state of Georgia. The need for action was evident, and DSC&I was the only defendant before the Court, and thus the only one to become subject to its injunction power. The Court declared that though the present evidence made it "impossible . . . to ascertain with certainty the reduction in the sulfur content of emitted gases necessary to render the territory of Georgia immune from injury," the state was entitled to relief. The company now had to pay for the folly of refusing to settle with the state when it could.[27]

The Court issued an injunction limiting DSC&I production in two ways. The company "shall not permit the escape into the air of fumes carrying more than 45 percent of the sulfur contained in the green ore." Next, "It shall not permit escape into the air of gases the total sulfur content of which shall exceed 20 tons during one day from April 10th to October 1st of each year or exceed 10 tons in one day during any other season." The vote was 6 to 3. Justice Charles Evans Hughes dissented, joined by Chief Justice Edward Douglass, who questioned whether "the evidence justifies the decree limiting production as stated." Justice Oliver Wendell Holmes, author of the 1907 opinion in Georgia's favor, also joined in the dissent. In his earlier opinion, Holmes gave warning of his reluctance about an injunctive remedy at Ducktown, and now, in 1915, he made good on the warning in his dissent.[28]

The Court addressed the statistical weakness of the case by appointing Dr. John T. McGill, a Vanderbilt University chemist, to conduct a six-month study of sulfur emissions by DSC&I and to assess their effect on vegetation. McGill conducted the most detailed study to date in the region, using a variety of approaches including analysis of production statistics, systematic air sampling, and field studies of vegetation damage. He substantiated the state's arguments about the persistent escape of gas. Daily tests documented that the company was generally compliant with the new limits. He made no attempt to allocate historical damage between the two companies. He did leave an intriguing remark about the current wave of private smoke suits from Georgians. After comparing the list of suitors to the locations claimed to be within the vicinity of the DSC&I plant, he concluded "that no visible injury of any consequence was done to vegetation on these lands by the smoke of the Ducktown Company's smelter during the period of our investigation." The Court accepted McGill's report and then slightly increased the daily limits from twenty tons to twenty-five tons during the growing season, and from forty tons to fifty tons in other seasons.[29]

The English company now found itself at a competitive disadvantage, one that it could have avoided had it elected to settle with the state. The Tennessee Copper Company gambled on business and won. The Ducktown Sulphur, Copper & Iron Co. gambled on the law and lost—heavily. The court-imposed limitations on it were much more severe than the limits TCC accepted in its voluntary settlement with the state of Georgia. The TCC settlement agreement obligated it to match the level of smelting activity to the capacity of its acid conversion plants during the three and a half months of the growing season. There was no fixed limit on the *amount* of sulfur emissions during the summer, and so the company could legally double production so long as it increased the capacity of its acid operations in proportion. TCC was free to operate outside the growing season without any limitations at all. By comparison, the Supreme Court decree placed Ducktown under year-round limitations on the amount of sulfur emissions with one fixed limitation for the growing season and another for the remainder of the year. In effect, DSC&I could not increase production unless it invented some new technology that harvested more than 45 percent of sulfur fumes. The company already employed state-of-the-art techniques; it could not operate any more efficiently than it already was. Tennessee Copper was free to grow larger, even if that meant greater releases of sulfur fumes, so long as it honored the proportionality rule during the summer. Production at the Ducktown Company was capped and would remain so unless and until the Supreme Court altered the injunction.

In 1916 the Tennessee Copper Company was free from the threat of an injunction but not free of political troubles in Georgia. Renewal of its settlement agreement with the state was in serious doubt because of continued complaints of smoke damage and anger about the claims arbitration system. The state's victory against DSC&I increased the risk that failure to renew the agreement would leave it exposed to a Supreme Court injunction against itself. The 1913 settlement at first met with great approval, especially regarding the terms relating to the arbitration program and the $16,500 paid annually by the company to fund it. Claimants were generally satisfied with the size of the awards issued in the program's first year. Most were in the range of $100 to $500, with a surprising number tending to the higher end of the range. The payments matched or exceeded the amounts that farmers could expect to recover at trial in the Polk County Circuit Court. Plaintiffs' attorneys were also happy because the governor consented to send arbitration checks directly to the lawyers rather than the clients, making it easier for the lawyers to collect their fees. The trouble came in the following years when the amount and number of awards plunged. Claimants were especially upset when the arbitrators failed to allot the entire annual settlement fund of $16,500.[30]

The anger among smoke claimants toward one of the arbitrators, J. J. Brown, was at a level that jeopardized renewal of the agreement. The arbitration system provided for three arbitrators, one for Tennessee Copper and one for the smoke claimants, with Brown filling the third slot as umpire to resolve differences between the other two. J. F. Holden, a Fannin County banker, wrote Governor Nat Harris on the eve of the 1916 General Assembly advising that "settling damages by arbitration is an absolute farce." Holden added, "Umpire Brown has used his position from beginning to end to give the Copper Company the long end of the deal." It was a difficult political problem for the governor. His predecessor appointed J. J. Brown to the panel because of Brown's stature as the current state president of the National Farmers Union. He was supposed to be politically and socially aligned with the farmers, especially the farmers in the highly organized NFU chapters in Fannin and Gilmer counties. The farmers rejected their spokesman and wanted Harris to fire him. Adding to the difficulty was J. J. Brown's status as Tom Watson's protégé. Watson was perhaps Georgia's most powerful politician and a man whom elected officials offended at their peril. Every political figure in the state knew that Hoke Smith won the governorship in 1906 with Watson's backing. They also remembered that Smith then lost in 1908 when Watson transferred his support to an opponent after becoming miffed over Smith's refusal to help a personal friend. Watson was the kingmaker. To make matters worse, Brown was running for commissioner of agriculture with Watson's backing.[31]

Will Shippen did all he could to raise the level of political heat by engaging in a noisy exchange of column-length letters in the *Atlanta Constitution* and other Georgia newspapers. He published a column that initially lauded Brown as the editor of Watson's populist magazine, the *Jeffersonian*. That said, Shippen then criticized Brown's performance as umpire in the smoke arbitration, alleging that he was in the pay of Tennessee Copper Company. Brown retaliated that Shippen was maneuvering to take his umpire job. Shippen countered by boasting that "everyone who knows me knows that I am not a $125-a-month man, the known salary that the copper company pays Mr. Brown each month." He added, "Mr. Brown evidently is a $125-a-month man, as shown by the grim hold he retains on his present little position." This was to suggest that Shippen's reputed wealth placed him above monetary motives. Brown retaliated with evidence from the recent Supreme Court proceedings establishing that Shippen had received $50,000 from Tennessee Copper, and so it continued, tit for tat.[32]

At the local level, one account reported that "a mass meeting of farmers and citizens" was held in the Fannin County seat at Blue Ridge where a resolution in favor of renewing the settlement agreement and retaining Brown as umpire prevailed on a vote of 171 to 1. Homer Legg of nearby Morganton responded with a letter to the editor alleging that the meeting was "a huge joke" because "the farmers of Fannin County knew practically nothing of any such meeting." Instead, the meeting "was reported to have been composed almost entirely of employees of the Tennessee Copper Company who were rushed into Blue Ridge by the car loads to dominate this meeting." Legg added that "the company not only paid their railroad fare, but their time as well." In Gilmer County, the local chapter of the National Farmers Union petitioned Governor Harris for Brown's removal on the argument that "there is a time when patience is no longer a virtue." Continuing, they said, "our mountain people are as independent and spirited as any in our good state, and we fail to understand why our chief executive should delay a moment in affording us the relief," which the Supreme Court "has repeatedly said we are entitled to receive." Even the Gilmer County Board of Education condemned Brown.[33]

In the General Assembly, attorney John D. Little lobbied for TCC while J. A. Drake and Lamar Hill acted in their many capacities as special counsel for the attorney general, as counsel for the National Farmers Union, and as private counsel for scores of private farmers (and NFU members) to make the argument against renewal. On the floor, A. H. Burtz of Gilmer County aligned with Shippen against the measure, and W. Y. Gilliam of Fannin spoke for renewal on behalf of TCC. More charges of interest were exchanged. Gilliam challenged Burtz for representing Shippen in a large smoke damage suit. Burtz retaliated by noting that Gilliam was vice president of the Bank of Copperhill, which handled the payroll account for TCC.[34]

In the executive offices, Attorney General Felder wanted to end the settlement agreement. He wrote Governor Harris on June 26, 1916, advising that notice of termination be issued to TCC. Governor Harris was probably tempted to let the unpopular settlement agreement die, rather than renewing it at the price of firing Brown. He instead chose a subtler approach. He made an inspection tour to Ducktown to demonstrate his interest and took care to review petitions from both sides. Then, in his annual message to the General Assembly, he acknowledged public dissatisfaction with the agreement and said he was leaving it to the General Assembly's disposal, though not without hinting in favor of renewal. Without an agreement, he warned, Georgians "will be left to their remedies in the Tennessee courts." The state would have to move for an injunction against TCC with all the risk of economic disaster to Georgia mine workers and the fertilizer industry if TCC was to be shut down. As to the umpire issue, Governor Harris knew that Brown was in the race for commissioner of agriculture and was confident that he would win it (which he did). If so, the duties of Brown's new office would force him to resign from the arbitration panel, leaving Harris free to pick a replacement without angering Tom Watson. As the political dust settled, the General Assembly renewed the Tennessee Copper contract, allowing the threat of injunction against the company to quietly die another death. For all of their hostility to the settlement agreement, farmers fully participated in the arbitration process by submitting 1,040 claims to the arbitrators at the end of the 1916 growing season, for amounts ranging from $10 to $3,000.³⁵

World War I may have been another element tilting toward renewal of the settlement agreement with the Tennessee Copper Company. By 1916 England, France, and Russia had been at war with Germany for two years, and signs were growing that the United States would become involved as a supplier of war goods, as a combatant, or as both. The two local copper companies produced two kinds of valuable war materiel. Their copper was used in its pure form as wiring for electrical and communication equipment. As brass, copper became bullet casings, artillery shells, and uniform belt buckles. As bronze, copper was cast into propellers that moved ships and their cargoes to the Allies. Sulfuric acid, Ducktown's second major contribution to the war, was key to the manufacture of explosives. All of this meant that demand and prices for Ducktown products would be high. On August 20, 1915, the *Copper City Advance* happily declared, "Big War Orders for the South—Copper Basin Gets 17 Million Dollars' Worth—Every Portion of the South Busy Making Contracts and Filling Orders."³⁶

The sky was the limit, unless contracts and court orders dictated otherwise. In 1915 the Du Pont Company needed 1 million tons of acid to make explosives. Tennessee Copper wanted to fill it but was bound by its long-term acid contract with the International Agricultural Company. The contractual restriction enabled IAC to benefit from the Du Pont order by reselling TCC acid at a handsome markup. IAC enjoyed a sixfold profit on TCC acid. The

New York Times reported that "the contract calls for a price of $4.81 a ton to be paid ... by the International," which the latter "then sold again above $30 a ton." IAC ran its acid operations through its Atlanta office, so some of the Du Pont money was bound to flow into the state.[37]

Tennessee Copper was determined to reap profits for itself. It built another large acid plant in 1916 to increase production from 210,000 tons of acid per year to 300,000 tons. More than 5,000 new people moved into the basin to work on all the new construction. A salesman named Virgil Hyatt rejoiced: "There is one of the biggest building booms going on up there you ever saw." "Everybody is excited," he said, "just like the old gold fever days." Officials at TCC explored another way to benefit from the wartime market. They believed they found a loophole in the contract to the effect that IAC had the right to acid originating only from ore mined by TCC, not from ore that it processed for other mining companies. If so, then TCC could bypass IAC by selling acid made from the other ore directly to the open market and reap the profits without the interference of a middleman. The Tennessee company knew that its aggressive contract interpretation would lead to litigation with IAC, but it was willing, as always, to take a calculated business risk. As for the copper, TCC was free to produce and sell all it could make, subject only to its agreement with the state to reduce production during the growing season.[38]

Over in Isabella, the Ducktown Sulphur, Copper & Iron Company suffered in two ways. Like TCC, it was subject to the IAC contract for acid. Worse, it was bound by the production limits imposed by the Supreme Court. DSC&I could not increase production to serve the profitable war effort without violating the terms of the injunction. The injunction also prevented it from expanding production, as its rival had done. With the restrictions in place, W. H. Freeland, the DSC&I general manager, was legally helpless to return his company to a competitive posture vis-à-vis TCC. Nothing would change unless the terms of the injunction were modified by the Court or by an agreement with the state of Georgia.

Back in London and much closer to the Great War being fought across the English Channel, the DSC&I directors now realized the enormity of their earlier refusal of settlement. They were now desperate to undo their error by belatedly asking the state for an agreement similar to the one they rejected three years earlier. They were too late in asking: the recent Supreme Court victory removed any motivation for the state to negotiate. Georgia officials were content with the injunction, and so were the voters. The new attorney general, Clifford Walker, inherited a victory obtained by his predecessors. He had nothing to gain politically, and much to lose, if he voluntarily relaxed the terms of the injunction. The same was true for Governor Nat Harris and the General Assembly.

There being no chance for an agreement in Atlanta, the Ducktown Company's lawyers, W. B. Miller and James A. Fowler, tried and failed to have the Supreme Court dismiss the case. They then began to work the political side of the matter by asking Tennessee's Governor Tom C. Rye and Attorney General Frank M. Thompson to intervene in the case. The request came with petitions signed by many people from the mining community in Polk County. The motion to intervene was more than a little ironic. In 1903, at the outset of Georgia's case, Tennessee's Governor Frazier disclaimed any legal authority to compel abatement of smoke from the copper companies. His state then remained outside of the litigation for the next twelve years. Now, in 1916, Georgia's attorneys asserted that "if the state of Tennessee ever had any right to intervene, it has forfeited such right," because "it has waited until the court has handed down a final decree before taking any action in the matter." As J. A. Fowler conceded, "it is rather difficult for the state to set forth ... legitimate grounds to become a party in the litigation." The Supreme Court agreed: on April 16, 1917, it entered a one-sentence order denying the state of Tennessee's motion to modify the injunction. Georgia's injunction decree remained intact for the moment.[39]

The Supreme Court issued its latest order in the case only ten days after the United States declared war on Germany. Though disappointed by the Court's decision, Ducktown's lawyers realized that America's entry into the Great War would allow the company to employ a new approach and a new theme to the smoke litigation: patriotism. The argument was simple. American industry needed to mobilize for the war effort. President Woodrow Wilson created the War Industries Board, led by Bernard Baruch, to coordinate the efforts of a wide array of industries to that end. Copper and sulfuric acid were needed for the war effort; there was thus a patriotic duty to maximize production of both commodities to fight the Germans.[40]

Baruch brought sulfuric acid within his scope through the actions of the Advisory Commission of the Council of National Defense, Committee on Chemicals. On August 11, 1917, the committee surveyed sulfuric acid supplies and then reported "evidence of approaching shortage increases," adding, "Plant capacity [is] already insufficient and all should be operated all the time." The committee observed, "Prior to 1917, over 60 percent of the acid in the country had been made from Spanish ore." Plants that depended on foreign ore now faced declining production as German submarines torpedoed and sank cargo vessels at an alarming rate. The copper companies of the Ducktown Basin were not dependent on foreign ore because they made their acid from locally mined ores. The success of enemy U-boats thus heightened the importance of the southern copper and acid industry.[41]

W. B. Miller realized that the War Industries Board provided a new layer of influence and a new national agenda that could prove useful in freeing DSC&I from the injunction. He had finally realized that political power was a force equal to or greater than the law alone. His letter of January 9, 1918, to NFU counsel Jesse Drake revealed the change in his methods. Instead of arguing the law, Miller devoted the first four paragraphs to a description of his meetings with Tennessee's congressional delegation, the War Industries Board, and the Commission on Car Service (which controlled the movement of railroad tank cars for acid). Not until the fifth paragraph did he mention, "the Government is making such heavy demands of this Company—we should show the War Department . . . the possibilities of increasing our output if restrictions were removed." He closed by saying, "I would rather approach the matter from a purely patriotic viewpoint." Miller's February 7 letter to Georgia's Attorney General Walker showed further advances in his newfound political techniques. He mentioned a visit to the Fuel Administration and then to the secretary of the navy: "The possible modification of the injunction was laid out before the Navy Department, supplemented by explanatory letters from our two Senators. It seems the Government is in extreme need of acid, and of course requires copper."[42]

Miller's Washington lobbying resulted in a letter dated February 14, 1918, from Secretary of the Navy Josephus Daniels to Georgia's Attorney General Clifford Walker. The secretary made his intentions clear with his first sentence: "It has been brought to my attention that the country's production of sulfuric acid and copper will be materially increased if arrangements can be made so that the Ducktown Sulphur, Copper & Iron Company, Ltd. and the Tennessee Copper Company can operate at their maximum capacities." The secretary added, "anything which can reasonably be done to increase production cannot fail to be a distinct national service." Walker sent the letter to Governor Hugh M. Dorsey on several occasions, the second time with a new threat that the federal government would intervene in the copper case, as "a war measure."[43]

It was not a bluff. The solicitor general, John W. Davis, filed an amicus curiae (friend of the court) brief in the Ducktown case in May. The brief contained the text of a letter revealing a three-way correspondence between Secretary of the Navy Daniels, Bernard Baruch, and the nation's attorney general, all to the end of securing modification of the injunction. Davis ended the brief with a request to the Court that if Georgia objected to the requested modification, "it should be overruled, because any possible damage that could result from the slight modification asked ought not to weigh against the need of the Nation, in its hour of peril, to use its resources for the common good and the safety of the country."[44]

W. B. Miller's skillful Washington lobbying transformed the case of *Georgia v. Tennessee Copper Co.* in much the same way that the state of Georgia's Supreme Court strategy had transformed the private nuisance actions in the Tennessee courts. On both occasions, the status quo was upended by inserting new themes and new players into a story that moved to new venues. Ducktown smoke litigation began as a mismatched contest in the Tennessee courts between a group of a dozen small mountain farmers and two powerful corporations. The copper lawyers, P. B. Mayfield, James G. Parks, and Howard Cornick, befuddled the mountaineers in the courts and in the Tennessee General Assembly until time and persistence taught the claimants and plaintiffs' lawyers how to file successful lawsuits. That is not to say that the farmers were powerless. The three smoke injunction cases gathered under the caption *Madison v. Ducktown Sulphur, Copper & Iron Co.* nearly shut the industry down until eventually overturned. Georgia entered the smoke wars in 1903 at a time when the cases were local affairs tried as private actions for nuisance. Governor J. M. Terrell and Attorney General Hart wielded the sovereign powers of the state to move the focus from Polk County to the U.S. Supreme Court in Washington. Hart and his assistant, Ligon Johnson, then aligned the case with the values of the national conservation movement, so that the case became an issue of natural resources as opposed to a damages issue concerning reduced tax revenue and related property claims. W. B. Miller accomplished another transformation in 1917 by recasting sulfur production as a crucial component of the war effort. He then brought the case to the attention of some of the nation's most powerful leaders. Governor Dorsey and Attorney General Walker knew they were unlikely to prevail in the Supreme Court against the triumvirate of Bernard Baruch, Josephus Daniels, and John W. Davis. Georgia's insistence upon preserving the terms of the DSC&I injunction began to look petty and unpatriotic.

News of the letter from the secretary of the navy found its way to the newspapers, alerting all stakeholders to a potential modification of the injunction. The gist of the modification was to allow Ducktown Sulphur, Copper & Iron to sign a contract like that made with the Tennessee Copper Company. Will Shippen had his lawyer write an eighteen-page brief for Governor Dorsey and Attorney General Walker advising them that they lacked the authority to agree to a modification of the injunction decree. Walker agreed and advised that "the applicants be referred to the Legislature if they desire to take any action before the convening of the Supreme Court." Neither he nor the governor was eager to decide the matter on his own.[45]

Will Shippen wrote yet another column in the papers, this time arguing that the federal government should be thanking Georgia instead of pressing for modification of the injunction. He reasoned that "not one atom of commercial sulfuric acid would doubtless ever have been made at these copper furnaces" if Georgia had not acted "to compel them to stop the wholesale destruction of vegetation." DSC&I demonstrated its newfound skill at public relations with its own column in response. Its new general manager, W. F. Lamoreaux, made two simple points. The first was that TCC was now able to act at highest efficiency for the war effort, free of the injunction constraints upon DSC&I. The second was a somewhat disingenuous assertion: "The request for modification . . . was made under solicitation of the War and Navy departments to the United States Supreme Court, and now awaits the affirmation of the Governor and Attorney General of the state of Georgia." Though true, his statement failed to mention the substantial lobbying conducted by W. B. Miller and the Tennessee congressional delegation, much less the company's refusal to accept settlement in 1913 that made it all necessary.[46]

The issue returned to the 1918 General Assembly where it considered separate bills for and against modification of the injunction. The Fannin and Gilmer chapters of the National Farmers Union waged a battle of petitions. Citizens sent mail bags full of letters, often written in pencil on lined paper mailed from rural post offices such as Higdon and Oasis. Business people and professionals spent the extra dollars necessary to send stacks of telegrams. All of the familiar arguments for and against the injunction were made once again. The more insightful supporters of the injunction realized that they needed to respond to the argument of patriotism injected by DSC&I counsel. Linda Pack, self-identified as a "country girl about 16 years old," reframed the issue for the benefit of the farmers: she insisted that the injunction remain intact to "help make food for the boys who are fighting for me." As she saw it, food from mountain farms was even more important to the war effort than copper and acid from mountain mines. She urged the governor "to do all you can to help keep off the smoke so we can help the boys who are fighting and striving and doing their best."[47]

The arguments of the federal government proved more persuasive to legislators than the plea of a teenage girl from a mountain farm. The Committee on Mines and Mining passed the resolution for modification of the injunction and authorized settlement on a vote of 28 to 10; it then passed in both houses. On August 20, 1918, Ducktown signed the settlement agreement it had refused five years earlier. It agreed to reduce production in the growing season to a level commensurate with the capacity of its acid plant, with a daily cap on sulfur emissions set at forty tons during the growing season. The cap on summer emissions was peculiar to the DSC&I contract; there was no counterpart to it in the Tennessee Copper deal. Its presence represented a compromise on the Supreme Court injunction decree by preserving an upper limit during the growing season, though at a higher level of forty tons per day compared to the twenty-five-ton limit set by the Supreme Court. During the rest of the year, the company was free to smelt copper without the quantitative or qualitative limitations set by the Court.[48]

The company also agreed to participate in the arbitration plan. Being the smaller firm, its annual funding was set at $8,500, about half of TCC's $16,500. That meant that the two companies set aside a combined $25,000 per annum for claims. The funds were enough to cover the small farmers, provided that the arbitrators were actually willing to award the money. There was a safeguard. Arbitration was voluntary for the claimants; if the system failed, they retained the right to avoid the procedure and pursue their claims in the Polk County Circuit Court. The state added one more clause to benefit the claimants. The agreement provided that "it shall not be necessary for the claimant to specify or prove which of said companies is legally responsible" for alleged damage. Essentially, the Ducktown Company waived its long-standing argument that the larger Tennessee Copper was to blame for most of the sins of the area's copper industry. The clause simplified matters for claimants because on proof of loss, the companies agreed to pay the claim together on a split of one-third by DSC&I and two-thirds by TCC.[49]

The Supreme Court approved the settlement and continued the case for the three-year term of the agreement. As a point of law, Georgia reserved the right to seek further injunctive relief if conditions deteriorated. It never again did so. The parties only returned to the Court as necessary to file renewals of the settlement agreements and to request further continuances. With the series of cyclical renewals, the case remained open on the docket until 1937, when the Supreme Court requested the parties to stipulate to a dismissal with prejudice to bring the case to a permanent end. Yet, for all practical purposes, the case of *Georgia v. Tennessee Copper Co.* ended in 1918.

The case had certainly seemed, as Attorney General Clifford Walker observed, to be a case that "we have with us always." Active litigation in the Georgia case began in 1904 and lasted until 1918, a span of fifteen years, actually sixteen, if events leading to the General Assembly's initial 1903 resolution are counted. The private smoke wars involving individual citizens or lumber companies against the copper companies lasted much longer; individual cases came and went, but as a whole the great tide of several hundred suits lasted more than a quarter century before receding.[50]

The settlement redirected the energy of smoke suitors from the courtroom into the arbitration system created by the settlement agreements between the state of Georgia and the two copper companies. The annual reports of the arbitrators reveal that in 1921, a group of sixty-one property owners submitted claims. A few were from Copperhill, Tennessee, and the rest were from Georgians throughout the Ducktown Basin, from communities such as Epworth, Blue Ridge, Fry, Chestnut Gap, Cad, and Mineral Bluff. Each of the claims was on a preprinted form with blanks for damage to corn, wheat, rye, oats, hay, peas, beans, tomatoes, potatoes, apples, peaches, and timber. There was no line for cotton because it was not a local crop. The arbitrators made awards on thirty-nine claims (64 percent), with most of the farmers receiving $5 each, and six others receiving somewhat larger amounts ranging from $10 to $20.[51]

The meager payouts discouraged claimants in most of the subsequent years. In 1922 the arbitrators received only nineteen claims and paid an aggregate of $75 to thirteen farmers. No awards at all were paid on the seven claims submitted in 1923; of the four claims submitted in 1926, only one received payment, being in the amount of $10. There were only two claims in 1927, neither of which were honored. The onset of the Great Depression caused numbers to spike in 1929, with sixty-nine claims and a total payout of $621 spread over forty-seven recipients. It did not last: in 1930 there were just seventeen claimants, and three awards of $5 each. By 1941, only one farmer bothered to file a claim; it too was unsuccessful.[52]

Even though they lost confidence in the arbitration system, farmers did not attempt to resume the smoke wars in the local courts. The continuity required to maintain a campaign of litigation no longer existed as the state of Georgia and the NFU withdrew from the fray. Georgia's elected leaders were tired of the case and glad to see it end. The case had been handled by five Georgia attorneys general and nine gubernatorial administrations. Their Supreme Court successes in *Georgia v. Tennessee Copper Company* compelled the copper companies to implement pyritic smelting and acid condensation, but having done that, there was nothing to gain with further litigation. The current state of mining technology had little else to offer toward the reduction of smelter fumes, and the state had no desire to destroy the industry.

The only remaining matters for the state were the periodic renewals of the settlement agreements and the claims arbitration system. Future governors occasionally received letters from complaining citizens but never enough to stir a renewal of the dormant Supreme Court case against the copper companies, or rather the company: the Tennessee Copper Company swallowed DSC&I in 1936. Governors disposed of routine complaints by simply referring aggrieved citizens back to the arbitrators. Governor Eugene Talmadge followed this course when responding to a 1935 letter from Mrs. Birdie Suit about her low arbitration award, and added, "I had nothing to do with the making of the awards and for that reason I am unable to give you any information."[53]

State leaders were content to allow the Supreme Court case to die a quiet death. In 1937 Governor E. D. Rivers and the legislature readily consented when the Court requested final dismissal of the case. The state extended the now thirty-year-old pattern of settlement agreements by entering into a new one that obligated TCC to continue funding the arbitration system at the rate of $5,000 per year and limiting sulfur dioxide emissions during the growing season (April 1 to October 1) to a generous 88 tons per day at its Copperhill facilities and 40 tons per day at its Isabella plant (the old DSC&I works), for a total of 128 tons per day.[54]

The 1937 settlement and dismissal occurred without objection from the National Farmers Union because it had long ago faded from the scene. The farmers of the NFU chapters in Fannin and Gilmer counties relied upon the organization's political and legal backing in every courtroom and legislative battle from 1907 to 1918. Their reliance on the state and national leadership came to an end when NFU strength plummeted in Georgia and the other southern states soon after World War I. Low postwar cotton prices and the ravages of the boll weevil discouraged membership throughout the cotton belt. During the years of the smoke litigation, a Georgian, Charles S. Barrett, served as the organization's national president from its headquarters in Union City, just south of Atlanta. With his departure in the early 1920s, the organization effectively abandoned Georgia and the South to instead serve grain farmers of the Midwest and Great Plains. The headquarters are now in Denver, Colorado, twelve hundred miles from its former Georgia home.[55]

Will Shippen, a central figure in almost every aspect of the Supreme Court case, ended his fight soon after the 1918 settlement between the state and Ducktown Sulphur, Copper & Iron Company. James G. Parks, DSC&I trial counsel, had been right: Shippen was "a wiry, untiring, and resourceful chap." He had injected more energy into Georgia's federal litigation than any other person. Yet by 1922, the company he fought so hard to protect was in receivership. He liquidated his Gilmer County holdings in 1926 and moved to Atlanta.[56]

In a 1933 letter to the editor of the *Ellijay Times-Courier*, Shippen recited his lumber career and his many years as a leader in the Ducktown smoke wars. He claimed credit for the General Assembly resolution that launched the Georgia case, saying, "I got busy and put a bill through our legislature that appointed a commission of which I was chairman to investigate this fume damage" and then "drew the report of this commission." He "personally solicited and obtained U.S. Government experts," took them to his "Ellijay home and into the woods, and qualified them to testify for the state." He also "got up over 4,500 affidavits and employed counsel to aid the state."[57]

As for his business, he said, "When I first started my fight on the Copper Companies, some of my good friends told me that I did not realize just what I was going up against" and was warned that "they were most powerful and would eventually put us out of business." The warnings came true when the Tennessee Copper Company "bought stock in the Lumber Company and in 1915 united with others and voted my brother and self out of the active management." (He failed to mention that he and Frank were the ones who brought TCC into the business by selling a large portion of their personal shares for $175,000, over and above the $50,000 paid to the company to settle its smoke suit.) Then, like many another deposed corporate leader, he lamented, "I knew well that they were going to wreck the company."[58]

As the Shippen brothers, the National Farmers Union, and the state of Georgia turned their attention away from the great controversy over Ducktown smoke, the population of the basin adjusted to present realities. The smoke wars had lasted for almost a quarter century; a new generation acknowledged that the Ducktown Basin was now the Copper Basin. Those who could farm under the conditions established in the settlement agreements continued to do so. Those who could not moved on. Mountain farming was a dying way of life, with or without the problems of sulfur smoke. The national rail network and new means of food processing and preservation transformed farming from a local farm-to-market business to a nationwide concern dominated by growing agribusinesses. Small, hardscrabble farms found it hard to compete with products from large mechanized farms in the Midwest and elsewhere. Though owners of the best mountain farms often remained in business, more and more of the marginal farmers settled into a pattern of part-time farming, supplemented by wage labor, or left farming altogether. The pool of potential smoke claimants slowly shrank with the decline of the farming population.[59]

Most of the people who remained in the basin were tied to copper as mine workers or as the merchants and contractors that served the industry. Life in the basin eventually bent around the industry, and most were content to leave it that way. It was not a bad way of life for workers who were used to it. The pay was unusually good and steady in a region lacking sufficient employment at a living wage. Residents enjoyed the close-knit mining community. They shared pride in doing a hard job well and in making their life in a bizarre but strangely beautiful landscape. Locals enjoyed the ever-changing light as it shone across the barren red hills. They also took a puckish glee in being free from the mosquitoes, chiggers, and snakes so common in the farms and forests of the southern mountains. It was not an ecologically sensitive viewpoint, but it did reflect their pride of place.[60]

Ducktown smoke litigation also faded away because, as an effort to combat air pollution, it was several generations ahead of its time. The cases, at both the state and federal level, were initiated by farmers, loggers, and political leaders, not by environmental agencies with their battalions of air-quality specialists and in-house environmental lawyers. The legal and bureaucratic infrastructure now operated by the federal Environmental Protection Agency and its state counterparts would not be created until the 1970s. The Ducktown smoke suits were grounded in the ancient law of nuisance, not in the modern array of air pollution statutes and regulations that now shape environmental law. Comprehensive air pollution laws would not be enacted until well after World War II. Before then, the political leaders and jurists who struggled over Ducktown smoke did so without the regulatory framework that would come later. The modern environmental agencies are sophisticated scientific entities with rule-making authority to set pollution standards and the authority to enforce them. In Ducktown, the only rule making during the twenty-five years of litigation was by the Supreme Court in 1915, when it imposed quantitative production standards upon DSC&I. The Court knew it was acting in a field beyond its institutional competence and found it necessary to hire a Vanderbilt University chemistry professor to provide some guidance.[61]

When compared to modern practice, the successes achieved in Ducktown three or four generations earlier become all the more impressive. The state of Georgia successfully extended the nuisance law to establish a state's sovereign right for injunctive remedy against interstate pollution. Attorney General Hart, Ligon Johnson, and their successors made maximum use of conservationist expertise from federal scientists. Georgia's success in the Supreme Court forced the copper industry to new technology. Acid condensation was far from perfect but did remove millions of tons of sulfur dioxide from Ducktown skies for conversion into beneficial by-products. The state implemented an arbitration program to resolve smoke damage claims without litigation. And to the relief of the thousands who relied upon the Ducktown Sulphur, Copper & Iron Co. and the Tennessee Copper Company for their livelihoods, the industry remained viable and continued to operate for most of the twentieth century.

They accomplished all of this without modern environmental laws, modern environmental agencies, modern environmental science, and indeed, without the modern environmental paradigm that now underlies law and science. From a legal perspective, it was a remarkable outcome for litigation started by mountaineer farmers a decade before Henry Ford produced his first Model T car. Viewed a different way, in terms of the landscape, the damage had been done. The Ducktown Desert was now the most prominent feature of the basin's landscape and would remain so for the rest of the twentieth century.

9 POWER DAMS, WHITEWATER RAFTING, AND THE RECLAMATION OF THE DUCKTOWN DESERT, 1916–2010

The Ducktown Desert remained after the smoke suits ended. Ducktown smoke litigation was never about restoring the badlands. Farmers and loggers sued for damages to their crops and timber. The state of Georgia sued for injunctive power to force reduction in the amount of smelter smoke. None of the litigants sued for restoration of forests to the naked hills, nor did the courts suggest it. The settlement agreements between the state and the copper companies, providing only for arbitration of smoke claims and a modest limitation of sulfur releases during the growing season, said nothing about reclamation. So far as the state, the farmers, and the timber owners were concerned, the Ducktown Desert was a fait accompli for which there was no practical remedy under the law. What was done was done.[1]

The enormous scale of the destruction staggered the imagination. Fifty square miles of Southern Appalachian hardwood forest had been leveled for the mining industry. In many places, even the stumps had been removed for firewood, leaving nothing to retain the soil. Decades of sulfur dioxide smoke killed remaining old growth and suppressed the new. Logging and smeltering in an environmentally susceptible area left the landscape bare and exposed to the powerful erosive force of sixty annual inches of rainfall. Poor grazing practices compounded the damage. In the center lay an area of thirty-six square miles of red, treeless, badly eroded land—an area roughly one and a half times the size of New York's Manhattan Island.[2]

The damage occurred in a pattern of concentric rings centered upon the smelting complexes at Copperhill and Isabella. Richard A. Wood, a researcher with the Tennessee Valley Authority (TVA), provided a typical description in his 1942 report, "Erosion Control and Reforestation of the Copper Basin." The outer ring contained the area of least damage; it was "covered with a thin stand of sedge grass and occasional scrub oaks and other inferior hardwoods." The outer ring had few gullies, but "a considerable amount of sheet erosion" was nonetheless occurring because of the lack of mature ground cover and humus. The middle ring was "deeply cut with numerous gullies" and the ridges between them were "patches of sedge grass which are gradually becoming smaller each year." The totally denuded center area was the worst: eleven square miles of heavily gullied land "characterized by the absence of any effective vegetal cover" and the total absence of topsoil. Another researcher, Kenneth Seigworth, wrote that the Ducktown Desert (also known as the Copper Basin Desert) was a "a copper-red gully-scarred area" which "has suffered as have few areas anywhere else in the world."[3]

The geography of the desert, and the pattern of ownership within it, effectively limited the number of parties with a legally recognizable property interest in its reclamation. Few Georgians or North Carolinians owned a portion of it, because great as it was, almost all of it lay in the Tennessee portion of the basin floor. Only a small portion extended up the watershed of North Potato Creek into North Carolina, and even less across the Ocoee River into Georgia. The copper companies owned more than half of the portion within Tennessee. In 1942 the TVA determined that the Tennessee Copper Company owned 54 percent of the badlands, or 12,500 acres (TCC acquired the land and assets of Ducktown Sulphur, Copper & Iron Company in 1936). The private owners who held the remaining 10,500 were "chiefly nonresidents." The TVA contacted a number of them and learned that they had "no interest in the surface but only speculative interest in potential ore deposits underneath." This was a realistic hope because the industry continued to lease private lands and to open new mines well into the twentieth century.[4]

The state of Georgia also lacked a significant stake in reclamation, especially because the Ocoee River carried the acid-laden silt away into Tennessee. Georgia's leaders had accomplished their twin goals of smoke reduction and industry continuity by encouraging the copper companies, under threat of a Supreme Court injunction, to extract sulfurous fumes for conversion into useful sulfuric acid. The industry prospered and the skies over the Ducktown Basin, if not totally smoke free, had improved compared to conditions at the turn of the century when smelter fumes hid the noonday sun and men tied bells to their wagon teams to avoid collisions with other teams as all traveled blindly through dense clouds of dark, odorous smoke. Having met its goals, the state had no interest in starting another wave of copper litigation to compel reforestation.[5]

Contributing to the inertia was the lack of comprehensive laws at either the federal or state level to require restoration of land degraded by mining. The first state laws, aimed at strip mining by the coal industry, were passed in the late 1930s and early 1940s. Congress eventually enacted nationwide legislation with the passage of the Surface Mining Control and Reclamation Act in 1978. In 1980 Congress passed another major reclamation act, the Comprehensive Environmental Response, Compensation, and Liability Act (commonly known as the Superfund law) to give the Environmental Protection Agency authority to clean up toxic waste sites and to impose liability upon potentially responsible parties. Superfund would become a factor in Ducktown at the end of the twentieth century but not before then. In the meantime, if the naked gullied hills were ever to be covered by forests again, the initiative would have to come from a new direction, from new parties seeking to protect new interests.[6]

The new direction and initiative came from Tennessee's nascent hydroelectric industry. The great natural resources of copper, sulfur, iron, and other minerals found in the Copper Basin were matched by the great potential of the Ocoee Gorge for hydroelectric power. It had all of the necessary features: a fast, dependable stream fed by high annual rainfall evenly distributed throughout the year, a narrow gorge that facilitated dam construction, a location near cities eager for electricity, and a high vertical drop that allowed the power of gravity, in the form of falling water, to be transformed into electricity. The Toccoa-Ocoee River began in the Blue Ridge Mountains of Georgia at an elevation of 2,025 feet, and then crossed the Ducktown Basin, where it made a gentle curve around TCC's Copperhill complex and passed several miles through the Ducktown Desert before entering the Ocoee Gorge. Within the gorge, the river fell 705 feet over the course of 10 miles before entering the Tennessee Valley.[7]

The Eastern Tennessee Power Company and its successor, the Tennessee Electric Power Company (TEPCO), built the Ocoee No. 1 dam where the river exited the gorge at Parksville. It was a large project, being a concrete gravity dam 840 feet wide and 110 feet high, fitted like a plug into the gap between Sugar Loaf and Bean Mountains. Built in only fifteen months, it was nonetheless built well and remains the oldest dam still in service for the Tennessee Valley Authority. The dam began operation in 1912 to great public acclaim. Polk County historian, Roy Lillard, wrote that a group of leading citizens watched a young woman, Miss Virginia Lockett, throw the switch at the power dam to send electricity forty miles away to Chattanooga, where, on top of the Hamilton Bank Building, an array of electric light bulbs spelled out the word "Ocoee" in five-foot-high letters to shine across the city.[8]

TEPCO built Ocoee No. 2 immediately after finishing the first dam. The second dam had two distinct units: a powerhouse located about 10 miles upstream from Ocoee No. 1 and a diversion dam 5 miles further up the river. The diversion dam shunted the river into a huge wooden flume that carried the flow 4.5 miles on an almost level plane to just above the powerhouse, where it then fell 250 feet to drive the turbines below. The flume itself was an engineering marvel: it was a structure made of 9 million board feet of Georgia pine assembled to form a box-shaped tunnel 14 feet, 2 inches wide and 9 feet, 9 inches high. A railroad track ran along the top to move workers and equipment between the diversion dam and the powerhouse. When finished in 1913, the new dam joined Ocoee No. 1 to send power across the region from Knoxville, Tennessee, to Rome, Georgia. Its electricity lit homes, powered factories, propelled streetcars, and refined aluminum at the new Alcoa plant in Maryville, Tennessee.[9]

As soon as the dams went into service, TEPCO began to experience waterborne trouble from the Ducktown Desert. The barren hills of the desert lost soil at the rate of 39 acre feet per square mile per year, compared to the rate of only 0.44 acre feet experienced in much of the watershed, and most of the sediment eventually washed downstream to the dams. Silt filled in the upper portion of Lake Ocoee, the reservoir behind Ocoee No. 1 and completely filled the 10-acre retention pond behind the Ocoee No. 2 diversion dam. The damage extended into the inner workings of the power plants. Francis E. Pray of the TVA noted that "sand and acid in the water coming from the Copper Basin created all sorts of difficulties with the flume, the tainter gates, and the water wheels." Dam operators found it necessary to replace the water wheels for the electric turbines every other year "because of erosion, corrosion, and cavitation."[10]

The repair costs and service interruptions caused by the silt and acid led TEPCO to file a suit for damages against the copper companies in 1916. The Tennessee Copper Company paid $50,000 to end the suit in 1921. It also agreed to reduce acidity in the river by treating the water of Davis Mill Creek as it flowed through the Copperhill complex just before joining the Ocoee River. The treatment plant functioned by directing the flow of acid water from the creek over a bed of limestone to counterbalance the acidity of the water. It was the first acid neutralization plant in the Ducktown Basin, the predecessor of other and more effective plants to come later.[11]

The TEPCO lawsuit was significant as a harbinger of the themes that would govern the reclamation of the Ducktown Desert. The earlier smoke litigation focused on air quality, specifically the toxic effects of sulfur dioxide on vegetation. The TEPCO suit shifted the focus to water quality, specifically the impact of siltation and acidity upon hydroelectric facilities. The conduct of the copper industry in an environmentally susceptible landscape transformed Southern Appalachian hardwoods into the Ducktown Desert. The impact of the Ducktown Desert upon the environmentally susceptible hydroelectric industry forced the copper companies to begin the work of transforming the Ducktown Desert back to the semblance of a forest.

The TVA inherited TEPCO's Ducktown problems when President Franklin Delano Roosevelt signed the Tennessee Valley Authority Act into law on May 18, 1933. The act gave the TVA a broad mandate to "improve the navigability and to provide for flood control of the Tennessee River; to provide for reforestation and the proper use of marginal lands in the Tennessee Valley; to provide for the agricultural and industrial development of said valley; [and] to provide for the national defense." Section 15 of the act authorized the TVA to purchase TEPCO's power facilities, including the Ocoee No. 1 and No. 2 dams.[12]

Conditions at Ducktown, now more commonly called the Copper Basin, intersected with some of the legislative goals. Damage caused by acid and silt to the hydroelectric facilities frustrated the TVA's core strategy of providing abundant low-cost power to attract industry and improve living conditions in the region. The massive loads of silt that washed down the river impaired flood control by reducing the storage capacity of the reservoirs behind the power dams. In 1936 Stuart Chase, a nationally prominent advocate for soil conservation, observed silt deposits at the head of Lake Ocoee, the reservoir formed by the Ocoee No. 1 dam. There he found "a thing that belongs in no lake ... a chocolate-colored tongue of shaking mud, half a mile long" and many "other tongues and trembling islands," all about six feet high. "The whole upper end of the reservoir is full of them" and the banks "have changed from good honest mud to this forbidding red-brown jelly."[13]

These concerns were heightened in 1941 when the TVA began construction of the Ocoee No. 3 dam at the head of the gorge to provide power for nearby defense plants in anticipation of World War II. Engineers considered the silt problem and designed two large sluices for its removal. Though the dam worked as intended, the reservoir became clogged with silt, losing its storage capacity in less than thirty years. The TVA reported that "since this reservoir is essentially filled in, most of the sediments from upstream are passed on downstream to Ocoee No. 1," where they added to the islands of "forbidding red-brown jelly" noticed by Chase.[14]

Beyond these concerns, the naked gullied hills of the Ducktown Desert presented a dramatic challenge to the TVA's conservationist mandate "to provide for reforestation and the proper use of marginal lands in the Tennessee Valley." David Lilienthal, the TVA's first chairman, called the basin "the dead land" where "every sign of living vegetation was killed and the soil became poison to life." He argued that the Ducktown Desert was an especially dramatic example of the devastation that occurred throughout the South because of its status as a "colonial region." The colonial economy deprived the South of wealth because the income earned for its resources paled against the value earned and added by manufactures outside the region. Worse, the lack of a broad manufacturing base in the South led to the single-minded exploitation of certain marketable resources at the cost of others. He wrote that in Ducktown, "copper had been developed. But all its other resources had been destroyed in the process." The same dynamic occurred with other southern products. "If a region depends ... almost entirely upon the income from cutting the lumber or growing the cotton" without also manufacturing products from those resources, "then the pressure to 'mine' the fertility of the soil, to devastate the forests for lumber, to deplete ... the coal reserves becomes very great indeed." The answer lay in the diversification of the southern economy, primarily by expanding local manufacturing.[15]

The chairman's comments about Ducktown reflected his desire for cleaner industries. He favored manufacturing jobs in electrically powered factories over environmentally destructive mining, logging, and cotton agriculture. Manufacturing of this sort would, he argued, reduce harm to natural resources while boosting the southern economy by the local production of value-added products. He warned that "if the industry is only exploitative, if it does not sustain the productivity of the resources upon which all of us depend, industry can exhaust a region and hurt its people's chances of security and happiness." To avoid that fate, he declared, "the unity of nature's resources must not be disregarded."[16]

Lilienthal sounded important themes during the Great Depression, when America suffered from the Dust Bowl and other conservation nightmares. Yet, his comments failed to acknowledge the copper industry's long-standing role as the provider of the industrial wage labor that the TVA generally encouraged. Copper mining in the Ducktown Basin was not a cut-and-run proposition like the logging industry; it was almost a century old and had been, for most of that time, one of the greatest employers in the Southern Appalachians. By 1934 ten thousand people lived in the Copper Basin, "practically all of whom" derived "their living from the copper and chemical companies operating the Basin." Thanks to stable employment, the Ducktown mining community endured the Great Depression in better shape than most of the South's millions of tenant farmers in the cotton belt.[17]

The Tennessee Copper Company was more than an extraction industry. When Lilienthal published his comments in 1944, the company was well advanced in its own transformation from an extraction industry into a sophisticated manufacturer of value-added chemical and mineral products essential to southern agriculture and manufacturing. The transformation began in 1908 when TCC and DSC&I began production of sulfuric acid from formerly wasted sulfur dioxide smelter gases. The next advance came with the adoption of flotation milling to extract other products out of Ducktown ore. The process began by crushing ore into a powder, mixing it with water to form slurry, and then agitating it within a large tank. The addition of different chemical reagents into the slurry caused the particles to separate by type, allowing copper, iron, sulfur, zinc, and trace metals to be isolated and skimmed off without the use of blast furnaces. DSC&I built the first local flotation plant in 1920, followed by TCC in 1924. The process allowed both companies to begin selling iron, in the form of iron sinter, to the steel industry in Birmingham, Alabama. Ducktown iron became Birmingham steel.[18]

Chicago meatpackers boasted that they used every part of a hog except its squeal. The flotation process allowed the copper companies to accomplish an analogous feat with a ton of ore. By 1939 TCC produced 700 pounds of iron, 500 pounds of sulfur, 23 pounds of copper, and 17 pounds of zinc from every ton of ore. Even the slag found marketable uses as an additive for Portland cement, in roofing shingles, in cast stone, and as ballast for railroad tracks (the gravel on which tracks are laid). Company chemists devised scores of new products derived from copper, sulfur, and zinc. The copper company was now a gigantic chemical company that produced several grades of sulfuric acid, oleum (a form of sulfuric acid used for the manufacture of explosives), copper sulfate, zinc sulfate, many types of copper-based insecticides, fungicides, and dyestuffs. TCC even harvested iron oxide dust from the iron roasters to produce Ferri-floc, a ferric sulfate compound used as a precipitating agent in sewage treatment plants.[19]

The progress of the copper industry was impressive, but the Ducktown Desert continued to provide Lilienthal, Chase, and others reason to criticize the mining practices of the past. In 1932 the *Chattanooga Times* wrote that "the 'man-made desert' becomes a 'desert of folly' when we find it came into existence as the result of continued cutting of timber and the escaping copper fumes from smelting plants in the Ducktown Basin." Officials at TCC and DSC&I were sensitive to the criticism, knowing that the very existence of the desert had been a substantial factor in the success of the smoke suits against them. In 1917, when DSC&I was desperately seeking to free itself from the 1916 Supreme Court injunction, the company created a demonstration farm to prove that crops could be grown on the barren ground, with the tacit suggestion that conditions were not as bad as complained. An employee later admitted that the farm received greater care than could be expected in normal farming, and that "their secret for preventing pollution damage was to sprinkle with water whenever the gas came down." The company performed another demonstration project in 1928 when it built a nine-hole golf course. Overall, such efforts were too costly and labor intensive for application to the entire desert.[20]

Revegetation took a more serious turn in 1929 when DSC&I engaged C. R. Hursh, a forest ecologist with the U.S. Forest Service, to conduct a study and make recommendations. The plan, formulated with the assistance of the U.S. Department of Agriculture and Extension Service foresters at the University of Tennessee, involved the planting of three thousand disease resistant Chinese and Japanese chestnut trees on the eastern edge of the desert near Angelico Gap and the North Carolina border. The Tennessee Copper Company began a project of its own in 1931 in cooperation with the Extension Service and the Appalachian Forest Experiment Station. The project tested various native species, with local pine trees proving the most successful.[21]

The initial revegetation efforts were limited in scope but often productive. When the TVA evaluated the results, it concluded that "the work, which involved elaborate soil preparation and the addition of many tons of organic material and fertilizer, was successful but expensive." Reforestation of the entire thirty-six-square-mile Ducktown Desert and especially its eleven-square-mile barren central zone would require new low-cost techniques together with a great deal of money, labor, and time. It was an argument for intervention by powerful government agencies such as the Tennessee Valley Authority, the Soil and Conservation Service, and the Forest Service.[22]

The TVA found a willing partner in the Tennessee Copper Company. Simple prudence dictated the company's cooperative attitude: having lost the TEPCO suit twenty years earlier, TCC leaders knew that an intransigent posture would only provoke the new owner of the dams into filing another successful suit. Company leaders respected the resources and power of the TVA, and were aware that it enjoyed the personal support of Franklin Roosevelt. The authority demonstrated its legal clout by winning a Supreme Court battle when TEPCO and seventeen other utilities attempted to overturn the 1933 act on constitutional grounds.[23]

The Tennessee Copper Company had other reasons to join with the government's reclamation plan. The company hoped to create a managed forestry program on its own lands with the help of the TVA and other agencies. If successful, the company could then meet its annual requirement of 750,000 thousand board feet of lumber without having to buy it from outside vendors. The company also hoped to reverse a peculiar problem for such a rainy land: the lack of water. The naked hills lacked the vegetation and humus needed to retain water. Rainfall instead washed quickly down the bare slopes to contribute to the serious flooding that occurred on the Ocoee River once or twice a decade. R. E. Barclay, who served as chairman of the Polk County Commission in addition to his duties with TCC, told a reporter that "water often ran off the denuded hills so rapidly that basements in Copperhill were flooded." The lack of vegetation also accelerated evaporation of whatever water had been absorbed into the soil. The downcutting of gullies deep into the hillsides further lowered the water table. Together, the processes caused major streams to dwindle, springs to dry up, and wells to fail. This was a serious problem for the many workers and their families in the scattered mining villages who, for lack of plumbing, depended upon these sources for their water. The copper companies adjusted more easily by piping water directly from the Ocoee River.[24]

The water problems within the Ducktown Desert corresponded with the stream-flow theories propagated by George Perkins Marsh in his famous book, *Man and Nature* (1864). His doctrines had provided the theoretical foundation for the forest conservation movement. They were also of critical importance to Attorney General John C. Hart and his colleague Ligon Johnson when they presented their arguments before the Supreme Court in *Georgia v. Tennessee Copper Company*. They now proved true, at least within the desert, to the discomfort of its inhabitants. A joint effort with the TVA in the reclamation of the desert might reverse the conditions that dried up local springs and flooded nearby basements.[25]

TCC officials realized that cooperation would also reposition the company in a more favorable light in the national campaign for better soil conservation practices. Soil conservation was an old topic but became a hot political issue when the horrors of the 1930s Dust Bowl forced a national response during Franklin Roosevelt's first term. As noted by Donald Worster, a leading Dust Bowl historian, "it took the May 1934 dust storm to make the plains visible to

Washington. As dust sifted down on the Mall and the White House, Roosevelt was in a press conference promising ... a new Great Plains relief program." Roosevelt appreciated that the dust storms were not just the result of a drought but were the result of a history of unwise farming practices. The administration responded in 1936 by creating the Soil Conservation Service (SCS) to address current problems and prevent their recurrence by educating the public on land use methods. A network of SCS offices opened to press the message across the country, including in Tennessee.[26]

The Tennessee Copper Company understood that if the Dust Bowl served as the iconic rationale for soil conservation in the West, the Ducktown Desert served the same function in the East thanks to Stuart Chase, David Lilienthal, and others. The company also recognized that the mission of the SCS aligned perfectly with the TVA's statutory mandate "to provide for reforestation and the proper use of marginal lands in the Tennessee Valley." That being so, the company decided it was wiser to join the cause than to resist it.

Company officials knew they could best shape their new conservationist message by collaborating with the SCS as it communicated its own. The service broadcast a radio series, *Fortunes Washed Away*, "a series of dramatizations of better land use" to highlight soil conservation issues around the country and scheduled a segment on the Ducktown Desert for broadcast for January 4, 1941. The SCS sent the script to TCC for evaluation, to which R. E. Barclay responded with the company's approval and minor editorial suggestions. In his cover letter to the service, he wrote: "The opinion here ... was that it is just another unnecessary dramatization of our already overadvertised district.... Your purpose, however, is most worthy, and all here are in support of every attempt at soil conservation."[27]

The program began with the show's standard introduction of an organ fanfare, the sound of the howling winds of a dust storm, and voices decrying "Black blizzards in the Dust Bowl. Wind, choking dust," followed by "Fire in the great pine forests. Havoc, destruction, waste," and then "Floods rushing down the great valleys. Floods, drowning, killing, wasting." Next was a thunderclap and the words of the announcer, "The Badlands of Tennessee: The 141st consecutive episode of *Fortunes Washed Away*." The core of the script presented a series of dialogues by characters from different times in Ducktown's history. The first scene, set in 1891 just after the railroad reached Ducktown, was a conversation of "smart Englishmen" discussing the advantages of local open heap roasting of copper ores: "We pile the raw ore on the ground, cover it with cordwood, of which you will agree there is an inexhaustible supply—and roast it until it gives up its copper. It will be like gold in our palms." Each element of the scene distanced the modern Tennessee Copper Company from the destructive smelting and logging practices of the past. The somewhat jingoistic reference to Englishmen harkened to the now defunct English-owned Ducktown Sulphur, Copper & Iron Company. The 1891 date pointed to a time eight years before TCC entered the basin. The evils of roasting and deforestation were thus the result of abusive and exploitative practices by greedy foreign investors. The reality was different: the American-owned TCC roasted ore on an even larger scale than DSC&I.[28]

The next scene, concerning the smoke litigation, cast the smoke suitors in a negative light while presenting the copper companies as reasonable corporate citizens and technological heroes. The dialogue involved Freeland (W. H. Freeland, manager of DSC&I), Adams (Randolph Adams, manager of TCC), and a claimant's attorney discretely named Smith. The lawyer approached the managers to announce, "There's a hundred farmers who'll sue you for damages, and they'll collect too." Then to emphasize his point, he said, "Why, the smoke and fumes are so bad that our children get lost on the school grounds.... If it weren't that the teacher rings the school bell to guide 'em, they'd fall down the mountainside." The managers argued the sympathetic response of their companies, noting that they "paid more damages out of court than in." Freeland then chastised Smith for his belligerent attitude, to which Smith retorted, "And I aim to keep on being belligerent. Our claims are in. . . . your mines can either pay up or close up." Exeunt Smith. Adams then lamented to Freeland: "Those unappreciative rascals. And that double-crosser Smith. Why our mines are the life-blood of the basin." Their conversation then turned to the exciting development of the sulfuric acid business. Freeland closed the scene with the hopeful thought: "Someday they may say: 1907 was the year that put an end to the deadly smoke clouds."[29]

There followed a musical transition, "Organ: mournful music behind . . ." and the announcer intoned, "But the damage had been done," followed by a description of the desert. The announcer asked his colleague, Charles: "You don't propose to blame anyone, then, for the destruction of land and vegetation in the Copper Basin?" Charles answered with the key corporate messages: "Certainly not—because you can't indict a whole people. The wastage of timber, and minerals and soil in the Copper Basin has been no more of a crime than similar destruction elsewhere. It is only more complete and obvious." He continued, "There are 10,000 residents in the basin who are a little tired of outsiders who ridicule them because of the bleakness of their home sites. And they'd like outsiders to know that they are doing what they can to bring vegetation back into the basin." Thus, attacks on the copper company became generalized as attacks on the ordinary people of the mining community. They were not evil people. Their sin, if it was a sin, was shared with the entire nation, and was from a time when everyone acted without the guidance of modern soil conservation principles promoted by the Soil Conservation Service.[30]

There followed four more pages of dialogue reporting details about the extensive cooperative effort between the copper company, the Forest Service, and the SCS to revegetate the desert. TCC appeared as a corporate citizen intent on reversing old wrongs committed by previous generations in the uninformed past. The announcer then summarized: "Well, that's a good report... makes it look as if, within a few generations, the great Copper Basin will cover up—heal over the scars of soil erosion." The program closed with another musical cue: "Organ: up and out."[31]

It was a brilliant corporate message, one that bore the endorsement of respected federal agencies committed to a noble conservationist cause. It also happened to be largely true. In 1940 the Tennessee Copper Company and its successors began a comprehensive reclamation program, in partnership with the TVA, the SCS, the Forest Service, and other federal and state agencies, that continued into the twenty-first century. The radio announcer was correct: it would take several generations to heal the scars.

The joint effort to reclaim the twenty-three thousand acres of the Ducktown Desert began with preliminary discussions in 1940, one year after the TVA acquired the Ocoee No. 1 and No. 2 dams. Hitler invaded Poland the year before, so the prospect of another war in Europe heightened the need for electricity from the two older dams and the new Ocoee No. 3 dam to make aluminum for military aircraft at the nearby Alcoa plant. With this in mind, the TVA noted, "the silt carried into these reservoirs by the Ocoee River is ample justification for the expenditure of public funds for the control of erosion in the Copper Basin.[32]

The TVA marshaled its resources as if for war. It analyzed the battlefield with new maps of the desert terrain, supplemented by careful field surveys. One map set forth the three zones of denudation, showing that the worst of it centered on a triangle defined by TCC's Copperhill complex in the south, the community of Ducktown in the northwest, and the DSC&I complex at Isabella in the northeast. Other maps delineated stream drainages within the basin and the location of earlier test plantings. A fourth map documented the areas where the sulfur content of the soil was highest, a calculation made by chemically analyzing the stems of cat briars, one of the few plants tough enough to grow in the barren central zone.[33]

The TVA then assembled the troops, specifically a unit of young men from the Civilian Conservation Corps (CCC). Franklin Roosevelt created the CCC in 1933 to accomplish a number of goals: it put unemployed, and often underfed, young men to work during the Great Depression; it enabled them to earn money to send home to their families; and it provided labor for a wide array of conservation projects in the national parks, national forests, and other sites. The CCC also had the not incidental benefit of preparing youths for the army by strengthening their bodies with ample food and physical outdoor work, and by exposing them to military regimentation under the supervision of the army officers who ran the camps. Camp F-10 on nearby Big Frog Mountain was typical of the seventy-seven CCC camps in Tennessee. The youths arrived in 1933 and at first lived in tents and ate from military mess kits until wooden barracks were finished. Army officers ran the camp with the help of sergeants. The young men spent the days in the forests building fire watchtowers, fighting fires, improving timber stands, and patrolling against poachers. At night, army education officers taught illiterate youths by "giving courses in reading, writing, cipher, and other academic subjects."[34]

In the Copper Basin, the CCC focused upon the reforestation of eleven hundred barren acres behind the town of Copperhill. The young men built check dams to slow the movement of silt in the gullies, planted trees and ground cover, and fought fires. Though a tractor broke open the hardened ground, most of the work of tree planting was by hand. As was true so often in Ducktown's history, the lack of available wood hindered their efforts. The TVA noted the "difficulty of obtaining materials" for the check dams, adding that the treeless job site "made it necessary to go from five to eight miles to obtain brush, posts, stakes, and poles."[35]

The CCC left Ducktown in 1942 when the government closed the camps nationwide and sent the now healthy young men to war. Their year of labor nonetheless allowed the TVA and the Tennessee Copper Company to evaluate test plantings of various species in the different zones of the desert. They tested four species of native pines—loblolly, shortleaf, Virginia, and pitch—along with black locusts and other species. Of the four pines, loblolly (the most desirable lumber tree) grew fastest and the pitch pine was the hardiest. They made no attempt to plant the magnificent eastern white pine because it was too susceptible to the level of sulfur dioxide still emitted at the copper works. Black locusts succeeded in the gullies, but were a short-lived tree intended to stabilize the ground only until pines and ground covers grew in their place. They tested twenty-two different grasses and groundcovers from around the world. Lespedeza and South African weeping love grass proved the most successful. Tests also proved that all plantings required soil preparation, fertilizer for nutrients, and lime to counteract high levels of soil acidity. Plants quickly died without these measures.[36]

The TVA and the Tennessee Copper Company continued planting on a reduced basis throughout the war years. As the war ended, the TVA made a detailed assessment of the progress to date. The work demonstrated that reforestation was possible with select species on well-prepared ground, but the pace was slow: of twenty-three thousand acres in the desert, less than fifteen hundred acres had been reforested. The constant erosive force of rain and frost, plus the pernicious effects of open range grazing and the deliberate burning of sedge grass, caused the desert to expand faster than the copper company and the government could replant it. The TVA lamented that erosion was "becoming progressively more extensive and the native grass cover ... slowly diminishing." The law finally changed in 1946 to put an end to open range grazing in Polk County.[37]

The scope of the remaining work was daunting, as were the environmental factors that hindered planting in the desert. There was no topsoil and, in many areas, no subsoil either. Sixteen vertical feet of soil and subsoil had washed away in the areas of the worst erosion. The remaining subsoil was "incredibly low in plant food and organic matter." It was also highly acidic, with pH levels ranging from 3.8 to 5.0, and averaging 4.0. (The pH scale is based upon a neutral value of 7.0; the smaller numbers are more acidic, thus battery acid has a pH of 0 compared to 7.0 for distilled water.) Sulfur dioxide emissions from the copper works, though significantly lighter than in the past, were "nevertheless injurious in the vicinity of the smelters under certain weather conditions." The situation restricted the choice of species "to a relatively few resistant kinds."[38]

Acid condensation was successful as a commercial venture but failed to completely eliminate the release of sulfur fumes into the basin's atmosphere. When the Supreme Court issued its 1915 injunction against the Ducktown Company, it considered the atmospheric release of 45 percent of the sulfur liberated from green copper ores to be acceptable. Efficiency improved over later decades, but sulfur gases remained a hindrance to reforestation. Copper workers had their own way to detect atmospheric acid: in 1979 an employee told a reporter that if he wanted to see its effects to "just look at the new cars around here after three months." Many employees made it a point to commute to work in worn-out cars rather than expose the shiny finish of the family car to the effects of airborne acid.[39]

Revegetation efforts continued after the war despite the difficulties. The TVA wanted to accelerate the pace of the work, and recommended the use of a "labor force of 300 to 700 and averaging 500" to finish the work in five years. It was hoped that local people would do the work to eliminate the cost and administrative burden of setting up camps. The proposal bespoke a longing for the prewar days when the TVA could command the cheap labor of strapping youths in the CCC. It was not a realistic idea in the postwar economy when workers had better opportunities and owned automobiles to commute to better jobs. Instead, the TVA and the Tennessee Copper Company carried on the work at a slower pace with a smaller workforce. The most successful reforestation occurred in the outer zones where there were some humus and grass cover to help new plants. A pattern soon developed in which the Ducktown Desert shrank from the outside toward the still barren hills in the center.[40]

The TVA, the Soil Conservation Service, the Forest Service, and researchers at the universities of Tennessee, Georgia, and North Carolina provided technical advice, conducted experiments, and published dozens of technical papers evaluating species and planting methods. They continued to favor loblolly pine because of its fume resistance and commercial value, along with pitch pine because it quickly established roots in the worst terrain. They also experimented with the use of Japanese kudzu to function as a "gully stopper," only to realize that the aggressive vine engulfed more desirable trees just as it did everywhere else in the South. It was also a fire hazard during the winter when it became dormant and dry.[41]

The Tennessee Copper Company and its successor companies provided funding and labor for the project. (TCC merged into Cities Service Company, a major oil corporation, in 1963; Cities Service then sold the operations to Tennessee Chemical Company in 1982; more corporate changes would happen when Tennessee Chemical declared bankruptcy in 1989.) By the 1970s the rate of planting increased to a half-million trees per year. The efforts were impressive; the results were less so. Until the early 1970s, the survival rate of the pine seedlings was a dismal 30 to 40 percent; two trees died for every one that survived. Many areas were planted several times over with the same sad results. Seedlings often washed away before they could establish roots.[42]

Cities Service achieved better survival rates when it began to plant each seedling with a slow-release fertilizer pill to support growth through the first three years. More improvements followed when the company entered into a five-year research study with the U.S. Forest Service Southeastern Forest Experiment Station near the University of Georgia in Athens and another study with the Soil Conservation Service–University of Tennessee Plant Materials Committee. Researchers devised the technique of subsoiling, using a crawler tractor with long ripping claws to break the hardpan subsoil to the depth of two feet. This improved water retention and made space for root growth. Lime and fertilizer were then mixed into the loosened soil. The scientists also learned that pretreatment of seedlings with *Pisolithus tinctorium* fungus improved health and longevity. The new techniques and continued use of the time-release fertilizer pill improved survival rates to 70 percent in the 1970s. They rose even higher when foresters with the Soil Conservation Service learned that treating seedling roots with water-holding gel "helped the tender root systems withstand an extended dry period." Once established, young pine trees tended to become chlorotic, that is, to turn yellow for lack of chlorophyll, which in turn pointed to the lack of essential nutrients. This occurred about three years after planting, when the fertilizer pill lost its potency. Cities Service met the problem with periodic applications of fertilizer spread from the air by a helicopter. The use of genetically improved seedlings also proved successful. Overall, the survival rate of loblolly pines climbed to 92 percent, a testimony to biological science and to persistence.[43]

The new methods allowed tree planters to regain the initiative against forces of erosion, and with time the progress became obvious to the many columnists and feature writers who published articles every year or so to track the progress. Captions for the early articles were mostly prospective: "The Great Copper Desert Will Bloom Once More" (1940) and "TVA to Reclaim Copper Basin" (1941). Postwar progress prompted captions with a tone of celebration and amazement: "Copper Basin Miracle" (1950), "Man and Nature Are Slowly Reclaiming the Tennessee Sahara at Copperhill" (1962), and "Copperhill Is Again an Area of Green Land" (1962). By the late 1970s, progress advanced to a point where many realized that the familiar badlands would eventually disappear. New captions sounded a note of nostalgia. Jimmy Townsend, a Ducktown boy and later columnist for the *Atlanta Constitution*, wrote of "Ducktown's Desert Beauty" (1978). Dan George, a writer for the Associated Press, wrote an article that appeared in a number of papers under such captions as "Barren Hills Are Beautiful to Copper Basin Natives," "Historic Copper Basin's Natives Find Beauty in Bare Red Hills," and similar phrasings. George quoted David Beckler, a native son and board member for the new Ducktown Basin Museum: "People pay to go out West and see something like this, and say it's beautiful. But when we have it here (they say) that's ugly. I've never seen it as ugly." Other locals sympathized with him, and a movement arose to set aside a portion of the barren hills as a permanent memorial. The TVA eventually honored the request by leaving a representative tract of denuded land near the museum's location on Burra Burra Hill.[44]

While many understandably lamented the loss of a landscape central to their sense of place, others worried about the decline of Ducktown's mining and chemical industry. A series of expensive investments by Cities Service management to modernize and expand facilities failed to yield an acceptable return. In 1976 Cities Service Company announced that it lost money on its Copperhill operations because of declining demand for its sulfuric acid. The price of copper fell below the price of production because of competition from Chile, Zaire, and Zambia. As demand and prices fell, the costs of production continued to increase. Underground mining was an expensive business at the best of times, and the costs rose as miners dug ever deeper in pursuit of ore. Shaft A at the Calloway Mine was 3,000 feet deep: it was 1,640 feet above sea level at the surface and 1,400 feet below sea level at the bottom. The costs of complying with new federal and state air pollution and water pollution laws enacted in the 1960s and 1970s further reduced profits. Another blow occurred with the failure of the company's $180 million investment in a new iron pelletizing plant. Cities Service shut the plant down in 1979 after it failed to produce pellets of the expected amount or quality. Four hundred workers lost their jobs when the unit closed. Another 170 workers lost their jobs in 1981 when the company cut production in other product lines. News of the second layoff occurred only a week after the blue jeans maker, Levi Strauss, announced the loss of three hundred jobs with the closing of its garment plant in nearby Blue Ridge, Georgia. It was a portent of worse things to come in the Copper Basin.[45]

Cities Service came to view its Tennessee operations as a drain on its otherwise profitable oil and gas business, and put the Copperhill facilities up for sale in 1981. The property remained on the market for a year and a half until purchased in 1982 by Tennessee Chemical Corporation (not to be confused with the former Tennessee Copper Company). Tennessee Chemical immediately experienced financial difficulties of its own because the nation suffered a major economic recession. Then, a year after the purchase, an underground collapse forced the new company to close the Calloway mine, its richest source of ore. The combination of calamity and market conditions put mining jobs at risk. The unions agreed to a pay cut to keep the company afloat, but the concessions merely delayed the inevitable.[46]

An *Atlanta Constitution* headline summarized the impact of the industry's declining fortunes: "Copper Slump Hurts 2 States." In Tennessee's Polk County, government leaders worried about the fate of Tennessee Chemical because property taxes on the copper industry normally funded 45 percent of the county budget. Tennessee Chemical challenged its assessed tax value of $101.5 million on the grounds that it paid only a third of that amount to buy the assets from Cities Service. County workers saw their hours cut and paychecks delayed while the valuation fight dragged on in the courts. The Polk County Board of Education halted the school buses as a desperate cost-saving measure to avoid an early end of the semester. In Georgia, the local economy relied on the paychecks issued to the large number of Georgia workers in the copper industry. The declining payroll led to reduced business among local merchants with a consequent reduction in sales taxes paid to state and local government. The mayor of McCaysville, Georgia, said that business was off because "people are holding back. They don't know when they might have to leave town."[47]

In 1987 Tennessee Chemical announced the end of mining operations in the Copper Basin, together with the loss of nine hundred more jobs. It was the end of an era, the end of almost a century and a half of hard rock mining in the shadow of Little Frog Mountain. Frank Russell, an environmental engineer in charge of the reforestation project for Tennessee Chemical, had seen it coming: "A wise old fellow told me a few years ago that all mining towns wind up as ghost towns." The closing ended a way of life. In better times, as one reporter noted concerning veteran miner Larry Welch, "A job and a union card gave all the security he needed to work near family and friends" the way it had "for generations of mountain people." Those days were now over. Some unemployed miners moved to find work; others began a life of lengthy commuting to jobs outside of the mountains. Frank Collis, local chairman of the machinists union said, "It don't matter if I have to drive one hundred miles each way to my next job.... When it comes bedtime, I'm going to sleep in these mountains." One saddened miner spoke for many when he said, "There will be a Copperhill in the future, but it will only be for tourists." Time would soon prove those words to be prophetic.[48]

The end of mining impacted every aspect of life and activity in the basin, including the ongoing effort to complete revegetation of the Ducktown Desert. Tennessee Chemical continued to operate the chemical facilities at Copperhill with a reduced workforce. Nonetheless, local confidence in the industry had been shattered, and the TVA worried that its fifty-year partnership with the copper industry in the reforestation effort might also come to an end. Tennessee Chemical upset the relationship in 1984 when financial straits forced it to cut all nonessential expenditures, "including the planting of 300,000 seedlings." As a company spokesman stated, "This all might be a blessing in disguise, because it makes everyone understand that this is a separate company that must stand on its own. There is no big oil money behind us now."[49]

Jack Muncy worked to reaffirm the TVA's relationship with Tennessee Chemical. There was still a great deal of work to do in the Ducktown Desert: as of 1986, 11,267 acres (18.2 square miles) "lacked complete vegetation," of which 1,412 acres (2.2 square miles) were totally denuded. Viewed another way, a little more than half of the desert, as measured in 1945, had been revegetated and the remaining half, consisting of the most sterile, acidic, and deeply gullied areas, continued to shed copious amounts of toxic silt into the Ocoee River. It was Muncy's task to devise a plan to finish the job with proven methods at a reasonable per acre cost that would not overly burden the struggling Tennessee Chemical Company. To its credit, the Tennessee Chemical Company resumed the work and reforested another 2,000 acres before it filed for bankruptcy in 1989.[50]

Muncy then found it necessary to revise and renew the plan with a different partner, a Swiss company called Boliden Intertrade AG (BIT). The desert now consisted of 9,074 partially vegetated acres and 528 denuded acres, for a total of 9,602 acres. A new problem, that of divided ownership, entered into the planning. Tennessee Chemical held 8,500 acres when it entered into bankruptcy, but BIT purchased only 1,624 acres when it acquired the Copperhill facilities. The trustee in bankruptcy held the remainder for disposition in smaller tracts by sale and auction. Boliden's limited holdings and its situation as a foreign-owned purchaser of a failed venture might have led it to resist the TVA; it chose instead to resume the work as the latest in the line of companies stretching all the way back to 1929 when the Ducktown Sulphur, Copper & Iron Company, soon followed by the Tennessee Copper Company, began to reforest the desert. BIT's positive response reflected a spirit of corporate citizenship and perhaps a stronger awareness that the work begun seventy years ago under one set of laws now continued within an entirely new legal landscape.[51]

Until the 1970s, reforestation of the Ducktown Desert happened within the legal context of common law nuisance. The 1916 suit by the Tennessee Electric Power Company taught the copper companies that they could be successfully sued in nuisance for damage to hydroelectric facilities caused by acid and silt washed into the Ocoee River. The TVA assumed TEPCO's legal posture when it acquired the Ocoee dams in 1939. For the copper companies, cooperative reforestation was a rational means of forestalling future lawsuits by the TVA about the power dams; never mind the additional benefits of an improved corporate image, an improved local water table for the mining community, a potential renewable source of timber for the industry, and a genuine desire to do the right thing to reverse the abuses of earlier generations. Though the conservationist mandate of the TVA and the Soil Conservation Service was broad, the number of decision makers concerning reforestation of the Copper Basin remained small. Planning could be accomplished primarily between the copper companies and the TVA with the helpful technical assistance of the Forest Service, the Soil Conservation Service, and other government agencies.

The 1970s brought a new set of laws and new parties. Congress changed the legal landscape by enacting a series of environmental laws, among them the Clean Air Act (1970) and the Federal Water Pollution Control Act (1972). The Clean Air Act established federal emission standards for air pollution and required the states to enforce them. The Federal Water Pollution Control Act (FWPCA) established national standards for toxic water pollutants and required federal wastewater permits for all sources of water pollution. President Richard M. Nixon then created the U.S. Environmental Protection Agency (EPA) to enforce both measures. Unlike earlier antipollution measures, the rationale and goals of the new environmental laws reflected a deliberate embrace of ecological science. Water quality in the Ocoee River would now be measured for its impact on aquatic biology, not just for its impact on hydroelectric operations. The new laws had an immediate impact upon the copper industry by imposing fixed pollution standards backed by the enforcement power of the EPA and its environmental counterparts at the state level. New laws, new standards, and new parties would permanently alter the comfortable working relationship between the TVA and the copper companies.[52]

The new air pollution standards imposed under the Clean Air Act forced Cities Service Company to install a system of expensive air scrubbers and monitors. The same standards left the company exposed to regulatory enforcement actions whenever a release of sulfur dioxide or other gas occurred, as happened in May and June 1973. When the gas drifted over the Ocoee River into McCaysville, Georgia, the *Atlanta Constitution* published articles under the headlines "Sulfur Poisoning Town?" and "Copper Plant Freeing Excessive Sulfur Gas." Georgia citizens wrote Governor Jimmy Carter demanding action. He responded with the customary assurances to the constituents and then delegated the complaints to air quality specialists in the Georgia Department of Natural Resources, Environmental Protection Division, who then collaborated with their counterparts at the EPA and at Tennessee's environmental agencies. Everyone worked with baseline ambient air quality standards established by scientists, though Georgia's standards were somewhat tougher than the standards set by the EPA. Cities Service Company cooperated with the investigation and resolution of the matter, for reasons both altruistic and practical. It had no other viable choice under the new environmental laws. The company paid settlements to citizens for damage to their trees, shrubs, and gardens. A Mrs. Waters told reporters, "A man from the plant was here today and said the company would send us a check next week." The company also installed an alarm to warn the citizens of McCaysville of future discharges.[53]

The new environmental legislation governed water quality in the Ocoee River. In 1975 the Tennessee Division of Water Quality Control tested the river and its tributaries pursuant to the mandate of the Federal Water Pollution Control Act. It noted the high levels of industrial wastes discharged into Davis Mill Creek at Copperhill: "Process wastes contain varying amounts of sulfate, sulfide, copper, iron, lead, and zinc. . . . Most effluents are acidic; pH below 2.0 is common." Several local creeks contained untreated sewage. Water quality officers also considered nonpoint runoff from the desert. In harsh terms, they wrote: "Reforestation efforts in the central basin have been notably and consistently unsuccessful. The Ocoee carries an enormous burden of silt after every rain." Three years later, they noted the reforestation efforts by Cities Service, but added, "Runoff from the denuded lands and the industrial mining and processing areas" continue "to cause severe and consistent violation of stream standards." They found that contamination from acid- and mineral-laden silt was so bad that unless abated, it was "doubtful whether stream standards can ever be attained in North Potato Creek or the Ocoee River" or that a "balanced indigenous aquatic community" could be reestablished.[54]

The reports demonstrated several news truths about reclamation in the Ducktown Desert. The issue was no longer simply a matter of protecting the power dams. Desert silt was now a toxic pollutant that violated federal and state clean water standards. The issue could no longer be handled solely between the TVA and the copper companies; both had to answer to the water quality control authorities if regulatory sanctions were to be avoided. In 1978 the EPA filed suit against Cities Service for violating approved standards for the discharge of copper, zinc, arsenic, cyanide, and other impurities. For its part, the TVA soon came under regulatory scrutiny for its practice of sluicing accumulated silt out of the Ocoee No. 3 dam. The dam, being the one closest to the desert, served as a silt trap to protect the two dams further downstream. It was nonetheless necessary to periodically flush silt away from the intake to the tunnel that carried river water to the Ocoee No. 3 powerhouse. This was done by opening the gates at the bottom of the dam for several days to several weeks, often with pulses of water from the Blue Ridge Dam upstream to increase the effect. The process flushed massive amounts of foul-smelling toxic sediments downstream that scoured the river channel, coated the rocks, and killed what little life there was in the river. TVA had been doing this without regulatory hindrance ever since completing the dam in 1942; it now had to answer to water control authorities.[55]

The TVA soon learned that the whitewater community presented another constraint upon its activities. The Ocoee whitewater industry developed by accident in 1976 when the TVA shut down the flume at Ocoee No. 2 for repairs. For sixty years, the river had been largely diverted into the wooden flume which carried it five miles to the No. 2 powerhouse. The same thing happened further up river between the No. 3 dam and the No. 3 powerhouse, with the exception that the river traveled through a two-mile tunnel drilled into the mountainside instead of through an exterior flume. In both stretches of the river, the diversions reduced the natural flow in the riverbed to a mere trickle, not enough to float a kayak, much less a large seven-person rubber raft. This changed, to the joy of whitewater enthusiasts, when the closure of No. 2 flume returned the river to its natural channel.[56]

News spread instantly through the whitewater community. New companies formed to provide rafting trips to visitors. The stretch below the No. 2 dam gained a reputation as one of the top ten whitewater streams in the East, but if the TVA had its way, the pleasure would last only for the year or two needed to repair the flume. The whitewater community insisted that it remain open, especially in the warmer months. Rafters noted that the TVA was willing to spend $26 million to repair the flume and argued that it should have proportionate concern for recreation. From the TVA's perspective, the No. 2 and No. 3 dam systems had always been single-purpose hydroelectric dams, built and maintained without reference to recreation. A return of water to the river meant the loss of salable power. Negotiations broke down, followed by a lawsuit. Recognizing the recreational and economic benefits of whitewater tourism, Tennessee state officials and members of its congressional delegation soon arranged an affordable compromise whereby the TVA agreed to allow the river to run freely for about a hundred days annually in exchange for congressional funding and modest user fees collected by the rafting companies.[57]

Ocoee whitewater enthusiasts inevitably got drenched and ingested water as they ran the rapids, so water quality downstream from the Copper Basin Desert was now a problem of human exposure, not just a concern for hydroelectric installations. A 1985 article in the *Atlanta Constitution* stated the theme: "Whitewater Mecca Is an Industrial Sewer." The article was actually somewhat retrospective: a river runner said that when he kayaked the river in 1979, "every time my head went into the water, my nose, eyes and mouth burned. Now, that doesn't happen any more." Jack McCormick, the water quality officer for Cities Service, acknowledged that the National Wildlife Federation once rated Davis Mill Creek as "the third most polluted waterway in the country." Under the compulsion of clean water laws, the company limited industrial discharges so that "the creek and the river are no longer purple or red."[58]

The aesthetic quality of Ocoee water to humans grew in importance as the whitewater business matured into a significant industry with several dozen companies serving hundreds of thousands of visitors annually. The popular river became a scheduled stop on the whitewater competition circuit, leading to its selection as the venue for whitewater events in the 1996 Atlanta Olympics. The U.S. Forest Service boasted that it was "the only Olympic whitewater site in the world located on a natural river." It was true that the course ran over a natural riverbed, unlike other venues with totally artificial courses, but there was otherwise little that was natural about it. The Ocoee course was on a stretch that normally ran dry because of the diversion of the river at Ocoee No. 3. As with the public whitewater area below the Ocoee No. 2 dam, the river ran through the Olympic site only when the TVA opened the tap. The organizers then engineered the riverbed to suit the needs of the events. They narrowed the channel by 50 percent and sculpted the riverbed to improve the rapids, and even inserted permanent artificial rocks.[59]

The river was not natural, nor was it alive. The Olympics whitewater events occurred on a biologically dead river flowing between two federally protected wilderness areas. A 1981 study by Tennessee authorities determined that the river was a "biological desert" between Copperhill and Lake Ocoee. A decade later, the TVA determined that the reservoirs downstream from Copperhill "provide poor aquatic habitat because of high concentrations of heavy metals." Copper, lead, and zinc were all at levels toxic to phytoplankton and zooplankton, causing the destruction of the aquatic food chain.

The physical effects of sluicing from the No. 3 dam made matters worse by releasing toxic sediments to coat every surface in the riverbed. The situation remained largely unchanged above Lake Ocoee when researchers conducted fieldwork for the 1994 environmental impact statement prepared for the Olympics whitewater events. Using standard electrofishing techniques, biologists sent shocks into the water to stun the fish for counting purposes, and then sent divers into the deeper pools to conduct a visual survey. They found almost nothing. They concluded that "the fish community currently residing in the upper Ocoee River could be described as nearly non-existent." If not safe for aquatic life, it was safe for the competition because water quality in the river exceeded "requirements for water contact activities."[60]

Water quality concerns for fish and rafters downstream now influenced planning for revegetation of the Ducktown Desert upstream. Muncy's 1986 revegetation plan with Tennessee Chemical Corporation reflected the new priorities when stating the call for new approaches to accelerate the work because of "the need to improve water quality in the Ocoee River, reduce the rate of sedimentation into Ocoee No. 1 reservoir, and to reduce the impact of sediment loads on hydro-power operations." Water quality now trumped power operations. His 1991 plan with BIT stated, "Siltation and heavy metals pollution resulting from erosion of Copper Basin lands" was now "one of the high priority water resource problems in the Tennessee Valley."[61]

TVA and BIT (later renamed Intertrade Holdings) continued the work of revegetation through the 1990s and into the next decade, even while industrial operations at Copperhill continued their steady decline. The company continued to operate the acid facilities, but with the closure of the mines, it no longer processed chemicals from local ore. Instead, the company processed sulfur shipped into the Copper Basin from other sources until the year 2000, when the company ended ninety-three years of acid production with yet another round of employee layoffs. The last train of acid left the Copper Basin on March 6, 2001. A Brazilian firm bought the No. 6 acid plant and had it dismantled for shipment to South America.[62]

In the meantime, a new business-government relationship developed between the EPA and Glenn Springs Holdings, Inc., the environmental reclamation unit of Oxy USA, Inc., which was, in turn, the domestic subsidiary of Occidental Petroleum Corporation, an international oil and gas company. Neither Glenn Springs nor Oxy ever owned or conducted a business in the Ducktown Basin, but Oxy became potentially exposed to liability for the costs of remediation under the broad sweep of the 1980 Comprehensive Environmental Response, Compensation, and Liability Act (Superfund). The intent of the law was to place the financial burden of hazardous waste site remediation upon what the law deemed "a potentially responsible party" rather than upon the taxpayers. The broad scope given to the phrase enabled the EPA to pursue current owners, any previous owner, or, in the present situation, a party that had acquired the liabilities of a previous owner. With Tennessee Chemical bankrupt and all of the other prior owners defunct, Oxy stood as the only potentially responsible party that was both still in business and economically viable. It acquired its status by virtue of its acquisition of Cities Service in December 1982, three months after the latter sold its Copperhill operations to Tennessee Chemical. When Oxy assumed the liabilities of Cities Service, it also acquired the latter's exposure as a potentially responsible person for the reclamation of the Ducktown Desert, even though it had never produced a gallon of sulfuric acid or processed a ton of copper ore there.[63]

Oxy denied that it was a successor in interest to Cities Service for the purposes of the Superfund law, but nonetheless elected to fully cooperate with the EPA in the Copper Basin. It had already gained expensive familiarity with Superfund when a federal court ruled it to be a responsible party for the 1978 Love Canal disaster by virtue of its acquisition of Hooker Chemical Company in 1968. The crisis began in Niagara Falls, New York, when a cluster of cancer cases led to the discovery that a residential neighborhood and elementary school had been built upon an abandoned canal filled with thousands of tons of toxic chemical waste dumped by Hooker in the 1940s. The company eventually filled in the canal and covered it with grass. A local school board, desperate for land to build a school, approached Hooker to buy the tract, but the company refused after disclosing the hazards posed by the buried waste. The school board persisted and, in 1953, the company transferred the property with a deed containing added language advising of the hazard and disclaiming liability. All of this happened many years before the 1968 purchase of Hooker, but Occidental nonetheless became the prime financial target when the crisis broke in the late 1970s to become the greatest national environmental story of the day. The crisis prompted Congress to pass the Superfund law, which then led to a Superfund lawsuit filed by the Department of Justice against Occidental. The financial exposure from the lawsuit and the impact of the worst possible publicity forced the company to eventually settle by paying the government $129 million.[64]

Oxy's work in Ducktown through Glenn Springs Holdings began in a much more positive context, without litigation and with the full cooperation of the EPA and the Tennessee Department of Environment and Conservation (TDEC). Glenn Springs assumed management of the reclamation effort in 1996 by devising a multifaceted strategy to improve water quality, to enhance safety around abandoned mine sites, and to prepare the land for beneficial use by the community. The water quality agenda was the most costly and complex. Two main watersheds, Davis Mill Creek and North Potato Creek, flowed through the old mine sites, across the barren remnant desert, and around the chemical plants before entering the Ocoee just below Copperhill. Untreated water from both creeks added enough metal to the river to make two automobiles every day. Davis Mill Creek contained so much iron hydroxide that it coated every living thing with a slick orange sludge. Both creeks were highly acidic, with pH values of 2.5 to 3.5.[65]

Under the direction of Frank Russell, an environmental engineer and the site manager for Oxy, Glenn Springs installed or improved wastewater treatment plants to extract the metals and lime plants to reduce acidity. In just two years, the treatment plants removed almost six million pounds of iron, zinc, manganese, copper, lead, and cadmium. Water acidity dropped to a neutral pH 7.0. Another phase of the work made use of an old open-pit mine. In the past, the flow of North Potato Creek had been diverted into Davis Mill Creek in order to dig the South Open Pit Mine, in the old creek bed. Glenn Springs redirected North Potato Creek back to its original channel, which allowed the two-hundred-foot-deep South Pit to function as a gigantic settling pond to capture minerals and sediments. In the Davis Mill Creek watershed, engineers installed a polyethylene pipeline sixty-three inches in diameter and three thousand feet long to redirect the clean headwaters of a feeder stream, Bell Creek, around the most contaminated areas of the watershed. This maintained water quality in Bell Creek while reducing the flow to be processed in downstream wastewater treatment plants.[66]

In other projects, workers pumped contaminated water out of old mines to protect nearby groundwater sources. Toxic mine wastes, including lead from the old acid chamber plants, were removed and covered using environmentally sound methods. Glenn Springs also supplemented the revegetation work of the TVA and BIT-Intertrade by planting trees and grasses across a broad expanse of powdery tailings from the London Flotation Mill. This stopped the winds from carrying mining wastes across the northern part of the basin. In one of the more dramatic acts of reversal, Glenn Springs engineers transformed one of the infamous roast yards into a biological passive flow treatment system to further cleanse water after leaving the treatment plants. Acres of green wetlands now existed where hundreds of burning heaps of roasting ore once sent clouds of sulfur dioxide directly into the skies.⁶⁷

All of the work was subject to a legally binding memorandum of agreement signed on January 11, 2001, by representatives from Oxy, Glenn Springs, the EPA, and TDEC. The agreement provided that Glenn Springs agreed to perform the remediation work ordered by the two agencies according to their standards and subject to their approval. The agencies agreed to provide supervision and technical assistance. All the parties agreed to work cooperatively to inform the public concerning project status and future plans. Though Oxy continued to deny liability as a potential responsible party under the Superfund law, it nonetheless agreed to perform the work in exchange for the EPA's agreement to "withhold the proposed listing of the site on the National Priorities List." The Superfund law required the EPA to maintain the list to designate the hazardous waste sites that it deemed the highest priorities for remediation. The parties were aware that the EPA considered adding the Copper Basin to the list, but all found it beneficial to instead enter into the memorandum of understanding. Voluntary remediation would allow the project to begin sooner and at less cost than might occur if listing entangled the project in a drawn-out lawsuit. Oxy also knew from its Love Canal experience that it wanted to avoid the stigma of association with a high-priority site. It was also aware that voluntary remediation of historical practices would produce substantial public relations benefits. Such agreements work only in a climate of mutual respect and trust. The EPA later acknowledged that Glenn Springs had a "long-standing reputation for responsible environmental stewardship and was willing to collaborate with EPA and TDEC." The Copper Basin agreement is now considered a model to be followed, where appropriate, at other hazardous waste sites.⁶⁸

The memorandum of understanding between Oxy, Glenn Springs, the EPA, and TDEC was the latest in a series of cooperative remediation agreements between government and industry that stretched back over sixty years to the 1939 pact between the Tennessee Valley Authority and the Tennessee Copper Company. The agreements reflected a growing awareness that conditions in the Ducktown Desert directly impacted matters downstream. Reclamation occurred because of what silt, acid, minerals, and chemicals from the copper industry did to property and life outside the basin. It was not impelled by a somewhat romantic desire to return sylvan beauty to a barren landscape.

Outside impacts prompted the response of outside parties: the Tennessee Electric Power Company, the Tennessee Valley Authority, and the environmental regulatory agencies. Each agreement reflected a keen appraisal by industry leaders of the legal and political strength of the outside parties. The initial struggle between TEPCO and the copper companies took place between relatively equal parties under the common law of nuisance. TEPCO's victory, though expensive, did not seriously impact the copper industry, so early reclamation efforts were at best tentative. The scope and pace of reclamation increased in response to the much greater power of the TVA. The balance altered again with the passage of comprehensive environmental legislation and the creation of the EPA, along with its state-level counterparts. Their ability to establish and enforce air and water quality standards gave them the power to govern the environmental practices of both the copper industry and the TVA. It would be overly cynical to assert that the copper industry engaged in reclamation only when compelled by a superior power, but it would be naive to believe that the industry acted solely for altruistic reasons without reference to its legal and political exposure under the Superfund law.

The reclamation agreements were more than the result of legal and political power. The TVA, the Soil Conservation Service, and the environmental agencies offered valuable technical assistance. Reforestation of the desert owed as much to research by government biologists and foresters as it did to industry money and labor. As with the smelter smoke, improvements in the desert and the river that flowed through it could not happen at a pace faster than the advance of science and technology. Early attempts at tree planting mostly failed until government researchers devised a cost-effective way to make seedlings grow in the sterile, poisoned soil of the naked red hills.

Their efforts were aided by the end of the mining and acid industries. Heavy industry left the Ducktown Basin, and so did most light industry. An economy that had for a century and a half rested upon the mineral resources of the basin now had to be rebuilt upon the recreational assets of the national forests, wilderness areas, and the Ocoee River. Vacation homes dotted those portions of the mountain rim in private hands, a phenomenon unthinkable in the days of heavy smoke. Copperhill and McCaysville made the transition from company towns to instead serve the needs of vacationers and daytrippers from Atlanta and other southern cities.[69]

Glenn Springs and its government partners proposed to transform the reclaimed copper lands into a resource "providing recreational ... improvements to encourage cultural, historic, and environmental tourism through enhancement of historic mining features." Its plans included hiking trails and mountain bike trails to connect with the nearby national forests and an equestrian center to make use of the open spaces in the basin. Plans also included a scenic railroad over Copperhill through the basin, around the Hiwassee Loop, and then to the Tennessee Valley. The words of the sad miner became true: "There will be a Copperhill in the future, but it will only be for tourists."[70]

EPILOGUE THE VIEW FROM THE MOUNTAIN

In 1860 Hardin Taliaferro stood on a mountaintop and saw the clouds of heavy black smoke rising from the heaps of roasting ores. With the eyes of a prose stylist, he described them as "huge columns of smoke ascending towards heaven, spreading out at top like vast sheaves" that combined to cover "the heavens with a smoky pall." Half a century later in 1915, Dr. John T. McGill stood on a mountaintop to observe smelter smoke at the request of the United States Supreme Court. By then, the smoke McGill saw no longer rose in sheaves from the roast heaps but instead pulsed from smokestacks as the dampers were "raised or lowered to regulate the amount needed for the sulfuric acid factory." In another eighty-five years, the sulfur smoke would be gone. No smoke rises from the dormant Tennessee Copper Company complex at Copperhill or from the abandoned ruins of the Ducktown Sulphur, Copper & Iron Company at Isabella. To the extent that there is air pollution in the basin, it is likely fumes from distant coal-fired power plants or automobile exhaust from Atlanta and other major cities.[1]

The local copper industry is gone, but the case it spawned remains potent as a legal precedent. *Georgia v. Tennessee Copper Co.* arose from the physical and political geography of the Ducktown Basin and then became a template for smelter litigation across the country. Its application then extended beyond the mining industry to serve as a vehicle for addressing other forms of transborder pollution. Finally, in 2007, the Supreme Court returned to its century-old decision in the copper case to serve as a crucial jurisdictional precedent for a landmark ruling on anthropogenic global warming.

Theodore Roosevelt and members of his administration grasped the significance of the 1907 victory achieved in the Georgia case by Attorney General John C. Hart and Ligon Johnson, his special counsel. Charles Bonaparte, Roosevelt's attorney general and Napoleon's grandnephew, hired Ligon Johnson to serve in a similar capacity on behalf of the federal government against other smelter companies for damage caused to national forests in the West. Johnson rescued the government's losing cause against Mountain Copper. He then pursued new actions in Montana against the Anaconda and Washoe smelters, and in northern California against Mammoth Copper, Balaklala Copper, Bully Hill Copper, and Engels Copper.[2]

In each case, Johnson demonstrated the same mastery of scientific evidence and mining technology that led to his victory in the Georgia litigation. He knew that in order to work effectively with the experts he had to acquire considerable knowledge of his own on a variety of subjects. As he explained, he became familiar with "plant pathology, entomology, agronomy, toxicology, veterinary medicine, and other technical matters." He knew, for example, that lead fumes from smelting caused horses to suffer a disease known as "roaring." He also knew that consumption of "too much dry food, particularly the first crop of alfalfa," would "paralyze the vagus nerve and produce conditions similar to fume poisoning." Thus, every claim of fume poisoning required him to first examine grazing conditions for livestock.[3]

Johnson put his knowledge to use in the Anaconda case. The company contended that it could neither successfully convert its furnace fumes into acid nor profitably sell the product because of high freight charges and the limited fertilizer market in the West. The company lobbied the White House on both points, arguing that it would be forced to close if compelled to install acid technology. Roosevelt was sufficiently concerned to withhold permission to file suit until persuaded that the industry was wrong on both points. He advised Attorney General Bonaparte, "My directions in connection with the Anaconda smelter matter are that we shall look carefully before we leap," because if industry contentions were correct, "tens of thousands of working men would be completely thrown out of employment, and half the State of Montana suffer seriously." Roosevelt wanted to know if the expense of smoke improvement technology would result in a shutdown, or "whether it merely represents a heavy necessary expense which can and should be borne which will allow the work to be done, but only under conditions that prevent its being noxious to the vegetation round about."[4]

Having answered similar questions in the Georgia case, Ligon Johnson prepared a detailed report for the president, demonstrating the technical and economic viability of four different approaches. He noted that abundant phosphate was near a local railroad. He then surveyed freight rates and determined the need for fertilizer in the West. With those factors in mind, he advised that Anaconda had four viable options. First, it could build an acid extraction plant like those used by the copper companies in Ducktown. Second, it could convert the fumes into another form of fertilizer, ammonium sulfate, as was now being done at the Mountain Copper smelter in California. Third, Anaconda could adopt the wet method of copper extraction as used in the Rio Tinto mine in Spain. The process, also known as flotation smelting, liberated sulfur from copper ore without using furnaces. Ore was instead pulverized and then oxidized by controlled exposure to air and water.

(The Ducktown Sulphur, Copper & Iron Company installed a flotation mill in 1920; Tennessee Copper followed with its own mill in 1923.) The fourth alternative was to move the Anaconda smelting to a treeless expanse in Montana where it would cause less trouble. With careful research and frequent citation to technical journals, Johnson's thirty-one-page report accomplished its goal. On February 4, 1909, Roosevelt advised, "in view of your report... we shall have to proceed with the suit to compel the Anaconda smelter people to do their duty along the lines you suggest."[5]

Johnson achieved a string of successful outcomes against the western smelting companies. In the Mountain Copper case, he salvaged a defeat suffered by earlier government counsel and won an agreement whereby the company paid damages, deeded a valuable piece of land to expand the Shasta National Forest, and "ceased the discharge of the injurious fumes." The company then relocated its operations to a site far from the national forest. In other suits involving Mammoth Copper, Balaklala Copper, the Bully Hill smelter, and the Engel smelter, the respective owners either installed smoke suppression devices or ended operations.[6]

Though successful, Johnson's work had a narrow geographic scope. In the absence of federal air pollution statutes (the first national Clean Air Act was not enacted until 1963), the government's interests in smelter problems were limited to the protection of its national forests. This narrowed the range of possible lawsuits to California and Montana, the only locations where smelters and national forests were in proximity at the time. Yet, to the extent success was achieved in the federal smelter actions, it occurred at the hands of Ligon Johnson, using the knowledge and tactics he gained in *Georgia v. Tennessee Copper Co.*[7]

Johnson eventually left government service and switched sides for a time to represent American Smelting & Refining Company and International Nickel Company, presumably for higher compensation than he received for his government work. He then left the field entirely to pursue a new phase of his legal career as a specialist in copyright and entertainment law on behalf of most of the major cinema companies, including Metro-Goldwyn-Mayer, 20th Century Fox, Warner Brothers, Paramount, and Disney. This was not a total surprise to those who knew him, because he represented some theatrical interests while still in Atlanta, and had always aspired to ever greater legal challenges in ever larger cities.[8]

The Georgia case became significant in pollution matters outside of the copper industry. The 1907 opinion by Justice Oliver Wendell Holmes established the jurisdictional precedent that allowed states to confidently assert Supreme Court original jurisdiction for other transborder pollution disputes. Additionally, the decision effectively created the federal common law of nuisance with its recognition that a state, as a quasi sovereign, had the right to protect its natural resources from invasive pollution.[9]

The Georgia case provided the jurisdictional basis for a 1908 Supreme Court water pollution case arising in the Northeast when New York sought to enjoin a New Jersey plan to divert sewage from the Passaic River via a tunnel to Upper New York Bay. Thirteen years later, the Court issued its decree rejecting the injunction. New Jersey's experts persuaded the Court that "the best obtainable sanitary engineers, chemists, and bacteriologists" designed the project.[10]

New Jersey filed its own transborder pollution suit against New York City in 1929 to stop the city from dumping garbage into the ocean, from whence it washed up on Jersey beaches. The Court issued a decree in 1933 ordering a stop to ocean dumping and requiring the city to build incinerators to dispose of its trash. The decree was extended and modified to allow the city to dump sewage sludge, though not garbage, into the waters.[11]

The principles and techniques developed in an interstate context in *Georgia v. Tennessee Copper Co.* next served as the framework for the international dispute between the United States and Canada regarding air pollution crossing over the border into Washington State from the Trail Smelter in British Columbia. The international commission formed to resolve the dispute consciously proceeded with the same emphasis upon technological solutions pursued earlier by John C. Hart and Ligon Johnson in the Ducktown Basin smoke controversies.[12]

Yet, for all of its importance in the New York–New Jersey cases, the Trail Smelter dispute, and similar cases, the holdings of *Georgia v. Tennessee Copper Co.* were applied with less frequency in later decades. The decline was in part due to the clumsiness of original jurisdiction proceedings before the Supreme Court. The justices did not like to function as a trial court, nor did they do so quickly. Eleven years elapsed between the initiation of the Georgia case and the injunction issued against Ducktown Sulphur, Copper & Iron Co. Thirteen years passed between the initiation of New York's injunction suit to stop the New Jersey sewage plan and the issuance of the Court's decree denying the requested injunction. In the latter action, the parties prepared their testimony between 1911 and 1913, but the Court did not set the matter for hearing until 1918. The Court then determined that the evidence was stale and ordered the parties to take additional testimony. The matter was not heard on the merits until 1921. The great delays reflected the low priority of the original jurisdiction docket compared to the Court's usual docket of appellate cases. For the litigants, the delays kept major public works projects in suspense, holding up designers, contractors, political administrations, and bond financiers, not to mention the intended beneficiaries among the taxpaying public.[13]

A more important reason for the declining use of *Georgia v. Tennessee Copper* as an antipollution weapon was that it was a decision ahead of its time. Though farsighted advocates fought pollution in earlier times, issues of environmental pollution had yet to seize public attention to the extent they did after World War II. The strength of the postwar environmental movement led state attorney generals to dust off the old Georgia case for use against a variety of polluters in the 1970s. The renewed interest reflected increased attention by states to their rights as quasi sovereigns to protect their natural resources.[14]

The Supreme Court was well aware that the newly powerful environmental movement promised an incoming tide of original jurisdiction cases unless it took action to stop it. It did so in *Ohio v. Wyandotte Chemicals Corp.* (1971), a case between a state and a corporation in another state regarding mercury contamination of drinking water. The facts were analogous to those in *Georgia v. Tennessee Copper Co.*, yet the Court nonetheless voted 8 to 1 to refuse original jurisdiction on the grounds that it should no longer function as the "principal forum for settling" controversies between a state and citizens of another state. In so ruling, the Court conceded that as an appellate tribunal, it was "ill-equipped for the task of factfinding" and that it had "no claim to special competence in dealing with the numerous conflicts between States and non-resident individuals that raise no serious issues of federal law." The same points could have been made three generations earlier to refuse original jurisdiction in *Georgia v. Tennessee Copper Co.*, but in an earlier age, the Court did not sense the burden of handling interstate pollution cases to the degree that was now apparent in the bloom of the environmental movement of the 1970s.[15]

The Court gained another rationale for refusing original jurisdiction pollution cases as Congress enacted a series of environmental bills in the 1970s. In only three years, it passed the National Environmental Policy Act (1970), Water Quality Improvement Act (1970), Federal Water Pollution Control Act (1972) (also known as the Clean Water Act), and the Endangered Species Act (1973). President Richard Nixon followed suit by creating the Environmental Protection Agency (EPA) to administer the new legal regime. Together, the new laws and bureaucracies went far toward filling the regulatory vacuum on environmental matters that existed at the time of the Georgia case.

The new laws gave the Supreme Court the opportunity to refuse application of federal common law to pollution cases. In *City of Milwaukee v. Illinois* (1981), the Court declared that the Clean Water Act created "a balanced and comprehensive remedial scheme" that preempted the Court from fashioning its own common law of pollution nuisance. Such cases would, in the future, be directed to the federal trial courts to be handled under environmental statutes passed by Congress. The Court soon extended the preemption doctrine to other aspects of pollution law.[16]

The doctrine of preemption involves complicated issues of federalism and congressional authority. The concept is nonetheless simple in its broadest strokes. There were no relevant pollution statutes when Georgia began its fight against the Tennessee Copper Company and the Ducktown Sulphur, Copper & Iron Company, Ltd. in 1903. In the absence of congressional action on air pollution, the Supreme Court had to create a body of pollution law of its own if the state was to have any remedy at all. Several generations passed before Congress established a comprehensive regime of pollution law, but when it did, the new statutes reduced the need for the Supreme Court and other courts to devise their own rules and remedies. Congress had finally spoken in a way that all but eliminated the need to apply federal common law nuisance cases such as *Georgia v. Tennessee Copper Co.*

The preemption law doctrine put an end to the Georgia case as an antipollution tool—or so most commentators thought. Then, in 2007, the majority opinion in *Massachusetts v. Environmental Protection Agency* demonstrated the continued relevance of the old case to pollution issues, this time concerning the problems of greenhouse gases and global warming. The new case involved a complaint filed by Massachusetts, together with other states and environmental organizations, alleging that the EPA had wrongfully refused to consider, much less to implement, regulations to control automobile tailpipe emissions in the fight against global warming.[17]

The substantive arguments of the plaintiffs involved the latest atmospheric science, but before they could be considered, the plaintiffs first had to establish standing—their legal right to bring their case to the Court. At least one of the plaintiffs had to be able to show, among other things, that the EPA's failure to regulate greenhouse gases presented "a concrete and particularized injury that is either actual or imminent." The EPA responded by arguing that global warming was a problem of such scale that it was difficult for any one plaintiff to establish an injury more specific, more "particularized," than that suffered by the public at large.[18]

Only one of the plaintiffs needed to establish standing. Massachusetts took the lead by relying upon the authority of *Georgia v. Tennessee Copper Co.* It argued that carbon dioxide and other gases caused global warming; that global warming led to rising ocean levels from the melting of glaciers and polar ice caps; and that rising sea levels threatened low-lying Cape Cod. The state then quoted from the 1907 Holmes opinion to assert that, as a "quasi-sovereign," it had the right to relief in the federal courts to protect the natural resources within its borders. The Supreme Court agreed, by a slim vote of 5 to 4. In the majority opinion by Justice John Paul Stevens, the Court quoted at length from the Georgia case. Stevens wrote, "Just as Georgia's 'independent interest ... in all the earth and air within its domain,' supported federal jurisdiction a century ago, so too does Massachusetts's well-founded desire to preserve its sovereign territory today." With standing established, the Court then ruled that carbon dioxide was a pollutant within the meaning of the Clean Air Act, and that the refusal of the EPA to regulate tailpipe emissions was arbitrary.[19]

The 1907 decision in *Georgia v. Tennessee Copper Co.* was the Supreme Court's first air pollution decision. The 2007 decision in *Massachusetts v. Environmental Protection Agency* was the Court's first global warming case. The earlier case created the jurisdictional and substantive rationale needed for states to combat smelter smoke and other regional forms of transborder pollution. The latter decision extended the rationale of the Georgia case to address the national and international problem of climate change. Thus, the Georgia case, a matter that owes its origins to smoke suits filed in the 1890s by mountaineers in the Ducktown Basin, proved to have legal ramifications far beyond the imaginings of any of its participants.[20]

The smoke suitors, the state of Georgia, and the copper industry functioned under very different legal and technological constraints from the situation in modern times. The legal remedies sought were confined to those of traditional nuisance law. The only truly effective technological remedy was acid condensation, which proved to be a partial cure at best. All of the parties operated in the absence of a comprehensive system of environmental law, much less the scientific and popular basis for it. There was no environmental bureaucracy to see things through. The litigants did what they could with the law and technology available to them.

There was one other alternative. Ducktown smoke could have been completely stopped, but only by destroying the copper industry. Georgia's legal and political leaders refused to let that happen. Though they wanted justice, and to a large degree obtained it in the Supreme Court, they never wanted the heavens to fall upon Ducktown's copper industry or upon the Georgians it employed. Time will tell how well, in the present day, citizens, industry, and government address the intersecting themes of law, environment, technology, economics, and employment in their attempt to meet the current challenges.

Notes

Abbreviations

DBM   Ducktown Basin Museum, Ducktown, Tenn.
GDAH Georgia Department of Archives and History, Morrow, Ga.
NARA National Archives and Record Administration, Washington, D.C.
NFU   National Farmers Union Collection, University of Colorado Archives, Boulder, Colo.
TSLA Tennessee State Library and Archives, Nashville, Tenn.

Introduction

1. Deposition of B. H. Sebolt (1914), Transcript, 265–72, *Georgia v. Tennessee Copper Co.*, U.S. Supreme Court, No. 1 Original, October Term, 1915. The docket numbers in this case are confusing because as an original jurisdiction action, it received a new docket number for each term in which the case came before the Supreme Court. The initial designation of the case when filed in 1905 was No. 13 Original, October Term, 1905, and when finally dismissed in 1937, it received its final designation, No. 1 Original, October Term, 1937. Documents filed in one term were frequently renumbered for use in subsequent terms. NARA archives the entire case file from 1905 to 1937 under its final 1937 designation. To avoid confusion, cited documents from the case are identified primarily by type and date rather than by docket number.
2. In most of the primary documents, writers used the old spelling "sulphur" instead of the modern, "sulfur." I use the modern spelling except in reference to the company name, Ducktown Sulphur, Copper & Iron Company and wherever the old spelling appears in titles of cited works. Note that Ducktown is also the name of a mining town (formerly known as Hiwassee) adjacent to the Burra Burra Mine; however, unless otherwise specified, references herein to Ducktown are to the region, such as the Ducktown Basin, the Ducktown Mining District, or, more generally, Ducktown. This was the common usage at the time of the smoke litigation.
3. McCallie, "The Ducktown Copper Mining District"; Chase, *Rich Land, Poor Land*, 49–53; Smallshaw, "Denudation and Erosion in the Copper Basin," 1; Clay, "Copper-Basin Cover-Up." Others acknowledged the devastation while appreciating the economic significance of the industry. See, e.g., Robards, "Tennessee's Wealthy Wasteland."
4. Barclay, "Information on Copperhill, Tennessee, for Inclusion in the American Guide Manual," Barclay Papers, box 1, folder 7, TSLA; Workers of the Writers' Program of the Works Progress Administration in Georgia, *Georgia: The WPA Guide to Its Towns and Countryside*, 472.
5. Burt, "Desert in the Appalachians"; Barnhardt, "The Death of Ducktown." For similar views, see Teale, "The Murder of a Landscape"; Ottewell, "There Are No Ducks in Ducktown."
6. McKinney, "Bad Lands of Copperhill"; Quinn, "Tennessee's Copper Basin: A Case for Preserving an Abused Landscape." Apart from local sensibilities, Quinn urged preservation of at least a portion of the badlands for their historical significance. The comments

about mosquitoes, etc., are from the staff of the Ducktown Basin Museum, most of whom are lifelong residents of the area. Georgia's Providence Canyon, another monument to landscape abuse, became a popular state park; see Sutter, "What Gullies Mean."

7   Quinn, "The Appalachian Mountains' Copper Basin and the Concept of Environmental Susceptibility."
8   Davis, *Where There Are Mountains*, 147–53, 157–59. Davis views Ducktown logging and charcoal practices against the background of similar practices in metal mines and iron forges throughout the Appalachians.
9   Most of the private cases were filed in either the district court or the chancery court of Polk County, Tennessee, though some of the larger cases were litigated in federal court in Chattanooga. The exact number of cases is difficult to determine because the original district court records were destroyed in a 1935 fire at the Polk County Courthouse. I derived the estimate of two hundred cases from appellate records archived at the Tennessee State Library and Archives and from docket sheets and attorney correspondence archived at the Ducktown Basin Museum. As is explained in chapter 3, a number of individuals filed multiple lawsuits during the course of the smoke wars.
10 *Georgia v. Tennessee Copper Co.*, 206 U.S. 230, 237 (1907).
11 Lazarus, *The Making of Environmental Law*, xv. For a recent overview of the case in the development of the federal common law of nuisance, see Percival, "The Frictions of Federalism."
12 *Georgia v. Tennessee Copper Co.*, 206 U.S. at 239.
13 Overviews of the conservation and environmental movements include Hays, *Conservation and the Gospel of Efficiency*; Hays and Hays, *Beauty, Health, and Permanence*; Nash, *Wilderness and the American Mind*; Shabecoff, *A Fierce Green Fire*.
14 The dates are from an annotated chronology: Dubé, *History of Tennessee Copper Company*.
15 Andrews, *Managing the Environment, Managing Ourselves*, ix. For a historical view of environmental nomenclature, see Rome, "Coming to Terms with Pollution." In addition to Andrews, a number of recent works emphasize law as a major factor of environmental history. A representative sample includes Brooks, Jones, and Virginia, *Law and Ecology*; Flippen, *Nixon and the Environment*; Keiter, *Keeping Faith with Nature*; McEvoy, *The Fisherman's Problem*; Merrill, *Public Lands and Political Meaning*; Steinberg, *Nature Incorporated*. For overviews regarding the impact of politics and economics on the development of American law, see Friedman, *A History of American Law*; Hall, *The Magic Mirror*; Horwitz, *The Transformation of American Law*.
16 *Worcester v. Georgia*, 31 U.S. (6 Pet.) 515 (1832).
17 For representative historical studies of smelter pollution, see MacMillan, *Smoke Wars* (regarding litigation in Butte, Montana); Charry, "Defending the 'Great Barbecue'" (lead smelting in Washington State); Church, "Smoke Farming" (Salt Lake area); Lamborn and Peterson, "The Substance of the Land" (also in the Salt Lake area). For studies of copper smelter pollution in Swansea, Wales, see Newell, "Atmospheric Pollution and the British Copper Industry"; Rees, "The South Wales Copper-Smoke Dispute." For the 1930s Trail Smelter dispute between the United States and Canada, see Wirth, *Smelter Smoke in North America*, 43, 119, 203 (influence of Tennessee Copper case upon the Trail Smelter dispute); Dinwoodie, "The Politics of International Pollution Control."
18 *Massachusetts v. Environmental Protection Agency*, 549 U.S. 497, 518–19 (quotation from *Georgia v. Tennessee Copper Co.*), 532 (greenhouse gases as pollutants within the mean-

ing of the Clear Air Act), 534–35 (EPA arbitrary and capricious), (2007) (5–4 decision; Roberts, dissenting, joined by Thomas, Scalia, and Alito).

19 For representative works, see Eller, *Miners, Millhands, and Mountaineers*; Eller, "Land as Commodity"; Ronald Lewis, *Transforming the Appalachian Countryside*. Other authors point to pre–Civil War industrialization in the mountains; Davis, *Where There Are Mountains*; Davis, "Living on the Land"; Dunaway, *The First American Frontier*.

20 Phillips, *Georgia and State Rights*.

21 Jeffersonian republican themes are explored on a national scale in McMath, *American Populism*; Sanders, *Roots of Reform*. For agrarian populism in Georgia, see Arnett, *The Populist Movement in Georgia*; Woodward, *Tom Watson, Agrarian Rebel*; Hahn, *The Roots of Southern Populism*; Shaw, *The Wool-Hat Boys*. For the National Farmers Union (NFU), see Barrett, *The Mission, History and Times of the Farmers' Union*; Fisher, *The Farmers' Union*; Crampton, *The National Farmers Union: Ideology of a Pressure Group*.

22 See Cobb and Dyer, "Economic Prosperity or Environmental Protection."

23 There is a rich body of work on the Ducktown area and its mining industry. Robert Edward Barclay laid the foundation with his three books, *Ducktown Back in Raht's Time*, *The Railroad Comes to Ducktown*, and *The Copper Basin, 1890–1963*, which are indispensable sources to anyone working on the area's history. His work is informed by the perspective he gained as a lifetime resident of the area and from his career as treasurer of the Tennessee Copper Company. For other published sources of local history, see Lillard, *The History of Polk County*; James Donaldson Clemmer Scrapbooks on Polk County, 1884–1934, Mf. 1525, TSLA; James W. Taylor, "Ducktown Desert"; Foehner, "Historical Geography." Lillard, a direct descendent of Polk County pioneers, incorporated Barclay's work on Ducktown proper within a broader work on the county's history. The Clemmer scrapbooks consist of thirty-seven volumes in an unusual format: each consisted of a Department of Agricultural yearbook into which he pasted local news clippings, programs of community events, letters, and other items of interest, though often without dates and sources. The oversight is forgivable because his clippings provide documents that are otherwise difficult or tedious to locate. Taylor and Foehner wrote useful historical geographies for their thesis projects. For surveys specific to Ducktown mining history, see Emmons and Laney, *Geology and Ore Deposits of Ducktown Mining District*; Dubé, *History of Tennessee Copper Company*. See also three collections of articles by the Polk County Publishing Company: *Polk County Scrapbook: A Tribute to the Miners*; *Polk County Scrapbook: 150 Years of Memories*; *Polk County News: Reprints of Articles Relating to the History of Polk County*. For Ducktown's place in Appalachian environmental history, see Davis, *Where There Are Mountains*. In addition to her previously cited articles, see Quinn, "Early Smelter Sites"; Quinn, "Industry and Environment in the Appalachian Copper Basin." See also Cobb, *Industrialization and Southern Society*. For historiographical essays on southern environmental history, see Stewart, "Southern Environmental History"; Morris, "A More Southern Environmental History."

Chapter 1

1 Barclay, *Ducktown*, 56–57.
2 Taliaferro, "Ducktown, by 'Skitt,' Who Has Been 'Thar.'"
3 The name Ducktown has been variously applied to the Cherokee village that preceded white settlement, to the mining town, formerly named Hiwassee, adjacent to the Burra

Burra mine, and to the copper mining district and to the geographic basin where it is situated. Copper Basin is the name appearing on U.S. Geological Survey maps and in the names of the local elementary and high schools, the Copper Basin Medical Center, and the Polk County/Copper Basin Chamber of Commerce. The older name is used by the Ducktown Basin Museum and some local businesses. Barclay marked the transition in the name of his three books: Barclay, *Ducktown*; Barclay, *Railroad Comes to Ducktown*; Barclay, *Copper Basin, 1890 to 1963*.

4 The description is derived from Barclay, *Ducktown*, 1–4; Emmons and Laney, *Geology and Ore Deposits of the Ducktown Mining District*, 1, 13; Safford and Tennessee Geological Survey, *Geology of Tennessee*, 469–71; LaForge, *Physical Geography of Georgia*, 114–16.

5 The stated elevations of local mountain peaks vary from source to source. The figures employed here are from topographical maps by the U.S. Geological Survey. Ducktown is at an awkward point in the USGS grid so several maps are required to encompass all of it. The following topographical maps were used: Defense Mapping Agency, "Eastern United States 1:250,000 Chattanooga, Tenn. Series V501p, Ed. 4, Sheet Ni 16-3"; Defense Mapping Agency, "Eastern United States 1:250,000 Rome, Ga. Series V501p, Ed. 4, Sheet Ni 16-6"; U.S. Department of the Interior, U.S. Geological Survey, "Epworth Quadrangle"; U.S. Department of the Interior, U.S. Geological Survey, "Mineral Bluff Quadrangle"; U.S. Department of the Interior, U.S. Geological Survey, and U.S. Department of Agriculture, "Ducktown Quadrangle"; U.S. Department of the Interior, U.S. Geological Survey, and Tennessee Valley Authority Mapping Services Branch, "Isabella Quadrangle."

6 The importance of the Ellijay Valley is described in LaForge, *Physical Geography of Georgia*, 100–101. For the wilderness designations, see *Wilderness Act*, Public Law 88-577, 78 *Stat.* 890 (1964). For more on the wilderness areas, see "National Wilderness Preservation System—Tennessee" (map and data). The foundational roles played by Aldo Leopold, Robert Sterling Yard, Benton MacKaye, and Bob Marshall in the wilderness movement and the passage of the Wilderness Act of 1964 are explored in Sutter, *Driven Wild*.

7 LaForge, *Physical Geography of Georgia*, 114–16.

8 The proper location of the Georgia-Tennessee border is a point of frequent controversy; see Coulter, "The Georgia-Tennessee Boundary Line"; Cadle, *Georgia Land Surveying History and Law*, 114–20. Georgia contends that in 1817, the surveyors laid the line a half mile south of the intended 35th parallel. In 1963 Georgia's Governor Carl Sanders contemplated a suit before the U.S. Supreme Court to return sixty square miles of land from Tennessee by restoring the border to its proper location. If done, this would have put the Tennessee Copper Company's valuable Copperhill complex on the Georgia side of the line, along with the town of Copperhill; see Max York, "A Quiet Little Border War," *Nashville Tennessean*, December 8, 1963. Nothing came of it then, but the issue revived in 2008. At stake was Georgia's desire to move the border north to tap the waters of the Tennessee River at Nickajack Lake near Chattanooga as a remedy to the Peach State's historic drought conditions. At this writing, the outcome of the initiative is in doubt; see Jeffrey Scott, "Georgia Town at the Heart of Border Dispute," *Atlanta Journal-Constitution*, February 8, 2008; Greg Bluestein, "Georgia Lawmakers Push for Border Change," *Atlanta Journal-Constitution*, February 21, 2008; Eric Schelzig, "Lawmaker: Tenn. Will Stand Up to Ga. on Border Dispute," *Nashville Tennessean*, February 22, 2008; Schaila Dewan, "Georgia Claims a Sliver of the Tennessee River," *New York Times*, February 22, 2008.

9 The constitutional issues created by transborder pollution at Ducktown foreshadowed the analogous problem of international law between the United States and Canada during the period 1927–41, when smoke from the Trail Smelter in British Columbia drifted over the border to cause harm in Washington State; see Wirth, *Smelter Smoke in North America*.
10 The Federal Road is described in Lillard, *History of Polk County*, 7–9; McLoughlin, *Cherokee Renascence*, 77–91. McLoughlin describes the maneuvering by the federal government to secure the right of way for the road from the Cherokees. The impact of the mountains on transportation and settlement in the region is discussed in LaForge, "The Geographic Control of Human Affairs," in *Physical Geography of Georgia*, 157–65. For the influence of Appalachian geography upon the course of settlement, see Salstrom, *Appalachia's Path to Dependency*, xiii–xix. Salstrom distinguishes three phases of settlement in Southern Appalachia: Older Appalachia (Virginia's Blue Ridge, the Shenandoah Valley, and the Tennessee Valley), Intermediate Appalachia (the Appalachian mountain systems of North Carolina, Tennessee, and North Georgia); and the New Appalachia (Tennessee's Cumberland Plateau and the highlands of Kentucky and West Virginia).
11 Perdue and Green, *Columbia Guide to American Indians of the Southeast*, 82–86; Persico, "Early Nineteenth-Century Cherokee Political Organization," 99–108. For the development of plantation slavery among the Cherokee, see Perdue, *Slavery and the Evolution of Cherokee Society*, 50–56.
12 For the motives and means of the civilization program, see Finger, *Eastern Band of the Cherokees*, 6–19; McLoughlin, *Cherokee Renascence*, 33–57; Perdue and Green, *Columbia Guide to American Indians of the Southeast*, 75–79.
13 Barclay, *Ducktown*, 4–6; Duggan, "Being Cherokee," 90, 105–13.
14 Tyner, *Those Who Cried*, 193–94; Duggan, "Being Cherokee," 108–19.
15 Cherokee presence in the Ducktown Basin before the Removal is discussed in Duggan, "Being Cherokee," 90–128. For discussion of Cherokees in Polk County, Tennessee, see Lillard, *History of Polk County*, 1–38. Cherokee land use patterns are discussed in Wilms, "Cherokee Land Use in Georgia before Removal."
16 Other works on the Removal, in addition to those already cited, include Anderson, *Cherokee Removal*; Ehle, *Trail of Tears*; Hoig, *The Cherokees and Their Chiefs*; King, *The Cherokee Nation*.
17 The ratified treaties are collected in United States and Kappler, *Indian Affairs: Laws and Treaties*, 8–11 (Treaty of 1785), 29–33 (Treaty of 1791), 140–44 (Treaty of 1817), and 177–81 (Treaty of 1819).
18 For Dahlonega gold and its impact on Georgia and the Cherokees, see David Williams, *Georgia Gold Rush*, 7–46.
19 Bartram, *Travels*, 350–51; Waselkov and Braud, *William Bartram and the Southeastern Indians*, 77, 80–81; Gilmer, *Sketches*, 247–50.
20 Phillips, *Georgia and State Rights*. For the 1802 cession, see United States, *American State Papers: Public Lands*, 1:113–14 (No. 68, Instructions to Land Officers, April 6, 1802, and No. 69, Georgia Cession, April 26, 1802). It was published separately as United States, *Articles and Agreement of Cession*. Georgia ratified it in Georgia, *Laws*, 1802. The agreement was part of an attempt to resolve the confusion arising from Georgia's Yazoo Land Fraud; see Coleman, *A History of Georgia*, 94–102; Lamplugh, "Yazoo Land Fraud."
21 Resolution, approved December 22, 1826, no. 249, 1826 *Ga. Laws* 227; Resolution, approved December 27, 1827, no. 236, 1827 *Ga. Laws* 236, 240.

22 Act, approved December 19, 1829, no. 93, 1829 *Ga. Laws* 98 (extending Georgia jurisdiction and nullifying Cherokee law); Act, approved December 2, 1830, no. 108, 1830 *Ga. Laws* 108, 154 (gold mines); Act, approved December 21, 1830, no. 98, 1830 *Ga. Laws* 127 (survey and lottery); Act, approved December 22, 1830, no. 88, 1830 *Ga. Laws* 114 (Cherokee assemblies, license and oath).

23 *Worcester v. Georgia*, 6 Peter (31 U.S.) 515 (1832); Norgren, *The Cherokee Cases*. For Worcester's translation work, see Worcester and Boudinot, *Cherokee Hymns*; Worcester and Boudinot, *Acts of the Apostles*.

24 *Worcester*, 6 Peter (31 U.S.) at 560–62.

25 Andrew Jackson, "State of the Union Address," *House Journal*, 21st Cong., 1st sess., December 8, 1829, 23–25.

26 United States and Kappler, *Indian Affairs: Laws and Treaties*, 448–51 (Treaty of New Echota). The motives of the Treaty Party are assessed in McLoughlin, *Cherokee Renascence*, 448–51; Perdue and Green, *Columbia Guide to American Indians of the Southeast*, 86–97. The story of Major Ridge's prediction and death appears in Wilkins, *Cherokee Tragedy*, 278.

27 Evans, "Fort Marr Blockhouse"; Thornton, "Demography of the Trail of Tears." For the late nineteenth-century presence of Cherokees in Ducktown, see Duggan, "Being Cherokee," 254–76; Lillard, *History of Polk County*, 20–21. Most of the Cherokees who evaded removal eventually coalesced to form the Eastern Band of the Cherokees that exists to this day in North Carolina.

28 Barclay, *Ducktown*, 12–13, 25; Foehner, "Historical Geography," 11.

29 Bird, "Exchange Networks," 51–53, 213; Goodman and Cantwell, *Copper Artifacts*, 7–9, 35–37, 70–74; Goad, "Chemical Analysis of Native Copper Artifacts from the Southeastern United States"; Hurst and Larson, "On the Source of Copper at the Etowah Site." For the Ontonagon Boulder, see Lankton, *Beyond the Boundaries*, 6–7.

30 Currey, *A Sketch of the Geology of Tennessee*, 72–74, quoted in Barclay, *Ducktown*, 44–45.

31 For the British copper and brass industry, see Day, *Bristol Brass*; Hamilton, *The English Brass and Copper Industries*. For the American copper and brass industry, see Brecher et al., *Brass Valley*, 1–6 (buttons, hardware, and watches); Hyde, *Copper for America*, 1–14 (sheathing, steam engines, early U.S. mines); Lathrop, *The Brass Industry in the United States*, 21–67 (clocks, lamps, engine parts, and photographic plates).

32 The account is in a letter from John Caldwell to Dr. R. O. Currey and C. A. Proctor, 1855, appearing in Safford, *A Geological Reconnoissance [sic]*, 61–62.

33 Barclay, *Ducktown*, 56–57.

34 Safford, *A Geological Reconnoissance*, 61–62.

35 Currey, *A Sketch of the Geology of Tennessee*, 76–77; Parmentier, Journal of Rosine Parmentier, 8, MSS 59, Cleveland (Tennessee) Public Library.

36 "The Mining Interest at the South." Contemporary corporate documents reveal the sources of investment: Charter and By-Laws of the Burra-Burra Copper Company (1861) (headquarters in New Orleans), Charter and By-Laws of the Polk County Copper Company (1860) (New Orleans), and the report of the Union Consolidated Mining Company (1857) (New York, with directors in Charleston and Savannah), all from Barclay Papers, box 1, folder 8, TSLA. Barclay asserts that the large majority of antebellum investors in Ducktown were American, not English. The extent of southern investment in the copper industry of antebellum Ducktown is in contrast to the usual New South themes of postwar northern capital for southern resource development;

see Eller, *Miners, Millhands, and Mountaineers*; Ronald L. Lewis, *Transforming the Appalachian Countryside*; Wright, *Old South, New South*. For pre–Civil War industrialization in the mountains, see Davis, *Where There Are Mountains*; Dunaway, *The First American Frontier*. Financing for Lake Superior mining came from Boston, New York, and Detroit; see Gates, *Michigan Copper*, 31–38; Hyde, *Copper for America*, 32–43. The comparative advantages of the Lake Superior and Ducktown districts are discussed in Gates, *Michigan Copper*, 3–5; Maury and Currey, *Polk County Copper Company*, 7, 14. For the early impact of Ducktown mining, see Barclay, *Ducktown*, 40–41 (the mines as a local market), 263–65 (copper haulers as aristocrats); J. D. Clemmer, "John S. Hutchins Tells of His Copper Hauling Days," *Polk County News*, February 2, 1938.

37 Currey, *A Sketch of the Geology of Tennessee*, 81–84. There are many published analyses of the Ducktown ores, the most thorough being Emmons and Laney, *Geology and Ore Deposits of the Ducktown Mining District*.

38 Maury and Curry, *Polk County Copper Company*, 4–7 (quotation). For the precipitation process, see American Bureau of Mines and Union Consolidated Mining Company, *Report*, 7; Barclay, *Ducktown*, 83, 256–57; Shepard, *Report*, 6.

39 The list of mines appears in American Bureau of Mines and Union Consolidated Mining Company, *Report*, 7. Olmsted, *A Journey in the Back Country*, 242–46.

40 American Bureau of Mines and Union Consolidated Mining Company, *Report*, 7; Barclay, *Ducktown*, 53; Clemmer, "John S. Hutchins Tells of His Copper Hauling Days" (freight rates circa 1870); "Copper Ore and Cotton: Dangerous Freight."

41 Shepard, *Report*, 4; Barclay, *Ducktown*, 150–51; Emmons and Laney, *Geology and Ore Deposits of the Ducktown Mining District*, 30–34.

42 Baker, "Charcoal." For discussion relating charcoal usage at Ducktown to the regional problem of deforestation arising from other metal industries, see Davis, *Where There Are Mountains*, 147–53, 157–59.

43 Currey's observation appears in Barclay, *Ducktown*, 53. The fuel statistics are from Maury and Currey, *Polk County Copper Company*, 11; Barclay, *Ducktown*, 262–63.

44 "The Mining Interest at the South," 446.

45 Gaussoin, *The Ducktown Copper*, 18; Taliaferro, "Ducktown, by 'Skitt,' Who Has Been 'Thar.'"

46 Taliaferro's sketches are collected in Taliaferro and Craig, *The Humor of H. E. Taliaferro*. Regarding the southwestern frame story, see Powers, *Mark Twain: A Life*, 19–21; Blair, *Native American Humor*, 90–101.

47 Barclay, *Ducktown*, 40, 77–83; Emmons and Laney, *Geology and Ore Deposits of the Ducktown Mining District*, 30–31.

48 Barclay, *Ducktown*, 186–205, 246–49.

49 For the importance of Ducktown to the Confederate war effort and the course of events during the war, see Barclay, *Ducktown*, 87–101; Donnelly, "Confederate Copper." For the mixed allegiances in the Southern Appalachians, see Noe and Wilson, *The Civil War in Appalachia*. For a local pro-southern view, see A. J. Williams, *A Confederate History of Polk County*.

50 The January 24, 1862, decree of sequestration was issued in *Confederate States v. Union Consolidated Mining Company* (District Court of the Confederate States for the Eastern Division of the District of Tennessee, 1862); a copy of the decree appears in the Barclay Papers, box 1, folder 9, TSLA. For an overview of similar actions, see Daniel Hamilton, "The Confederate Sequestration Act."

51 American Bureau of Mines and Union Consolidated Mining Company, *Report*, 8; Barclay, *Ducktown*, 130–33.
52 Burra-Burra Copper Company, "Shareholder Report, 1871"; Union Consolidated Mining Company, "Shareholder Report, 1866," 5, both from Barclay Papers, box 1, folder 8, TSLA. For the statistics, see Barclay, *Ducktown*, 262–63.
53 Act, approved February 26, 1876, no. 499, 1876 *Ga. Laws* 388; Act, approved October 13, 1879, no. 172, 1878–79 *Ga. Laws* 187. For representative works regarding the intrusion of class and economic status on common law rights or traditions of hunting, fishing, timber harvesting, and the use of waterpower for milling, see Hahn, *The Roots of Southern Populism* (stock laws as an element of Redeemer control over yeomen); Jacoby, *Crimes against Nature* (game regulations); Judd, *Common Lands, Common People* (game laws as a local initiative); Steinberg, *Nature Incorporated* (corporate control of New England rivers); Watson, "The Common Rights of Mankind" (impairment by slave owners of yeomen subsistence shad fishing).
54 Polk County Copper Company, "Report to the Board of Directors of the Polk County Copper Company at a Special Meeting Held October 20th, 1869," 4; Union Consolidated Mining Company of Tennessee, Report to the Stockholders, June 21, 1871, both from Barclay Papers, box 1, folder 8, TSLA; Barclay, *Ducktown*, 217 (Raht quotation).
55 Barclay, *Railroad*, 1–91; Polk County Copper Company, "Report to the Board of Directors, October 20, 1869," Barclay Papers, box 1, folder 8, TSLA.
56 For discussion of the industry's decline, see Barclay, *Ducktown*, 133–58.
57 Ibid., 159–84. He adds that Raht compelled miners to rely on commissary credit by paying wages quarterly instead of monthly or weekly (122–23).

Chapter 2

1 B. F. Perry, "The First Trip by the Hiwassee—the New Line to Knoxville," *Atlanta Constitution*, August 23, 1890, 3.
2 Barclay, *Copper Basin*, 1–48.
3 Barclay, *Ducktown*, 126–28. For illustrative works on Appalachian industrialization, see Eller, *Miners, Millhands, and Mountaineers*; Ronald L. Lewis, *Transforming the Appalachian Countryside*; Salstrom, *Appalachia's Path to Dependency*. For antebellum Appalachian industry, see Davis, *Where There Are Mountains*; Dunaway, *The First American Frontier*.
4 Henrich, "The Ducktown Ore-Deposits" quoted in Brief and Argument of the Ducktown Sulphur, Copper & Iron Company, Ltd. on Final Hearing, 19–20, *Georgia v. Tennessee Copper Company*, U.S. Supreme Court, No. 5 Original, October Term, 1906.
5 LaForge, *Physical Geography of Georgia*, 130–31. For conditions in Ducktown during the dormant years, see James W. Taylor, "Ducktown Desert," 45–46. For Appalachian droving practices, see Davis, "Living on the Land"; Dunaway, *The First American Frontier*, 218–21; Salstrom, *Appalachia's Path to Dependency*, 8–9, 37, 123.
6 Foehner, "Historical Geography," 11–24; James W. Taylor, "Ducktown Desert," 45–46.
7 For Ducktown grazing and fire practices, see Richard A. Wood, "Erosion Control and Reforestation," 7–8; Barclay, "Introduction to the Copper Basin."
8 Barclay, *Railroad*, 61–90.
9 The use of convict labor in Georgia is discussed in Elizabeth Taylor, "The Origin and

Development of the Convict Leasing System." For convict labor on southern railroads, see Scott Reynolds Nelson, *Iron Confederacies*; Lichtenstein, *Twice the Work of Free Labor*.

10 "The Biped Zebra," *Atlanta Constitution*, April 8, 1877, 4; "A Convict's Death," *Atlanta Constitution*, September 14, 1881, 8.

11 Barclay, *Railroad*, 104-7.

12 Brief and Final Argument of Ducktown Sulphur, Copper & Iron Co., *Ducktown Sulphur, Copper & Iron Co., Ltd. v. Barnes*, 60 S.W. 593 (Tenn. 1900), 20; Barclay, *Ducktown*, 146-47; *Georgia v. Tennessee Copper*, U.S. Supreme Court, No. 5 Original, October Term, 1906.

13 For heap roasting, see Barclay, *Copper Basin*, 4-5, 33-46; Barclay, *Railroad*, 179-80; Quinn, "The Appalachian Mountains' Copper Basin"; Smallshaw, "Denudation and Erosion in the Copper Basin," 4. The beginnings of the Tennessee Copper Company are described in American Institute of Mining Engineers, "A Brief Description," 1-15; Barclay, *Copper Basin*, 23-48.

14 For the advantages of pyritic smelting, see Quinn, "Industry and Environment in the Appalachian Copper Basin"; "The Ducktown Basin, Its Ore Deposits and Their Development." For the use of sulfur in phosphate fertilizer, see Lewis B. Nelson, *History of the U.S. Fertilizer Industry*. Regarding the DSC&I name, see Freeland and Renwick, "Smeltery Smoke as a Source of Sulphuric Acid."

15 The composition of Ducktown ores varied from mine to mine and from year to year. The numbers provided are from Emmons and Laney, *Geology and Ore Deposits of the Ducktown Mining District*, 41-54. For other experiments in pyritic smelting, see Barclay, *Railroad*, 180-81.

16 LaForge, *Physical Geography of Georgia*, 126-27; Smallshaw, "Denudation and Erosion in the Copper Basin," 5.

17 Transcript at 354 (deposition of G. W. Prince) and 387 (deposition of A. J. Bell), *Ducktown Sulphur, Copper & Iron Co. v. Barnes*, 60 S.W. 593 (Tenn. 1900). All of the transcripts to Tennessee appellate cases mentioned herein are archived in Supreme Court Trial Cases, 1796-1955, RG 170, TSLA. The scientific testimony is discussed in later chapters but includes the following: Testimony and Exhibits of the Complainant, the State of Georgia, upon Final Hearing, at 5 (first affidavit of J. K. Haywood), at 9 (second affidavit of J. K. Haywood), at 18 (affidavit of John M. McCandless, State Chemist of Georgia), *Georgia v. Tennessee Copper Co.*, U.S., No. 5 Original, October Term, 1906; Report of John T. McGill, *Georgia v. Tennessee Copper Co.*, U.S., No. 1 Original, October Term, 1914; Haywood, *Injury to Vegetation by Smelter Fumes*. See also Quinn, "Early Smelter Sites." Quinn situated the Ducktown experience within the historiography of acid precipitation.

18 Andrews, *Managing the Environment, Managing Ourselves*, 207-10. His book includes an extensive chronology of environmental law at 385-86.

19 Stradling, *Smokestacks and Progressives*, 2, 61-84. For a contemporary federal report on urban smoke abatement, see Flagg, *City Smoke Ordinances and Smoke Abatement*. For representative works on urban smoke pollution in the Progressive Era, see Grinder, "The Battle for Clean Air"; Gugliotta, "Class, Gender, and Coal Smoke"; Luconi, "The Enforcement of the 1941 Smoke-Control Ordinance"; Platt, "Invisible Gases"; Tarr, *The Search for the Ultimate Sink*; Tarr and Zimring, "The Struggle for Smoke Control in St. Louis." A contemporary view of urban smoke in Tennessee appears in Switzer, "Economic Aspects of the Smoke Nuisance." Switzer advocated new technology to increase

the efficiency of coal-fired industrial boilers, using a large textile mill in Knoxville as his example.

20 For the Butte ordinance and Deer Lodge Valley problems, see MacMillan, *Smoke Wars*, 25–46, 83–99. Other works on rural smoke pollution include Charry, "Defending the 'Great Barbecue'" (lead smelting in northeast Washington State); Church, "Smoke Farming" (Salt Lake area); Lamborn and Peterson, "The Substance of the Land" (also in the Salt Lake area).

21 *Ducktown Sulphur, Copper & Iron Co., Ltd. v. Barnes*, 60 S.W. 593 (Tenn. 1900).

22 The ethnic background of Ducktown miners appears in Heffington, "Tennessee's Copper Basin: An Ethnic Overview"; Barclay, "The Great Copper Basin." The Scots-Irish heritage shared by most of the smoke suitors and by the white Polk County citizens from which local juries were then drawn precluded the copper companies from exploiting nativist and ethnic tensions as was done against ethnic immigrant smoke suitors in the 1921 Northport smelter litigation in Washington State; see Charry, "Defending the 'Great Barbecue.'"

23 "Ordered to Leave at Once: Tennessee White Miners Threaten to Kill the Blacks if They Don't Leave," *Atlanta Constitution*, April 30, 1894, 1; Barclay, *Copper Basin*, 19–22. For Polk County slavery, see Historical Census Browser, "Tennessee: Total Slaves," Geospatial and Statistical Data Center, University of Virginia Library, ⟨http://fisher.lib.virginia.edu/collections/stats/histcensus/php/newlong3.php⟩ (accessed November 4, 2006). For Appalachian slavery, see Inscoe, *Mountain Masters*. Racial incidents occurred throughout the Southern Appalachians during the period; see Brundage, "Racial Violence, Lynchings, and Modernization in the Mountain South." Barclay, "The Great Copper Basin."

24 For local squatters, see Parish, *The Old Home Place*, 71–72; *Wetmore v. Rymer*, 169 U.S. 115 (1898) (lawsuit to remove squatters from privately owned forest tracts in the basin).

25 Residence data come from the pleadings in *Ducktown v. Barnes*.

26 Haywood, *Injury to Vegetation by Smelter Fumes*, 19.

27 Farm data, by county, are found in U.S. Bureau of the Census, *Report on the Statistics of Agriculture in the United States at the Eleventh Census: 1890*, tables 5, 8, 10, 12, 14, 16, 20, 22, and 24.

28 For historiography on the hillbilly construct, see Inscoe, "The Discovery of Appalachia."

29 "Ducktown Copper," *Atlanta Constitution*, March 29, 1876, 1; "Our Ducktown Letter," *Knoxville Journal*, April 27, 1874, 4.

30 Transcript at 153 (deposition of A. J. Bell), *Ducktown v. Barnes*.

31 Answers to Errors Assigned and Brief of Respondents, *Ducktown v. Barnes*.

32 The rise of pollution cases from large-scale industries led gradually to the relaxation of the standard of strict liability expressed by the maxim, as discussed in chapter 3.

33 *Ducktown v. Barnes*, 597–98.

34 James G. Parks to W. H. Freeland, August 12, 1905, DBM. Attorneys and clients of the period made frequent use of "law" as a verb, meaning to litigate, an apt usage this author regrets is no longer prevalent in the courts. Compare the litigation strategy of DSC&I and TCC to the extensive use of scientific evidence in the Northport smelter suit; see Charry, "Defending the 'Great Barbecue.'"

35 W. H. Freeland to P. B. Mayfield, June 1, 1899; P. B. Mayfield to W. H. Freeland, July 10, 1899, both DBM.

36 The labor troubles of 1899 received extensive coverage in the *Atlanta Constitution*: "Reese Shot from Ambush," June 6, 1899, 2; "Ducktown May Have Strike—Tennessee Companies Refuse to Employ Union Men," September 9, 1899, 1 (doctor issue mentioned); "Ducktown Strike Settled," September 16, 1899, 1 (doctor issue resolved); "Tennessee Miners on Strike—Asked an Eight Hour Shift with Nine and a Half Hours' Rate," October 29, 1899, 5; "Serious Trouble Is Brewing—Six Hundred Miners at Ducktown, Tenn. Are Armed," November 9, 1899, 1; "Trouble Now at Ducktown—Non-Union Men Will Be Put in the Mines and Mills Today," November 13, 1899, 2; "Deputies Going to Ducktown—Strikers Expected to Cause Trouble When Non-Union Men Work," December 2, 1899, 7; "Increased Miners' Wages," December 21, 1899, 8. The American Federation of Labor threat is mentioned in G. G. Hyatt to Mayfield, Son & Aiken, September 8, 1899, DBM. The Tennessee Copper Company experienced another major strike in 1939; see Simson, "Parades amid the Standoff in the Old Red Scar."

37 Clay, "Copper-Basin Cover Up" (cow quip); "Copper Companies Not at All Pleased with Their Tax Assessments," *Chattanooga News*, September 20, 1900, also found in James Donaldson Clemmer Scrapbooks, vol. 2, p. 252 (TSLA).

38 Navin, *Copper Mining and Management*, 304–6; Transcript at 52 (Margaret Madison demurrer), *Ducktown v. Barnes*.

39 This is the argument in Hahn, *The Roots of Southern Populism*, 1–10. Hahn asserts that eighteenth-century "republicanism gave populism its ideological force and political vitality." Yeoman republican values certainly surfaced among the largely Scots-Irish farmers of Ducktown, but the degree to which this was a conscious embrace of Jeffersonian thought as opposed to the common human desire to protect home and livelihood was uncertain, especially because many farmers readily embraced seasonal employment at the mines. For leading overviews of populism, see Goodwyn, *Democratic Promise*; Hicks, *The Populist Revolt*; Hofstadter, *The Age of Reform*; McMath, *American Populism*; Sanders, *Roots of Reform*. For works with a focus on Georgia and the South, see Arnett, *The Populist Movement in Georgia*; Ayers, *The Promise of the New South*; Bartley, *The Creation of Modern Georgia*, 75–103; Shaw, *The Wool-Hat Boys*; Woodward, *Tom Watson, Agrarian Rebel*. Cotton was a significant crop in the Tennessee Valley lowlands of Polk County, Tennessee, with 1,045 bales produced in 1899; see U.S. Census Office, *Twelfth Census of the United States, Taken in the Year 1900*, vol. 5: *Agriculture*, part 1 *Farms, Livestock and Animal Products*, table 10, "Acreage and Production of Cotton Fiber in 1899."

40 For equity jurisdiction, see 1877 *Tenn. Pub. Acts*, ch. 97, §§1–2, codified as *Tenn. Code Ann.* §16-11-102 (2005). Despite the modern trend toward the unification of law and equity, the distinction persists in Tennessee. Law courts and chancery courts remain separate institutions, and chancery is still generally barred from awarding unliquidated damages. *Ducktown Sulphur, Copper & Iron Co. v. Fain*, 109 Tenn. (1 Cates) 56, 64, 70 S.W. 813, 815 (1902). Unliquidated damages are defined in *Swift & Co. v. Memphis Cold Storage Whse. Co.*, 128 Tenn. 82, 158 S.W. 480 (1913).

41 Transcript at 10–28 (DSC&I bill of injunction) and 127–28 (Fortner complaint), *Ducktown v. Barnes*; see also opinion, *Ducktown v. Barnes*. The abolition of Tennessee's champerty law became a factor in Ducktown smoke litigation, as discussed in chapter 3.

42 The distinction between jurisdiction *in personam* and jurisdiction *in rem* is drilled into law students during their discussion of *Pennoyer v. Neff*, 95 U.S. 714 (1877), in the opening sessions of the typical class on civil procedure. The case has been superseded by later

expansion of jurisdiction principles but is still often assigned by professors because of its utility as a tool for Socratic hazing of new students. I still remember it almost thirty years after my law school days. The possible resort of Georgia suitors to the federal courts is discussed in chapter 3.

43 *Ducktown v. Barnes*, 595–96 (DSC&I Bill of Injunction); *International Shoe Co. v. Washington*, 326 U.S. 310 (1945).
44 Record on Appeal, 16, *Ducktown v. Barnes*.
45 Ibid., 86.
46 Ibid., 54–55 (Margaret Madison), 85–86 (William Madison).
47 Ibid., 555–63 (John Quintell).
48 Robert A. Barclay, *Ducktown*, 155 (early production figures). For later production figures, see Emmons and Laney, *Geology and Ore Deposits of the Ducktown Mining District*, 32; American Institute of Mining Engineers, "A Brief Description of the Operations of the Tennessee Copper Company."
49 Record on Appeal, 10–28 (DSC&I Bill of Injunction), *Ducktown v. Barnes*.
50 *Ducktown v. Barnes*, 598, 606, 607. As discussed in chapter 3, the court addressed multiplicity of actions more thoroughly in another smoke case, *Ducktown Sulphur, Copper & Iron Co. v. Fain*, 109 Tenn. (1 Cates) 56, 70 S.W. 813 (1902).
51 *Ducktown v. Barnes*, 599–607.
52 Ibid., 607.
53 P. B. Mayfield to William H. Freeland, Aug. 12, 1901, Mayfield Papers, TSLA; William H. Freeland to James G. Parks, Jan. 13, 1903, DBM.

Chapter 3

1 P. B. Mayfield to William H. Freeland, August 12, 1901, Mayfield Papers, TSLA; *Ducktown, Sulphur, Copper & Iron Co. v. Barnes*, 60 S.W. 593 (Tenn. 1900).
2 At the time of the Barnes litigation, the leading American treatise on the law of nuisance was H. G. Wood, *A Practical Treatise on the Law of Nuisances*. A new treatise was published in 1906, after the Tennessee Supreme Court had issued its key decisions on the Ducktown litigation; Joyce, *Treatise on the Law Governing Nuisances*.
3 For a definition of "laurel hell," see Montgomery and Hall, *Dictionary of Smoky Mountain English*, s.v. "laurel bed." The anecdotes appear in Kephart, *Camping and Woodcraft*, 24–25; Ellison, "Talking About Mountain Balds and 'Laurel Hells.'"
4 Drawing from all of these sources, I collated the cases into a searchable chart by names, dates of filing, and docket numbers to arrive at the figures herein.
5 *Ducktown v. Barnes*, 60 S.W. at 603 (maxim), 606 (damages).
6 For representative works on the development of nuisance law relating to industrial pollution, see Bone, "Normative Theory and Legal Doctrine in American Nuisance Law"; Brenner, "Nuisance Law and the Industrial Revolution"; Hoffer, *The Law's Conscience*; Kurtz, "Nineteenth-Century Anti-entrepreneurial Nuisance Injunctions"; McLaren, "Nuisance Law and the Industrial Revolution"; Provine, "Balancing Pollution and Property Rights"; Rosen, "Differing Perceptions of the Value of Pollution Abatement"; Rosen, "Knowing Industrial Pollution."
7 For the Chicago stockyards, see Cronon, *Nature's Metropolis*, 207–62.
8 For an example of judicial changes to nuisance law, see Kurtz, "Nineteenth-Century

Anti-entrepreneurial Nuisance Injunctions," 642–49, 663–69; Rosen, "Knowing Industrial Pollution," 573–86.
9 Steinberg, *Nature Incorporated*, 166–86, quotation at 186.
10 John Allen to P. B. Mayfield, January 25, 1899; P. B. Mayfield to G. G. Hyatt, February 4, 1899, Mayfield Papers, TSLA.
11 Act of April 17, 1901, ch. 139 (emphasis added), 1901 *Tenn. Pub. Acts* 246.
12 Transcript, 57 (demurrer and answer of Margaret Madison), *Ducktown v. Barnes*.
13 P. B. Mayfield to Tully R. Cornick, December 5, 1900, Mayfield Papers, TSLA.
14 P. B. Mayfield to W. H. Freeland, July 17, 1899 (demagoguery); P. B. Mayfield to W. H. Freeland, March 13, 1901 (open advocacy), Mayfield Papers, TSLA.
15 Tully Cornick to P. B. Mayfield, January 24, 1901; P. B. Mayfield to Tully Cornick, January 26, 1901 (Memphis strategy); P. B. Mayfield to Howard Cornick, March 13, 1901; P. B. Mayfield to W. H. Freeland, April 20, 1901, all from the Mayfield Papers, TSLA. For the enacted legislation, see Act of April 4, 1901, ch. 126, 1901 *Tenn. Pub. Acts* 197 (pauper's oath); Act of April 17, 1901, ch. 139, 1901 *Tenn. Pub. Acts* 246 (incidental benefits). State senator C. D. M. Greer of Memphis introduced both bills as shown in the index of bills in 1901 *Tenn. S. Journal*, 883.
16 *Ducktown Sulphur, Copper & Iron Co. v. Fain*, 109 Tenn. (1 Cates) 56, 59, 70 S.W. 813 (1902); Tenn. R. Civ. Pro. 23 (1971), Advisory Commission Comments to §23.07.
17 Transcript at 6 (bill for injunction), 57–70 (affidavits of nonresidents), *Madison v. Ducktown Sulphur, Copper & Iron Co.*, 113 Tenn. (5 Cates) 331, 83 S.W. 658 (1904).
18 Transcript at 117–28 (DSC&I demurrer), *Madison v. Ducktown*; "The Riches of North Georgia," *Atlanta Constitution*, May 12, 1902, 6.
19 Transcript, 123, 127 (demurrer and answer of DSC&I), *Madison v. Ducktown*. "Pabulum" is the Latin spelling that in the shortened form "Pablum" became a trademarked name for infant cereal in the 1930s.
20 "Smoke from Sulphur Works," *Atlanta Constitution*, August 15, 1901, 3; "Injunctions Dissolved by Judge M'Connell," *Knoxville Journal and Tribune*, August 21, 1901, 5.
21 W. H. Freeland to P. B. Mayfield, November 13, 1901; P. B. Mayfield to W. H. Freeland, November 15, 1901, both in Mayfield Papers, TSLA; James G. Parks to W. H. Freeland, February 18, 1903, DBM; Parks, *A Manual of the Law of Pleading*.
22 James G. Parks to Howard Cornick, February 13, 1903, DBM.
23 *Ducktown v. Fain* at 109 Tenn. (1 Cates) at 64, 70 S.W. at 815.
24 James G. Parks to W. H. Freeland, February 17, 1903, DBM.
25 Ibid.; *Tennessee Copper Co. v. Madison*, No. 287 (Chancery, Polk Co., Tenn., filed October 1902); *Ducktown Sulphur, Copper & Iron Co. v. Crofts*, No. 292 (Chancery, Polk Co., Tenn., filed December 21, 1902); James G. Parks to W. H. Freeland, February 17, 1903, DBM.
26 Howard Cornick to James G. Parks, September 8, 1902 (exchange of authorities), DBM; *Swain v. Tennessee Copper Co.*, 111 Tenn. (3 Cates) 430, 444, 78 S.W. 93, 96 (1903).
27 *Tenn. Code Ann.* §28-3-105(1) (statute of limitations).
28 Act of April 7, 1899, ch. 178, 1899 *Tenn. Pub. Acts* 321; Code of Tenn. §1781 (1858); *Hayney v. Coyne*, 57 Tenn. (10 Heisk.) 339, 342 (1872). Contingent fees are now expressly authorized: Tenn. Sup. Ct. Rule 8, Canon 1.5 (c) (2005). Work injuries remained tort actions in Tennessee until it adopted workers' compensation in 1919.
29 Transcript at 308 (deposition of N. B. Graham), *Ducktown v. Barnes*. Being Clerk of Court, Graham's willingness to testify for a party in a pending case was, at the least, a

dubious ethical choice. James G. Parks to J. K. P. Marshall, January 27, 1909; James G. Parks to W. H. Freeland, January 8, 1906 (shysters); James G. Parks to F. L. Mansfield, February 6, 1909 (blackmailers), all DBM.

30. James G. Parks to W. H. Freeland, November 6, 1903; James G. Parks to W. H. Freeland, January 24, 1908, DBM.

31. James G. Parks to W. B. Miller, February 1908, DBM.

32. Haywood, *Injury to Vegetation by Smelter Fumes*, 7–8; J. V. Kisselburg to Thomas S. Felder, July 5, 1912, RGS 9-1-1, box 4, folder 4, GDAH.

33. Transcript at 16 (dissatisfaction), 524 (Barnes employment), *Ducktown v. Barnes*; J. E. Mayfield to James G. Parks, Oct. 9, 1903, Mayfield Papers, TSLA; James G. Parks to W. H. Freeland, Jan. 8, 1906; W. H. Freeland to James G. Parks, Jan. 11, 1906, both DBM.

34. The debate over the extent of subsistence agriculture and the degree to which individuals engaged in public work has generated a substantial literature. Compare Eller, *Miners, Millhands, and Mountaineers*, xv–xxvi, 86–137, to Dunaway, *The First American Frontier*, 2–21, 123–28; see also Salstrom, *Appalachia's Path to Dependency*, 41–59 (part-time mountain farmers).

35. Bill of injunction, *Avery McGhee v. Tennessee Copper Co.*, No. 314 (Chancery, Polk Co., Tenn., filed July 13, 1903).

36. U.S. Constitution, art. 3; *Ladew v. Tennessee Copper Company*, 218 U.S. 357 (1910); *Wetmore v. Tennessee Copper Company*, 218 U.S. 369 (1910); Surrency, *History of the Federal Courts*, 103–12. Federal circuit courts functioned as one of the two main trial courts in the federal system until they were merged with the federal district courts in 1911. J. B. Mayfield to Edmund Crawford, July 18, 1913, Mayfield Papers, box 9, folder 4, TSLA.

37. Surrency, *History of the Federal Courts*, 35–64, 83–92. For the amount in controversy, see 15 *Moore's Federal Practice* 3d §§102App.04 [3–7] (2007); Act of March 3, 1887, 24 Stat. 552 ($2,000); Act of March 2, 1911, 36 Stat. 1091 ($3,000). The amount in controversy requirement rose to $75,000 in 1996.

38. Barclay, *Copper Basin*, 106, 148 (first car and Abernathy); Lillard, *History of Polk County*, 177 (Kimsey Highway).

39. E. M. Harbison, *Knoxville Journal and Tribune*, July 24, 1911, quoted in Barclay, *Copper Basin*, 107–8.

40. "How Injunction Would Work," *Washington Post*, January 20, 1902, 8; E. M. Harbison, *Knoxville Journal and Tribune*, July 24, 1911; "Disaster Threatens Ducktown," *Knoxville Journal and Tribune*, January 20, 1902, 1; Act of April 17, 1901, ch. 139, 1901 *Tenn. Pub. Acts*. 246.

41. Decree, January 17, 1902, *Farner v. Ducktown Sulphur, Copper & Iron Co. and Tennessee Copper Co.*, No. 261 (Chancery, Polk. Co., Tenn., filed January 10, 1902).

42. Transcript (Court of Chancery Appeals majority opinion), *McGhee v. Tennessee Copper Co.* The case is one of the three resolved in *Madison v. Ducktown Sulphur, Copper & Iron, Co.*, 113 Tenn. (5 Cates) 331, 83 S.W. 658 (1904).

43. Transcript (Court of Chancery Appeals dissenting opinion), *McGhee v. Tennessee Copper Co.*; Act of April 17, 1901, ch. 139, 1901 *Tenn. Pub. Acts* 246 (incidental benefits).

44. Milton H. Smith to J. Park Channing, October 14, 1903; Howard Cornick to James G. Parks, October 16, 1903; James G. Parks to W. H. Freeland, September 6, 1904; James G. Parks to Howard Cornick, September 6, 1904; Howard Cornick to James G. Parks, October 28, 1904; James G. Parks to W. H. Freeland, December 3, 1904, all DBM.

45 *Madison v. Ducktown Sulphur, Copper & Iron Co.*, 113 Tenn. (5 Cates) at 339, 363–66, 83 S.W. at 659, 666–67.
46 W. H. Freeland to DSC&I directors, November 28, 1904; DSC&I directors to W. H. Freeland, November 29, 1904; W. H. Freeland to James G. Parks, December 1, 1904, all DBM. The use of the balancing test in *Madison* is considered in Hoffer, *The Law's Conscience*, 155–67.
47 James G. Parks to Howard Cornick, February 13, 1903, DBM.
48 For the added term of court, see B. B. C. Witt to J. E. Mayfield, February 13, 1903; James G. Parks to Howard Cornick, March 29, 1903, both DBM; Act of April 15, 1903, c. 355, *Tenn. Pub. Acts* 1079; James G. Parks to F. L. Mansfield, February 6, 1909, DBM. Both the Senate and House versions of the special terms bill died in committee. H.B. 112, 56th General Assembly (Tenn., 1909); S.B. 148, 56th General Assembly (Tenn., 1909).
49 James G. Parks to Howard Cornick, February 13, 1903, DBM.
50 James G. Parks to W. H. Freeland, February 14, 1903; James G. Parks to Howard Cornick, February 13, 1903, both DBM.
51 Howard Cornick to James G. Parks, February 9, 1903; James G. Parks to W. H. Freeland, February 18, 1903, both DBM.
52 *Georgia v. Tennessee Copper Co.*, 206 U.S. 230 (1907); *Georgia v. Tennessee Copper Co.*, 240 U.S. 650 (1916).

Chapter 4

1 Bierce, *Devil's Dictionary*, s.v. "litigation"; James G. Parks, "Status of Smoke Litigation of the D. S. C. & I. Co. February 24th, 1903," DBM.
2 Corporate law firms routinely engage local counsel to avoid home cooking, even when it is not otherwise required by applicable licensure rules.
3 These issues regarding the decision rendered in *Georgia v. Tennessee Copper Co.*, 206 U.S. 230 (1907) are addressed in chapter 6. For the land lottery, see Gigantino, "Land Lottery System."
4 Senate Resolution 47, 1903 *Ga. H. Journal*, 436; 1903 *Ga. H. Journal*, 752.
5 Resolution of August 17, 1903, no. 47, 1903 *Ga. Laws* 691.
6 The biographical sketch is drawn from Jones, "The Administration of Governor Joseph M. Terrell," and from Cook, "Terrell, Joseph Meriwether."
7 "The Ducktown Copper Mine," *Atlanta Constitution*, August 20, 1903, 6.
8 Report of the 1903 Ducktown Commission to Governor J. M. Terrell, November 20, 1903, attached as Ex. A to Motion for Leave to File Bill of Complaint, 13–14, *Georgia v. Ducktown Sulphur, Copper & Iron Co.*, U.S. Supreme Court, No. 14 Original, October Term, 1903. Note: the caption of Georgia's initial Supreme Court filing reflected its confusion over the proper defendants, as is addressed more fully in chapter 5, note 2.
9 Newell's party is described in "Copper Fumes Killing Trees," *Atlanta Constitution*, October 29, 1903, A1.
10 Report of Wilmon Newell to Governor J. M. Terrell, October 28, 1903, attached as Ex. B to Motion for Leave to File Bill of Complaint, 15–18, *Georgia v. Ducktown Sulphur, Copper & Iron Co.*, U.S. Supreme Court, No. 14 Original, October Term, 1903 (herein Newell Report). The opening lines of the report reference the governor's role in his investigation: "In accordance with your instructions, I visited . . ."

11 Ibid., 15–16.
12 *Ducktown Sulphur, Copper & Iron Co., Ltd. v. Barnes*, 60 S.W. 593 (Tenn. 1900).
13 Newell Report, 17.
14 Ibid.
15 Ottewell, "There Are No Ducks in Ducktown" (man-made desert); Barnhardt, "The Death of Ducktown" (bona-fide desert); James W. Taylor, "Ducktown Desert."
16 For comparison of damage caused during the first and second eras of Ducktown mining, see Foehner, "Historical Geography," 9–49; James W. Taylor, "Ducktown Desert," 28–60.
17 Newell Report, 17.
18 Marsh, *Man and Nature*. For Marsh's place in the conservation movement, see Pisani, "Forests and Conservation"; Wynn, "'On Heroes, Hero-Worship, and the Heroic'"; Tyrrell, "Acclimatisation and Environmental Renovation"; Judd, "George Perkins Marsh: The Man and Their Man"; Marcus Hall, "Provincial Nature of George Perkins Marsh." For Marsh's life and work, see Lowenthal, *George Perkins Marsh*; Larson, "Marsh, George Perkins."
19 Dodds, "The Stream-Flow Controversy"; Newell Report, 17; "Georgia May Sue a Sister State," *Atlanta Constitution*, November 1, 1903, 9; "Furnaces Denude Georgia," *New York Times*, November 2, 1903, 9.
20 "Report Ready for Governor: Sulfur Fumes Destroy Vegetation in Fannin," *Atlanta Constitution*, October 14, 1903, 7; "Copper Fumes Killing Trees," *Atlanta Constitution*, October 29, 1903, A1; "Georgia Forests Withered by Sulfur Fumes," *Atlanta Constitution*, November 22, 1903, A5.
21 "Georgia May Sue a Sister State," *Atlanta Constitution*, November 1, 1903, 9; "State Gossip Caught in Capitol Corridors," *Atlanta Constitution*, November 2, 1903, 5; "Gossip at the Capitol," *Atlanta Constitution*, November 5, 1903, 6, and November 6, 1903, 6.
22 James G. Parks to W. H. Freeland, October 20, 1903 (hopes for successful appeal to the Tennessee Supreme Court), DBM. For correspondence regarding collaboration between the copper companies and the Louisville & Nashville Railroad, see Howard Cornick to James G. Parks, October 15, 1903; J. Parke Channing to Milton H. Smith, October 14, 1903; Milton H. Smith to J. Parke Channing, October 14, 1903; James G. Parks to W. H. Freeland, November 11, 1903, all DBM.
23 "Tennessee Copper," *Wall Street Journal*, October 29, 1903, 5.
24 Governor J. M. Terrell to Governor J. B. Frazier, November 25, 1903, container 8, file 8, Frazier Papers, 2, TSLA.
25 Georgia codified the maxim in *Ga. Code* 1863 §3017; it is now codified as *Ga. Code Ann.* §23-1-10.
26 Ibid.
27 Ibid.
28 Governor J. B. Frazier to Governor J. M. Terrell, December 14, 1903 (emphasis added), container 2, folder 11, Frazier Papers, TSLA. The dog bite defense is nicely demonstrated by a hotel clerk to Inspector Clousseau in Peter Sellers's comedic film, *The Pink Panther Strikes Again* (1976).
29 Ibid., 2.
30 "Frazier Says He Cannot Act," *Atlanta Constitution*, December 16, 1903, 3; Motion for Leave to File Bill of Complaint, *Georgia v. Ducktown Sulphur, Copper & Iron Co.*, U.S. Supreme Court, No. 14 Original, October Term, 1903.

31 Act of February 9, 1854, no. 363, 1853–54 *Ga. Laws*, vol. 1, 425. Other Ducktown railroad enactments include Act of October 24, 1870, no. 223, 1870 *Ga. Laws*, vol. 1, 340 (creating the Marietta & North Georgia Railroad and lending state funds to it); Act of August 27, 1872, no. 124, 1872 *Ga. Laws*, vol. 1, 179, 181 (authorizing line from Cartersville to Ducktown); Act of August 17, 1872, no. 212, 1872 *Ga. Laws*, vol. 1, 324 (railroad from Atlanta to the Tennessee border "at or near the Ducktown copper mines"); Act of August 27, 1872, no. 238, 1872 *Ga. Laws*, vol. 1, 360 (creating the North Georgia & Ducktown Railroad Co.); Act of February 25, 1876, no. 54, 1876 *Ga. Laws*, vol. 1, 40 (leasing convict labor to the M. & N. G. R. R.); Act of February 28, 1876, no. 271, 1876 *Ga. Laws*, vol. 1, 254 (M. & N. G. R. R. branch line to Ducktown); Act of October 7, 1885, no. 27, 1884–85 *Ga. Laws*, vol. 1, 671 (loan forgiveness). For pro-Ducktown river legislation, see Act of February 26, 1876, no. D, 1876 *Ga. Laws*, vol. 1, 388; Act of October 13, 1879, no. 246, 1878–79 *Ga. Laws*, vol. 1, 187.

32 "The Marietta and Ducktown Railroad," *Atlanta Constitution*, November 6, 1870, 3; "The North Georgia Railroad," *Atlanta Constitution*, July 11, 1874, 2; "The Governor's Trip," *Atlanta Constitution*, July 17, 1874, 2.

33 "The Ducktown Failure," *Atlanta Constitution*, July 21, 1878, 1; *Atlanta Constitution*, February 25, 1880, 2; see also "The Marietta and North Georgia," *Atlanta Constitution*, August 12, 1882; "The First Trip by the Hiwassee," *Atlanta Constitution*, August 23, 1890, 3.

34 For representative works on the New South, see Woodward, *Origins of the New South*, and Ayers, *Promise of the New South*. For intellectual and cultural approaches, see Gaston, *New South Creed*; Silber, *The Romance of Reunion*. For an economic approach, see Wright, *Old South, New South*, 156–97; Cobb, *Industrialization and Southern Society*, 1–26. For the New South in Georgia, see Coleman, *A History of Georgia*, 207–37, 252–54; Bartley, *Creation of Modern Georgia*, 45–102. For the boll weevil and other factors shaping cotton monoculture, see Daniels, *Breaking the Land*, 3–22, 91–133, 155–83; Geisen, "The South's Greatest Enemy?"

35 Wright, *Old South, New South*, 7 ("low wage region...").

36 "Tennessee Copper," *Wall Street Journal*, October 29, 1903, 5.

37 For production figures, see Emmons and Laney, *Geology and Ore Deposits of Ducktown Mining District*, 32; American Institute of Mining Engineers, "A Brief Description of the Operations of the Tennessee Copper Company." For Ducktown ore composition, see Weed, *The Copper Mines of the World*, 350.

38 Motion for Leave to File Bill of Complaint, 9, *Georgia v. The State of Tennessee, The Ducktown Sulphur, Copper & Iron Company, Ltd., and the Pittsburgh and Tennessee Copper Company*, U.S. Supreme Court, October Term, 1903; "Georgia Seeks to Repel Invasion by Tennessee," *Atlanta Constitution*, January 24, 1904.

39 Sherman's march unquestionably involved deliberate, extensive destruction, but the extent and nature of the destruction is a matter of debate. Compare Grimsley, *The Hard Hand of War*, 171–204, and Marszalek, *Sherman's March to the Sea*. For an environmental perspective, see Summers, "Desolation and War"; Royster, *The Destructive War*.

40 "Confederate Memorial Day in Georgia," (http://www.cviog.uga.edu/Projects/gainfo/confmem.htm) (accessed May 20, 2007). For a recent work on southern historical memory, see Cobb, *Away Down South*.

41 Transcript, 43 (deposition of J. H. Verner), *Madison v. Ducktown Sulphur, Copper & Iron Co., Ltd.*, Chancery Court, Polk Co., TN, No. 247.

42 For the EPA figures and discussion of sulfur dioxide in Georgia, see R. Harold Brown,

The Greening of Georgia, 209–23; Act of April 14, 1967, no. 433, 1967 *Ga. Laws*, vol. 1, 581–90.

43 *Vason v. South Carolina R.R. Co.*, 42 Ga. 631 (1871) (affirming denial of injunction); *South Carolina R.R. Co. v. Steiner*, 42 Ga. 631 (1871) (affirming denial of injunction, remanding for determination of damages); *Guess v. Stone Mountain Granite and Ry. Co.*, 72 Ga. 320 (1884). For other cases denying injunctions against railroads, see *Guess v. Stone Mountain Granite and Ry. Co.*, 67 Ga. 215 (1881); *Powell v. Macon & Indian Springs R.R. Co.*, 92 Ga. 209, 17 S.E. 1027 (1893); *Ga. R.R. and Banking Co. v. Maddox*, 116 Ga. 64, 42 S.E. 315 (1902) (grant of injunction reversed except on Sundays); but also see *Kavanaugh v. Mobile and Girard R.R. Co.*, 78 Ga. 271, 2 S.E. 636 (1886) (injunction proper where city authorization was invalid); *Coker v. Atlanta, Knoxville, and Northern Ry. Co.*, 123 Ga. 483, 51 S.E. 481 (1905) (invalid ordinance); *Ducktown Sulphur, Copper & Iron Co., Ltd. v. Barnes*, 60 S.W. 593 (Tenn. 1900).

44 *Mygatt v. Goetchins*, 20 Ga. 350, 358–59 (1856). For other anti-injunction rulings in favor of steam-powered factories, see *Cunningham v. Rice*, 28 Ga. 30 (1856) (flour mill); *Powell v. Foster*, 59 Ga. 790 (1877) (gristmill); *Knox v. Reese*, 149 Ga. 379, 100 S.E. 371 (1919) (cotton gin); *Holman v. Athens Empire Laundry Co.*, 149 Ga. 345, 100 S.E. 207 (1919).

45 For a description of hydraulic mining in Georgia, see Stephenson, *Geology and Mineralogy of Georgia*, 102–4.

46 For Georgia gold mining, see David Williams, *Georgia Gold Rush*; "There's Gold in Them Thar Hills, Gold and Gold Mining in Georgia," ⟨www.http://dig.galileo.usg.edu/dahlonega⟩ (accessed March 16, 2007); Act of December 13, 1858, no. 163, 1858 *Ga. Laws*, vol. 1, 157 (right of way); Act of December 11, 1858, no. 60, 1858 *Ga. Laws*, vol. 1, 72 (Yahoola).

47 The California case is *Woodruff v. North Bloomfield Gravel Mining Co.*, 18 F. 129 (Cir. Ct., N.D. Cal, 184). For a scientific overview of California hydraulic mining, see Gilbert, *Hydraulic-Mining Debris in the Sierra Nevada*. Gilbert estimated that hydraulic mining sent more than a billion cubic yards of sediment into the Sacramento and San Joaquin river basins and ultimately into San Francisco Bay, where it threatened navigation. The Cartersville cases are *Satterfield v. Rowan*, 83 Ga. 187, 9 S.E. 677 (1889); *Woodall v. Cartersville Mining and Manganese Co.*, 104 Ga. 156, 30 S.E. 665 (1898).

48 Weed, "Copper Deposits in Georgia"; "Copper in Georgia," *Atlanta Constitution*, November 9, 1904, 7; "Copper in Georgia," *Atlanta Constitution*, November 22, 1904, 6; Barclay, *Copper Basin*, 49–50.

49 Aldous Huxley, "Wordsworth in the Tropics," quotation no. 30,000, *The Columbia World of Quotations* (New York: Columbia University Press, 1996); ⟨www.bartleby.com/66/⟩ (accessed April 19, 2007).

50 For a detailed description of Ducktown topography and drainage, see LaForge, *Physical Geography of Georgia*, 114–16.

51 Ligon Johnson, "The History and Legal Phases of the Smoke Problem," 200.

52 Galloway, *Little Girl in Appalachia*, 15–16.

53 Georgia enacted the sales tax in 1929 and the modern income tax in 1931, Act of August 29, 1929, no. 427, 1929 *Ga. Laws*, vol. 1, 103–17; Act of March 31, 1931, no. 5, 1931 *Ga. Laws*, vol. 2, 24.

54 For Tate family history, see Lucius Eugene Tate, *History of Pickens County*; William Tate, *Documents and Memoirs*; Barclay, *Railroad*, 173–78.

55 "Supreme Court Hears Motion; Judge Hart and Ligon Johnson Present Their Case in Washington," *Atlanta Constitution*, January 26, 1904, 8.

Chapter 5

1 "Supreme Court Hears Motion; Judge Hart and Ligon Johnson Present Their Case in Washington," *Atlanta Constitution*, January 26, 1904, 8.
2 Motion for Leave to File Bill of Complaint, *Georgia v. Ducktown Sulphur, Copper & Iron Co.*, U.S. Supreme Court, No. 14 Original, October Term, 1903. Georgia's first Supreme Court case was initially filed against "The State of Tennessee; the Ducktown Sulphur, Copper & Iron Company (Ltd.); and the Pittsburg and Tennessee Copper Co." Georgia soon dismissed its claim against the state of Tennessee and also realized that the Pittsburg and Tennessee Copper Co. had been acquired by the New York–backed Tennessee Copper Co. before the institution of the suit. The state then filed a motion to amend the original bill to reflect the changes; the revised caption became *Georgia v. Ducktown Sulphur, Copper & Iron Co. (Ltd.), Pittsburgh and Tennessee Copper Co., and Tennessee Copper Co.* To avoid confusion, I refer to Georgia's first Supreme Court case as *Georgia v. Ducktown Sulphur, Copper & Iron Co.* The state's second Supreme Court case was filed in 1905 under the more familiar caption, *Georgia v. Tennessee Copper Co.*
3 Hart, *Second Annual Report of John C. Hart*, 5–8.
4 For Hart's life, see "Attorney General Hart and His Important Work," *Atlanta Constitution*, November 20, 1904, A-5.
5 For Johnson, see *The National Cyclopedia of American Biography*, s.v. Johnson, [Robert] Ligon. The railroad case is *Wright v. Louisville & Nashville R.R.*, 189 U.S. 512 (1903) (grant of certiorari), 195 U.S. 219 (1904) (reversed in favor of Georgia). His work on the railroad case is mentioned in "Georgia Seeks to Repel Invasion by Tennessee," *Atlanta Constitution*, January 24, 1904, 7.
6 For a comparison of the two methods, see Friedman, *A History of American Law*, 463–502. For a general overview, see Stevens, *Law School: Legal Education in America*.
7 For a modern example of the custom in Georgia, Griffin Bell continued to be called Judge Bell for the rest of his life after he resigned from the Fifth Circuit Court of Appeals in 1976 to serve as the nation's attorney general and later while in private practice at the Atlanta firm of King & Spalding.
8 "Georgia Wins First Point," *Atlanta Constitution*, February 2, 1904, A5; "Georgia Seeks to Repel Invasion by Tennessee," *Atlanta Constitution*, January 24, 1904, 7. For reaction in Tennessee, see "Georgia-Tennessee State Line Dispute," *Chattanooga News*, January 25, 1904; *Knoxville Tribune* quoted in "Talks about Atlanta," *Atlanta Constitution*, February 23, 1904, 6. The dichotomy is expressed in an aptly titled essay, White, "Are You an Environmentalist or Do You Work for a Living?"
9 U.S. Constitution, art. 3, §2. The cases are collected in Moore, *Moore's Federal Practice*, §402.02[1][e]. For the problem of river boundaries, see Twain, *Life on the Mississippi*, ch. 1; Steinberg, *Slide Mountain*, 21–51 (Missouri River).
10 Petitioner's Brief for Leave to File Bill of Complaint at par. nineteenth, *Georgia v. Ducktown Sulphur, Copper & Iron Co.*, U.S. Supreme Court, No. 14 Original, October Term, 1903.
11 For the Supreme Court's view of original jurisdiction purpose and history, see the majority opinion by Justice Shiras in *Missouri v. Illinois*, 180 U.S. 208 (1901).

12 *Chisholm v. Georgia*, 2 U.S. (2 Dall) 419 (1793); U.S. Constitution, art. III, §2, cl. 2 (emphasis added). The amendment was ratified in 1795 and formally added by presidential message in 1798; see Orth, *The Judicial Power of the United States*.
13 U.S. Constitution, amend. 11 (emphasis added); Moore, *Moore's Federal Practice*, §402.02[2][a]. See 28 *U.S.C.* §1251 (1948, amended 1978), which further narrows article 3 by making a distinction between the Supreme Court's original and *exclusive* jurisdiction (cases between states, as in boundary disputes) and its original but *nonexclusive* jurisdiction (cases by a state against citizens of another state). "Georgia Seeks to Repel Invasion by Tennessee," *Atlanta Constitution*, January 24, 1904, 7.
14 *New Hampshire v. Louisiana*, 108 U.S. 76 (1882) (action on bonds); *Louisiana v. Texas*, 176 U.S. 1 (1899) (quarantine); Note, *Virginia Law Review*.
15 See, e.g., the Demurrer of Tennessee Copper Company, par. second, *Georgia v. Tennessee Copper Co.*, U.S. Supreme Court, No. 13 Original, October Term, 1905; Ligon Johnson, "The History and Legal Phases of the Smoke Problem," 201.
16 Petitioner's Brief for Leave to File Bill of Complaint at pars fourth through seventeenth, nineteenth, *Georgia v. Ducktown Sulphur, Copper & Iron Co.*, U.S. Supreme Court, No. 14 Original, October Term, 1903. Wilmon's report appears as Exhibit B to the Brief. Marsh, *Man and Nature*. Marsh's influence upon the Newell Report is discussed in chapter 3.
17 Petitioner's Brief for Leave to File Bill of Complaint at pars. eighteenth and nineteenth. See U.S. Constitution, art. 1, §10, cl. 3 (power to wage war exclusively reserved to Congress); art. 4, §4 (guarantee to states).
18 These incidents, except Reconstruction, are surveyed in Phillips, *Georgia and State Rights*. *Chisholm v. Georgia*, 2 Dall. (2 U.S.) 419 (1793). See also United States, *Articles and Agreement of Cession*; *Cherokee Nation v. Georgia*, 5 Pet. (30 U.S.) 1 (1831); *Worcester v. Georgia*, 6 Pet. (31 U.S.) 515 (1832).
19 For their Civil War record, see Howe, *Oliver Wendell Holmes: The Shaping Years*, 80–175; Beth, *John Marshall Harlan*, 53–67; Highsaw, *Edward Douglass White*, 18–20.
20 Urbinato, "London's Historic Pea Soupers"; *EPA Journal* 20, nos. 1–2 (Summer 1994): 44; Halliday, *The Great Stink of London*.
21 U.S. Constitution, art. I, §7, cl. 3 (Commerce Clause); *Refuse Act of 1899*, 33 *U.S.C* §§407, 408, 411, 413 (2004); Young, "Criminal Liability under the Refuse Act of 1899." Jurisdiction problems in environmental law are considered in Lazarus, *The Making of Environmental Law*, 29–42.
22 Craig, *The Clean Water Act and the Commerce Clause*, 10–35, 110–18.
23 *Missouri v. Illinois*, 180 U.S. 208 (1901).
24 Ibid.
25 Ibid., 241. Missouri lost on the merits five years later when the Court denied its request for injunction, *Missouri v. Illinois*, 200 U.S. 496 (1906). The dispute continued under different captions until it ended when the Court ordered Illinois to build sewage treatment facilities. *Wisconsin v. Illinois*, 289 U.S. 385 (1933); Percival, "The Frictions of Federalism."
26 *Missouri v. Illinois*, 241; Snow, *On the Mode of Communication of Cholera*. For recent works on Snow, medical mapping, and cholera, see Vinten-Johansen et al., *Cholera, Chloroform, and the Science of Medicine*; Koch, *Cartographies of Disease*; Hempel, *The Strange Case of the Broad Street Pump*. For a more skeptical view of Snow and the pump handle, see McLeod, "Our Sense of Snow."
27 Transcript, 42–43 (deposition of J. H. Verner), *Madison v. Ducktown Sulphur, Copper &*

Iron Co., Chancery Court, Polk Co., TN, no. 247; deposition of H. A. Rogers, affidavit of Fred M. Kimsey, and affidavit of H. A. Rogers, all at *Madison v. Ducktown Sulphur, Copper & Iron Co.,* 113 Tenn. 331 (Transcript pp. 111-13, 333). For a history of industrial hygiene, see Sellers, *Hazards of the Job.*

28 Agency for Toxic Substances and Disease Registry, "Toxicological Profile for Sulfur Dioxide," 1-4, 13-42 (as paginated in the report).

29 "Georgia Wins the First Point," *Atlanta Constitution,* February 2, 1904, A5.

30 Barclay, *Ducktown,* 45, 61, 66-68, 84-86. For other views concerning the control of industrial fumes, see LeCain, "When Everybody Wins Does the Environment Lose?"; Uekoetter, "Solving Air Pollution Problems Once and for All"; Church, "Smoke Farming." LeCain phrased the alternatives as "the transformational techno-fix, the relocational techno-fix, and the delaying techno-fix." Uekoetter discussed the often-tense interaction between civic reformers, professional engineers, industry, and often-dubious inventors in the search for effective abatement of smoke from coal-fired furnaces. Church described bag-house technology used to remove arsenic, lead, and other metals from smelter smoke in Utah. For studies of copper smelter pollution in Swansea, Wales, see Newell, "Atmospheric Pollution and the British Copper Industry"; Rees, "The South Wales Copper-Smoke Dispute." Hart's goals are discussed in a multitopic column under the title, "Georgia Veteran Tells Why He Is for Simmons," *Atlanta Constitution,* April 17, 1904, 3.

31 The commission suggestions appear in "Georgia Forests Withered by Sulphur Fumes," *Atlanta Constitution,* November 22, 1903, A5.

32 Tarr, *The Search for the Ultimate Sink,* 7-35.

33 "Abate the Smoke Nuisance: A Scientific Plan for Controlling the Noxious Gases That Are Destroying Vegetation," *Ducktown Gazette,* ca. June 1902, collected in the James Donaldson Clemmer Scrapbooks, vol. 2, pp. 120-21, TSLA.

34 Barclay, *Ducktown,* 163 (August Raht quotation); Henrich, "The Ducktown Ore Deposits," 173-245, especially at 228-29.

35 The story of the discovery is told in Barclay, *Copper Basin,* 4-7.

36 W. H. Freeland testified that roasting ended there on August 16, 1902; see Transcript, 58 (deposition of W. H. Freeland, March 25, 1905), *Madison v. Ducktown;* "Tennessee Copper: New System of Smelting to Be Inaugurated," *Wall Street Journal,* October 29, 1903, 5; "Tennessee Copper: Surplus after Dividend Payments," *Wall Street Journal,* January 7, 1904, 5.

37 "To Eliminate Sulphur Fumes," *Atlanta Constitution,* January 13, 1903, 4; A. B. Dickey to James G. Parks, March 10, 1904, DBM; "Tennessee Copper: New System to be Inaugurated," *Wall Street Journal,* October 29, 1903, 5.

38 *Georgia v. Ducktown Sulphur, Copper & Iron Co.,* 194 U.S. 629 (1904) (granting motions for leave to file amended bill, to dismiss state of Tennessee, and for leave to file stipulation of settlement); "Will Not Press Suit of State," *Atlanta Constitution,* February 10, 1904, A3; "Ducktown Case Out of Court," *Atlanta Constitution,* April 19, 1904, 5.

39 James G. Parks to W. H. Freeland, December 3, 1904, DBM.

40 See Brooks, Jones, and Virginia, *Law and Ecology,* regarding how ecological science, public environmentalism, politics, and environmental law combined to create the modern environmental regulatory system.

41 James G. Parks to W. H. Freeland, April 21, 1903, DBM; Freeland, "Smelting of Raw Sulphide Ores at Ducktown."

42 James G. Parks to Howard Cornick, May 21, 1903, and Howard Cornick to James G. Parks, May 22, 1903, both DBM.
43 Barclay, *Copper Basin*, 36–37; compare Quinn, "Industry and Environment in the Appalachian Copper Basin," at n. 46.
44 The procedural history is found in *Madison v. Ducktown Sulphur, Copper & Iron Co.*, 113 Tenn. 331, 337–47, 83 S.W. 658, 659–61 (1904). My examination of court filings and docket sheets revealed that, as of 1904, Madison filed three actions for damages and was party to four different injunction actions as either a plaintiff or defendant. Farner had three claims for damages, and McGhee had two.
45 James G. Parks to W. H. Freeland, December 3, 1904, DBM.
46 "Tennessee Copper: Capacity Will Be Increased to 24,000,000 Pounds," *Wall Street Journal*, January 10, 1905, 5. The production figures are from the United States Bureau of Mines, quoted in James W. Taylor, "Ducktown Desert," 51.
47 Report of John T. McGill, January 1, 1916, *Georgia v. Tennessee Copper Co.*, No. 1 Original, October Term, 1914. Sources concur that the greatest period of destruction occurred after the industry's revival in 1891: Foehner, "Historical Geography," 25–50 (setting it from 1891–1946); James W. Taylor, "Ducktown Desert," 46–60 (from 1890 to 1904); Quinn, "Industry and Environment in the Appalachian Copper Basin."
48 Baes and McLaughlin, "Trace Elements in Tree Rings"; Meierding, "Marble Tombstone Weathering and Air Pollution in North America."
49 Barclay, *Copper Basin*, 7, 40, 80–81; Quinn, "Industry and Environment in the Appalachian Copper Basin," 590–91.
50 U.S. Forest Service, "Area Affected by Sulphur Fumes, Ducktown Region, Tennessee, Georgia, and N. Carolina," discussed in Quinn, "Industry and Environment in the Appalachian Copper Basin," 575–612, especially 602–3; Tennessee Valley Authority, Department of Forestry Relations, "A Proposal for Erosion Control" (1945), 1–5, 22.
51 Burt, "A Desert in the Appalachians"; Teale, "The Murder of a Landscape"; Hursh, *Local Climate in the Copper Basin*.
52 McGill Report, 26–27; Smallshaw, "Denudation and Erosion in the Copper Basin," 5.
53 Seigworth, "Ducktown—a Postwar Challenge"; Richard A. Wood, "Erosion Control and Reforestation," 7–8; Barclay, *Introduction to the Copper Basin*.
54 Quinn, "The Appalachian Mountains' Copper Basin."
55 Lankton, *Beyond the Boundaries*, 205–12.
56 Galloway, *Little Girl in Appalachia*, 5–10.
57 Ibid.
58 Ibid., 11; the sketch map appears as the frontispiece. Another Ducktown resident said there were two trees near Isabella; see Smallshaw, "Denudation and Erosion in the Copper Basin," 5. Either way, the celebration of a specific tree or pair of trees was remarkable in a once densely wooded land.

Chapter 6

1 A. B. Dickey to James G. Parks, August 7, 1905, DBM.
2 John C. Hart to Cornick, Wright & Frantz, August 22, 1904, DBM.
3 Harold Day Foster to John C. Hart, October 1, 1904; John C. Hart to Harold Day Foster, October 5, 1904, both from RG 9-1-1, box 1, folder 10, GDAH.
4 Klein, *Portrait of an Early American Family*; deposition of Will H. Shippen, August 20,

1914, 317–46, and deposition of Frank Shippen, August 21, 1914, 393–429, both in Transcript of Evidence, vol. 1, *Georgia v. Tennessee Copper Co.*, No. 1 Original, October Term, 1914 (herein Transcript [1914]). Will Shippen recounted his career in timber in "Mr. W. H. Shippen Will Maintain His Long Residence in Gilmer," *Ellijay Times-Courier*, January 6, 1939, 1.

5 Schenck, *Logging and Lumbering*, 97–110.
6 For their letterhead, see Will H. Shippen to Governor Jos. M. Brown, April 9, 1912, DBM; Mary Elizabeth Johnson, "Box and Container Industry."
7 Deposition of Frank Shippen, Transcript, vol. 1, 411, *Georgia v. Tennessee Copper Co.*, U.S. Supreme Court, No. 1 Original, October Term, 1915.
8 Schenck, *Logging and Lumbering*, 17–34, 61–88, 111–26. For industrial logging circa 1920 on the basin's southwestern rim, see Roper, "Logging the Cohutta Wilderness."
9 "A Great Country, Which Is Growing Greater Every Day," *Atlanta Constitution*, July 8, 1888, 14; deposition of Frank Shippen, Transcript (1914), 411.
10 For removal of timber suits to federal court, see Charles Seymour to James G. Parks, September 2, 1902; James G. Parks to Howard Cornick, April 1, 1904, both DBM; deposition of J. P. Vestal, Transcript (1914), 161–86.
11 For Wetmore, see *Marquis Who's Who on the Web*, s.v. George Peabody Wetmore (⟨marguiswhoswho.com⟩, accessed September 10, 2007); "Wetmore Estate Valued at $4,809,054," *New York Times*, September 22, 1922, 12. For Parmentier, see McCrary and Graf, "Vineland in Tennessee"; Landis, *You Have Stept Out of Your Place*, 200. For local timber companies, see Lillard, *History of Polk County*, 147–50.
12 Ward, *The Annals of Upper Georgia*, 382, 393.
13 Gennett, *Sound Wormy*.
14 James G. Parks to Howard Cornick, July 11, 1908, DBM.
15 Will H. Shippen to Warren A. Grice, October 29, 1914, DBM.
16 W. H. Freeland to James G. Parks, November 13, 1905 (unemployed miners); W. H. Freeland to James G. Parks, February 9, 1913 (settlement offer), both DBM. For the TCC settlement, see deposition of Will Shippen, Transcript, vol. 1, 329–35, 512–55, *Georgia v. Tennessee Copper Co.*, U.S. Supreme Court, No. 1 Original, October Term, 1915.
17 "Condemned Man to Gallows, Hilburn Seeks Liberty," *Atlanta Constitution*, January 28, 1905, 6.
18 "Wreck of the Old '97," public domain. For discussion of the song's public domain status, see Cohen, "Robert W. Gordon and the Second Wreck of the Old '97." The Kingston Trio borrowed the melody for its 1959 hit, "M.T.A."
19 "Condemned Man to Gallows, Hilburn Seeks Liberty."
20 Alfred K. Chittenden, "Report on a Preliminary Examination of the Effects of the Ducktown Sulphur Fumes on the Forests of Polk County, Tennessee," attached as Ex. F. to Original Bill of Complaint, 29–35, *Georgia v. Tennessee Copper Co.*, U.S. Supreme Court, No. 13 Original, October Term, 1905. The forestry unit of the U.S. Department of Agriculture experienced three name changes, beginning as the Division of Forestry in 1881, then the Bureau of Forestry in 1901, and finally the U.S. Forest Service in 1905. The latter two names straddle the time span of Georgia's smoke suit and are used interchangeably here.
21 Ibid., 31–34.
22 Ibid., 32, 34–35.
23 For the statistics, see Michael A. Williams, *Americans and Their Forests*, 118–20 (farm

clearing); Cox et al., *This Well-Wooded Land*, 111–12 (lumber production). The timber famine quotations appear in Clepper, *Professional Forestry in the United States*, 135–36. See also Pisani, "Forests and Conservation, 1865–1890" (for discussion of timber famine and the forest conservation movement).

24 Michael A. Williams, *Americans and Their Forests*, ch. 12 "Preservation and Management, 1870–1910," 393–424. The literature on American forests is extensive. Representative works, in addition to those already cited, include Hays, *Conservation and the Gospel of Efficiency* (presenting forest conservation as a model of progressive ideology and method); Nash, *Wilderness and the American Mind*, 122–81 (comparing John Muir's wilderness preservation to Gifford Pinchot's sustainable use); Bolgiano, *The Appalachian Forest* (a regional and social perspective). Histories focusing upon the Forest Service include Steen, *The U.S. Forest Service*, 3–102 (an institutional history with emphasis on the conservation movement); Hays, *The American People and the National Forests*. Useful essay collections on the Forest Service include Miller, *American Forests: Nature, Culture, and Politics*, and Steen and Forest History Society, *The Origins of the National Forests*. Critiques of the Forest Service include Hirt, *A Conspiracy of Optimism*, and Langston, *Forest Dreams, Forest Nightmares*.

25 Pinchot, *Breaking New Ground*; Steen, *The U.S. Forest Service*, 47–102; Miller, *Gifford Pinchot*. Michael A. Williams, *Americans and Their Forests*, 406–7. The state of New York provided the model for eastern forest reserves by creating the Adirondack Forest Reserve in 1885 to protect the Hudson River watershed; for a revisionist view of that project, see Jacoby, *Crimes against Nature*, 11–80.

26 Charles Dennis Smith, "The Appalachian National Park Movement"; Shands, "The Lands Nobody Wanted"; Eller, "Land as Commodity." For Georgia resolutions and enabling legislation for a southern national forest reserve, see Resolution of December 18, 1900, no. 182, 1900 *Ga. Laws*, vol. 1, 500–501; Act of December 18, 1901, no. 68, 1901 *Ga. Laws*, vol. 1, 84–85; Resolution of August 16, 1910, no. 288, 1910 *Ga. Laws*, vol. 1, 1279; Resolution of August 15, 1916, no. 338, 1916 *Ga. Laws*, vol. 1, 1045–46; Act of August 18, 1917, no. 51, 1917 *Ga. Laws*, vol. 1, 182–83; Act of August 17, 1918, no. 73, 1918 *Ga. Laws*, vol. 1, 206–8.

27 Charles Dennis Smith, "The Appalachian National Park Movement," 59–62 (discussion of the funding bill, H. R. 10538); U.S. Department of Agriculture et al., *Message from the President of the United States* (herein Wilson Report). Most of the contents were revised and republished as Ayres, Ashe, and Smith, *The Southern Appalachian Forests*.

28 Wilson Report, 32–34, 128–37; the quotation is at 129.

29 Ibid., 3–5, 25, 35–36, 38–40. The Weeks Act appears as Act of March 1, 1911, *U.S. Statutes at Large* 36, Part 1, Chap. 186, 961–63. The origin of the Cherokee National Forest appears on its official Web site at ⟨http://www.fs.fed.us/r8/cherokee/about/degraded.shtml⟩ (accessed April 27, 2010).

30 Affidavit of A. B. Patterson, September 24, 1906, and affidavit of Charles A. Keffer, September 18, 1906, both in Supplemental Evidence and Exhibits on Application for Restraining Order, *Georgia v. Tennessee Copper Co.*, U.S. Supreme Court, No. 5 Original, October Term, 1906. Patterson stated in his affidavit that he "acted in his official capacity and under the orders of the Forest Service." Hart acknowledged that the service detailed Patterson at the state's request; see Memorandum of Argument for Complainant upon Final Hearing, 8, *Georgia v. Tennessee Copper Co.*, No. 1 Original, October Term, 1906 (herein Georgia Brief on Final Hearing). Keffer acknowledged that he made his

report at the request of Charles Seymour, attorney for three great timber owners, Wetmore, Stephenson, and the Ocoee Timber Company. Note: the docket numbers are confused for this phase of *Georgia v. Tennessee Copper Co.*, which spanned the years of 1905 through 1907. Owing to the practice of reassigning docket numbers for each term of court, documents for this period bear the docket numbers given variously as No. 1 Original, No. 5 Original, and No. 13 Original. A docket number assigned to a document in an earlier term was often scratched through and reassigned a new docket number if the same item was used again in a later term. For the sake of clarity, all subsequent references to the case record in this chapter, regardless of the docket number given to a document at the time of filing, are to be understood as referring to the same case, which was decided as *Georgia v. Tennessee Copper Co.*, U.S. Supreme Court, No. 1 Original, October Term, 1906, with the opinion published as *Georgia v. Tennessee Copper Co.*, 206 U.S. 230 (1907).
31 Haywood, *Injury to Vegetation by Smelter Fumes*.
32 Ibid., 9–17.
33 Affidavit of J. K. Haywood, September 17, 1906; affidavit of J. K. Haywood, November 27, 1905.
34 Hart, *Fifth Annual Report of John C. Hart*, 12–13; "Governor Goes to Charlotte — Will Attend National Forestry Congress Today," *Atlanta Constitution*, March 3, 1906, 7; "Knotty Issue in Arbitration," *Atlanta Constitution*, October 16, 1905, 6. Johnson's forestry advocacy appears in a series of *Atlanta Constitution* articles: "Association Formed to Preserve Forests," December 8, 1907; "Save Forests — Roosevelt," January 14, 1908; "Governors Are Asked to Send Delegations to Fight for Forests," January 17, 1908; "Adamson's Plea for Anti-Jug Law," February 1, 1908 (bottom paragraph). For Johnson's article, see "Vast Importance to South of Forest Reserves," *Atlanta Constitution*, December 1, 1907, E3.
35 The scientific and legal arguments of the state are marshaled in its Memorandum of Argument for Complainant upon Final Hearing, filed February 25, 1907.
36 For *Atlanta Constitution* coverage on the resolutions, see "Ducktown Case to be Reopened — Resolution on Subject Goes to the Senate Today," July 14, 1905, 7 (Senate resolution); "Prohibition Bill Appears in House," July 15, 1905, 7 (Powell's House resolution); "To Investigate Ducktown Plant, Resolution Providing for Commission Introduced in Senate," July 15, 1903, 7. Whether Shippen knew the term "talking points" is not known, but he certainly understood the concept. For a near contemporary use of the term, see Sinclair Lewis, *Babbitt*, 42 (page citation is to the reprint edition).
37 Resolution No. 17, "In Relation to Injuries from Copper Mines at Ducktown, Tenn.," 1905 *Ga. Laws* 1250; "Gossip at the Capitol," *Atlanta Constitution*, August 17, 1905, 6; Report of the 1905 Legislative Commission, August 30, 1905, attached as Exhibit G to Plaintiff's Bill of Complaint, *Georgia v. Tennessee Copper Co.*, U.S. Supreme Court, No. 13 Original, October Term, 1905; "Will Look into Sulphur Ovens: Georgia Commission Departed Yesterday to Investigate Ducktown Problem," *Atlanta Constitution*, August 29, 1905, 7.
38 "North Georgia Is Devastated, Legal Steps Urged to Abate the Sulfur Nuisance," *Atlanta Constitution*, August 31, 1905, 7; Order of Governor J. M. Terrell, September 21, 1905, attached as Exhibit H to Plaintiff's Bill of Complaint, *Georgia v. Tennessee Copper Co.*, U.S. Supreme Court, No. 13 Original, October Term, 1905; "No Franchise, Says Board in the Bull Sluice Case — Discuss Ducktown Case," *Atlanta Constitution*, October 14, 1905, 6; "Knotty Issue in Arbitration — the Ducktown Case," *Atlanta Constitution*, October 16,

1905, 6. The case reached a formal, as opposed to effective, end when the Supreme Court dismissed it with prejudice on May 16, 1938, *Georgia v. Tennessee Copper Co.*, U.S. Supreme Court, No. 1 Original, October Term, 1937.

39 The quoted version of the adage is from Albert Gore Jr. in the *Washington Post*, July 23, 1982, quotation no. 1527, *The Columbia World of Quotations* (New York: Columbia University Press, 1996), ⟨www.bartleby.com/63/⟩ (accessed August 19, 2007).

40 Affidavit of John M. McCandless, November 1905 (the specific date was not provided in the original).

41 Affidavit of Will H. Shippen, April 13, 1906; affidavit of Frank E. Shippen, November 10, 1905; affidavit of L. D. Rogers, August 21, 1906. The extent to which they actually employed conservationist methods in the logging business is subject to question.

42 Fernow's testimony appears in the Brief and Argument of the Defendant Tennessee Copper Company on Final Hearing, 18–19 (herein TCC Brief on Final Hearing). Steep grassy Alpine pastures were the setting of Julie Andrews's title song in the film *The Sound of Music* (1965).

43 Schenck, *The Biltmore Story*; Jolly, "The Cradle of Forestry."

44 Schenck, *Forest Protection*, 141–52, 144.

45 Schenck's testimony is quoted in the Brief and Argument of the Ducktown Sulphur, Copper & Iron Co. Limited on the Final Hearing, 23–24 (herein DSC&I Brief on Final Hearing).

46 Ibid., 24; see also TCC Brief on Final Hearing, 34–36.

47 J. Parke Channing, quoted in the *Engineering and Mining Journal*, June 12, 1905, and discussed in Georgia Brief on Final Hearing, 62–63; affidavit of John H. Sussman, September 29, 1906; affidavit of J. Parke Channing, April 24, 1906.

48 Affidavit of R. F. Sams, September 1906; affidavit of D. B. Osborne, September 1, 1906; affidavit of H. A. Rodgers, September 1906; affidavit of A. A. Smith, September 25, 1906.

49 DSC&I Brief on Final Hearing, 27–37, 81–84; affidavit of W. H. Freeland, November 15, 1905.

50 Affidavit O. F. Chastine, November 17, 1905. For the number of affidavits, see Georgia Brief on Final Hearing, 10–11, 33–34.

51 Resolution from Harper Town Store, February 3, 1906; affidavit of F. M. Jones, September 10, 1906; affidavit of J. B. Witt, September 10, 1906; affidavit of James Akin, September 11, 1906.

52 For challenges to procompany local witnesses, see affidavit of J. A. Tipton, April 4, 1906; affidavit of W. A. Daves, April 3, 1906; affidavit of W. F. Hampton, March 19, 1906; affidavit of Sherman Martin, April 12, 1906.

53 For anti-injunction challenges, see affidavit of W. R. Early, September 22, 1906; affidavit of G. G. Hyatt, September 25, 1906.

54 Affidavit of Martin Vogel, October 1, 1906. Johnson's flyer was attached to Vogel's affidavit.

55 TCC Brief on Final Hearing, 82–101; DSC&I Brief on Final Hearing, 13–16.

56 *Louisiana v. Texas*, 176 U.S. 1, 24 (1899); TCC Brief on Final Hearing, 82–99; Georgia Brief on Final Hearing, 52–58. The procedural history appears in "Ducktown Case Ordered Filed," *Atlanta Constitution*, October 24, 1905, 9; DSC&I Brief on Final Hearing, 1–2, 4–16; TCC Brief on Final Hearing, 10–16, 21–26; Order overruling demurrers without prejudice, November 5, 1906.

57 *Ducktown Sulphur, Copper & Iron Co., Ltd. v. Barnes*, 60 S.W. 593 (Tenn., 1900).

58 TCC Brief on Final Hearing, 107–19 (quotation at 109); DSC&I Brief on Final Hearing, 73–81.
59 DSC&I Brief on Final Hearing, 3–4, 14–16.
60 *Madison v. Ducktown Sulphur, Copper & Iron Co.*, 113 Tenn. (5 Cates) at 339, 363–66, 83 S.W. at 659, 666–67; Act of April 17, 1901, ch. 139, 1901 *Tenn. Pub. Acts* 246.
61 *Mountain Copper v. United States*, 142 F. 625, 638, 640–43 (9th Cir., 1906).
62 Howard Cornick to James G. Parks, April 16, 1906, DBM.
63 Compare *Missouri v. Illinois*, 180 U.S. 208 (1901) (herein *Missouri I*) to *Missouri v. Illinois*, 200 U.S. 496 (1906) (herein *Missouri II*).
64 Percival, "The Frictions of Federalism," 4–15; *Missouri II*, 52–23 (quotations).
65 J. G. Gordon to James G. Parks, December 18, 1905, DBM; Hart, *Sixth Annual Report of John C. Hart*, 3–6.
66 "Ducktown Case with the Court—How Judge Hart Answered Question as to State's Right to Bring Suit," *Atlanta Constitution*, March 2, 1907, 7; James G. Parks to W. H. Freeland, March 7, 1907, DBM.
67 *Georgia v. Tennessee Copper Co.*, 206 U.S. 230 (1907).
68 Ibid., 236.
69 Ibid., 237.
70 *Missouri I*; *Louisiana v. Texas*, 176 U.S. 1 (1899).
71 *Georgia v. Tennessee Copper Co.*, 206 U.S. at 237.
72 Ibid., 238.
73 Ibid.
74 Ibid., 238–39.
75 Ibid., 239.
76 Ibid.
77 John H. Frantz to James G. Parks, May 20, 1907, DBM; Holmes, *The Common Law*, 1.
78 Holmes, *The Common Law*, 1.
79 *Georgia v. Tennessee Copper Co.*, 206 U.S. at 239.

Chapter 7

1 *Georgia v. Tennessee Copper Co.*, 206 U.S. 230 (1907).
2 Hart's compensation is discussed in "Attorney General Hart and His Important Work," *Atlanta Constitution*, November 20, 1904, A5.
3 *Georgia v. Tennessee Copper Co.*, 206 U.S. at 239.
4 The decree of the trial court is quoted in *American Smelting & Refining Co. v. Godfrey*, 158 F. 225, 228 (8th Cir. 1907). The court in *Godfrey* rejected the use of the balancing test employed by the Tennessee Supreme Court and others in the growing trend of American courts; see, e.g., Brenner, "Nuisance Law and the Industrial Revolution"; Hoffer, *The Law's Conscience*, 147–79; Kurtz, "Nineteenth-Century Anti-entrepreneurial Nuisance Injunctions"; McLaren, "Nuisance Law and the Industrial Revolution"; Provine, "Balancing Pollution and Property Rights"; Rosen, "Differing Perceptions of the Value of Pollution Abatement"; Rosen, "Knowing Industrial Pollution."
5 *Salt Lake City Tribune*, December 28, 1906, quoted in Lamborn and Peterson, "The Substance of the Land," 321; Church, "Smoke Farming" (discussing *inter alia* the use of baghouse technology to scrub heavy metals from smelter smoke after the *Godfrey* decision).
6 Coverage in the *Atlanta Constitution* on both matters appears in the following articles:

"Committees of House Pass Busy Afternoon," July 17, 1907, 6; "Large Delegation Here for Ducktown Hearing," July 17, 1907, 4; "Test Votes Tell Story in House—Prohibition Forces Shown to Be in Full Control of the Lower Branch," July 20, 1907, 1.
7. J. G. Gordon and Lewis Mortimer to W. H. Freeland, June 11, 1907, DBM.
8. John C. Hart to W. A. Daves, July 1, 1907; and to similar effect, John C. Hart to J. M. Clement, July 2, 1907, both RGS 9-1-1, box 1, GDAH.
9. Resolution of July 27, 1907, 1907 *Ga. Laws*, vol. 2, 991.
10. *Georgia v. Tennessee Copper Co.*, 194 U.S. 629 (1904) (granting leave to file stipulation of settlement); "Will Not Press Suit of State," *Atlanta Constitution*, February 10, 1904, A3; "Ducktown Case Out of Court," *Atlanta Constitution*, April 19, 1904, 5; Hart, *Sixth Annual Report of John C. Hart*, 3–6.
11. Kiefer, "Sulfuric Acid: Pumping Up the Volume."
12. Lewis B. Nelson, *History of the U.S. Fertilizer Industry*, 55–96.
13. Stoll, *Larding the Lean Earth*, 150–65, and specifically 160 (Hammond quotation). Stoll work is a survey of soil amendment strategies (or lack thereof) in antebellum America, with attention to their social and political implications. Edmund Ruffin devoted his career to the replenishment of southern farms before he became a notorious Confederate firebrand; see Mathew, *Edmund Ruffin and the Crisis of Slavery in the Old South*.
14. Blakey, *The Florida Phosphate Industry*, 1–12.
15. Lewis B. Nelson, *History of the U.S. Fertilizer Industry*, 11 (von Leibig quotation); Isenberg, *The Destruction of the Bison*, 123–63.
16. Affidavit of T. G. Hudson, Georgia Commissioner of Agriculture, September 1906 (specific date omitted in the original), *Georgia v. Tennessee Copper Co.*, U.S. Supreme Court, No. 1 Original, October Term, 1906.
17. Blakey, *The Florida Phosphate Industry*, 7, see generally 1–12; "A Visit to the Phosphate Wells of South Carolina," *Atlanta Constitution*, March 24, 1870, 1.
18. Lewis B. Nelson, *History of the U.S. Fertilizer Industry*, 84–88. Early articles on superphosphate and acid in the *Atlanta Constitution* include "The Atlanta Acid and Fertilizing Company," March 19, 1869; "Fertilizer—Georgia State Agricultural Society," December 28, 1873, 3 (price quotation); "State Agricultural Convention," August 11, 1875, 3; "A Big Bonanza—Fertilizers—How to Make Them," April 17, 1879, 1 (Dr. Pratt).
19. Floyd, *Tennessee Rock and Mineral Resources*, 50. For an earlier view of the Ducktown sulfur industry, see Wilbur Nelson, "The Manufacture of Sulphuric Acid in Tennessee."
20. Barclay, *Copper Basin*, 39–40; "Conservation of Injurious Gases of Ducktown Copper Plants Results in Enormous Benefits to Farmers of the South," *Atlanta Constitution*, August 2, 1911, 5.
21. "Tennessee Copper Co.," *Wall Street Journal*, May 14, 1907, 5; "$1,000,000 Spent by Tennessee Copper Company to Keep Sulphur Fumes from Injuring Vegetation," *Atlanta Constitution*, July 24, 1910, B7; "Conservation of Injurious Gases of Ducktown Copper Plants Results in Enormous Benefit to Farmers of the South," *Atlanta Constitution*, August 2, 1911, 5.
22. J. G. Gordon and Lewis Mortimer to W. H. Freeland, June 11, 1907; W. H. Freeland to J. G. Parks, June 12, 1907, both DBM.
23. J. G. Gordon and Lewis Mortimer to W. H. Freeland, June 11, 1907, 3, 8, DBM.
24. "Conservation of Injurious Gases," *Atlanta Constitution*, August 2, 1911, 5.
25. Howard Cornick to John C. Hart, September 3, 1907 (invitation and arrangements regarding Hart's visit); B. Britton Gottsberger, Asst. Mgr. to John C. Hart, Septem-

ber 27, 1907 (2,125 employees, of which 1,275 came from Georgia), both RGS 9-1-1, box 1, GDAH.
26 "Copper Mines Do Good Work," *Atlanta Constitution*, September 28, 1907, 4.
27 For a historical view of the flexibility of injunctive remedies, see Hoffer, *The Law's Conscience*. As discussed in chapter 8, such flexibility would be demonstrated by the Supreme Court in 1915 when it issued an injunction specifying limits upon different permissible amounts of sulfur dioxide releases during and after the growing season.
28 *Georgia v. Tennessee Copper Co.*, 206 U.S. 230, 237, 239 (1907).
29 Motion for Leave to Postpone Entry of Final Decree, October 21, 1907, *Georgia v. Tennessee Co.*, No. 1 Original, October Term, 1907. The order was granted on October 28, 1907.
30 *Madison v. Ducktown Sulphur, Copper & Iron Co.*, 113 Tenn. 331, 339, 666, 83 S.W. 658, 659, 666 (1904). One more attempt to obtain a Tennessee injunction surfaced in 1910 but was soon abandoned; see *Davenport v. Tennessee Copper Co. and Ducktown Sulphur, Copper & Iron Co.*, Polk County Chancery Court, no. 474; "Seek to Enjoin Copper Company—Two Hundred Georgia Citizens File Bill," *Atlanta Constitution*, September 3, 1910, 2.
31 *Padgett v. Ducktown Sulphur, Copper & Iron Co.*, 97 Tenn. 690, 37 S.W. 698 (1896); *Ducktown Sulphur, Copper & Iron Co., Ltd. v. Barnes*, 60 S.W. 593 (Tenn., 1900); *Ducktown Sulphur, Copper & Iron Co. v. Fain*, 109 Tenn. 56, 70 S.W. 813 (1902); *Jones v. Ducktown Sulphur, Copper & Iron Co.*, 109 Tenn. 375, 71 S.W. 821 (1902); *Swain v. Ducktown Sulphur, Copper & Iron Co.*, 111 Tenn. 430, 78 S.W. 93 (1903).
32 James G. Parks to Howard Cornick, February 13, 1903 ("exhausted this means"); James G. Parks to W. B. Miller, January 23, 1909 ("matter of routine"); James Parker to Howard Cornick, September 16, 1908; Howard Cornick to James G. Parks, December 21, 1908 (Parks settlement offer), all DBM.
33 James G. Parks to W. H. Freeland, October 22, 1908, DBM.
34 W. H. Freeland to James G. Parks, October 26, 1908; James G. Parks to W. H. Freeland, October 27, 1908, both DBM.
35 "Tennessee Copper—Profit and Loss Surplus Increased $447,000 in Past Calendar Year," *Wall Street Journal*, March 1, 1907, 7; James G. Parks to W. H. Freeland, January 24, 1908; James G. Parks to W. H. Freeland, September 10, 1908 ("a 'little something'"), both DBM.
36 Howard Cornick to James G. Parks, December 7, 1908. See letters from James G. Parks to A. B. Dickey dated July 4, 1908 (two of the same date), July 11, 1908, July 17, 1908, October 9, 1908, and October 30, 1908; and letters from A. B. Dickey to James G. Parks dated July 10, 1908, and July 15, 1908, all DBM. See also Howard Cornick to James G. Parks, July 13, 1908 (forwarding tax digests); James G. Parks to Howard Cornick, September 21, 1908 (inspectors); James G. Parks to Howard Cornick, October 9, 1908 (final demands); all DBM. The demand and recovery figures for each plaintiff combine the separate figures for the paired actions against TCC and DSC&I. Withrow is spelled Witherow in some sources.
37 James G. Parks to W. H. Freeland, January 24, 1908, DBM.
38 Howard Cornick to James G. Parks, June 3, 1908 (invoice); Howard Cornick to James G. Parks, June 9, 1908 (survey issues), both DBM.
39 C. A. Schenck to Howard Cornick, June 3, 1908; James G. Parks to W. B. Miller, February 1908 (specific date omitted in the original), DBM; Graves, *Forest Mensuration*, 202–9.
40 C. A. Schenck to Howard Cornick, June 18, 1908, DBM.
41 *Ladew v. Tennessee Copper Co.*, 218 U.S. 357 (1910); *Wetmore v. Tennessee Copper Co.*, 218 U.S.

369 (1910). James G. Parks to W. H. Freeland, January 18, 1908; James G. Parks to W. B. Miller, February 1908, 13, both DBM.

42 Haywood, *Injury to Vegetation from Smelter Fumes*; Howard Cornick to James G. Parks, November 9, 1907; also Howard Cornick to James G. Parks, October 22, 1907, both DBM. Haywood's other publications on smelter smoke include *Injury to Vegetation and Animal Life by Smelter Wastes*, "Injury to Vegetation and Animal Life by Smelter Fumes," and "Smelter Smoke."

43 Howard Cornick to James G. Parks, November 9, 1907, DBM. For the article, see Murrill, "A Serious Chestnut Disease."

44 "Wins $32,371 Verdict from Copper Company," *Atlanta Constitution*, February 26, 1911, 12.

45 See chapter 2.

46 McCandless, "Report of Dr. Jno. M. McCandless to Hon. Jno. C. Hart"; James G. Parks to John C. Hart, March 7, 1908, RGS 9-1-1, GDAH.

47 John C. Hart to Judge J. R. Chastain, March 19, 1908, RGS 9-1-1, box 2, GDAH.

48 J. M. Hackney to John C. Hart, July 9, 1908 (Fannin); J. T. Deweese to J. G. Hart, September 11, 1908 (Gilmer), both RGS 9-1-1, box 2, GDAH.

49 Barrett, *The Mission, History and Times of the Farmers' Union*, 45–48, 97–121; Fisher, *The Farmers' Union*, 8–18; Crampton, *The National Farmers Union*, 3–22, 55–60. Crampton demonstrates that with the end of Barrett's long presidency and the post–World War I decline of cotton prices, the Farmers Union faded in the South and established its current concentration among the grain farmers of the Midwest and Plains states. Note: the name National Farmers Union appears in the literature with and without an apostrophe after Farmers. The modern NFU eliminated the apostrophe, and its practice is followed herein.

50 Barrett, *The Mission, History and Times of the Farmers' Union*, 153–55, 213–22; Ward, *The Annals of Upper Georgia*, 399. Felton was appointed to fill the remaining term of Tom Watson upon his death. Her tenure lasted only one day before she was replaced by the newly elected, Walter F. George; see her obituary, *New York Times*, January 25, 1930, 9.

51 Barrett, *The Mission, History and Times of the Farmers' Union*, 16 (Watson quotation), 213–22 (Georgia details). Columns under the title "Farmers' Union Department," appeared in *Watson's Jeffersonian Magazine*, beginning with 1, no. 2 (February 1907) through 2, no. 2 (February 1908). The NFU presence at Union City is described in "Union City is Growing Daily," *Atlanta Constitution*, August 1, 1909, 7.

52 Barrett, *The Mission, History and Times of the Farmers' Union*, 49–53. For a description of farming and the economy in Gilmer and Fannin counties, see Ward, *The Annals of Upper Georgia*, 350–90; Parker, *Days Gone By*, 326–32. Barrett was a plenary speaker at the 1910 Southern Conservation Congress; see "First Meeting on Conservation," *Atlanta Constitution*, October 7, 1910, 1; beyond that mention, there is no extant mention of the smoke suits or conservation issues in general in early NFU records. Records from Barrett's long tenure were never transferred to NFU archives despite repeated pleas from the organization. In 1957 the NFU librarian wrote, "We have never been able to ascertain what records Mr. Barrett may have left behind him." See Phoebe F. Hayes, Librarian to Genevieve Pyle Demme, February 19, 1957, NFU I-1-2; C. E. Huff to Charles A. Barrett, October 8, 1957, NFU II-1-1.

53 Hicks, *The Populist Revolt*, 95, 405, 404–23; Woodward, *The Origins of the New South*, 175–205, 235–64; Hofstadter, *The Age of Reform*, 62, 82; Goodwyn, *Democratic Promise*, 537,

539; McMath, *American Populism*, 16–17, 40–42. For NFU in-house counsel, see Barrett, *The Mission, History and Times of the Farmers' Union*, 299 (Drake), 307–8 (Ladson).
54 The Hetch-Hetchy fight was one of the seminal events in the formation of natural resource policy in America. For an overview, see Nash, *Wilderness and the American Mind*, 161–81; Righter, *The Battle over Hetch-Hetchy*; Simpson, *Dam! Water, Power, Politics, and Preservation in Hetch Hetchy*.
55 *National Field*, January 1, 1914 (editorial), and February 12, 1914 (Pinchot's letter to the editors and, in the same issue, a laudatory article on Pinchot, complete with portrait).
56 The NFU's Union Phosphate Plants received regular coverage in the *Atlanta Constitution*: "Union Farmers Form a Fertilizer Company," November 19, 1907, 7; "Phosphate Plant of Union Opened," October 16, 1910, 5; "Farmers Saved Immense Sums—through Operation of Union Phosphate Co.," October 17, 1910, 5; "Enlarge Plants for Ga. Farmers—Union Phosphate Co. Will Develop Phosphate Holdings," June 21, 1911, 9.
57 The resolution received extensive coverage in the *Atlanta Constitution*: "Want to Close Copper Plants," July 17, 1908, 3; "Gases Still Ruining Crops," July 22, 1908, 9; "Ducktown Copper Mines before House This Week," July 27, 1908, 5; "Copper Plants in Status Quo," July 29, 1908, 7 (committee tally); James G. Parks to W. H. Freeland, August 10, 1908, DBM (House vote).
58 John C. Hart to Editor, *Blue Ridge Post*, September 12, 1908; John C. Hart to W. B. James, September 28, 1908, both RGS 9-1-1, box 2, GDAH.
59 John C. Hart to Editor, *Blue Ridge Post*, September 12, 1908; James G. Parks to John C. Hart, September 29, 1908; W. H. Freeland to John C. Hart, May 14, 1909; John C. Hart to Howard Cornick, October 8, 1908, all RGS 9-1-1, box 2, GDAH.
60 Progress on the plants appears in H. F. Wierum to John C. Hart, May 3, 1909; W. F. Freeland to John C. Hart, June 25, 1909, both RGS 9-1-1, box 2, GDAH. A published description of the DSC&I appears in Freeland and Renwick, "Smeltery Smoke as a Source of Sulphuric Acid." John C. Hart to Howard Cornick, July 1, 1909 (cabbage patch); Howard Cornick to John C. Hart, August 26, 1909; E. L. Worsham to John C. Hart, June 26, 1909 (beans), all RGS 9-1-1, box 2, GDAH.
61 W. T. Buchanan to John C. Hart, June 30, 1909; J. T. DeWeese to John C. Hart, September 4, 1909, both RGS 9-1-1, box 2, GDAH. For the suits, see C. T. Ladson to Cornick, Wright & Frantz, February 1, 1909, DBM; Jesse A. Drake to John C. Hart, October 27, 1909, RGS 9-1-1, box 2, GDAH; John C. Hart to W. H. Freeland, November 2, 1909; John C. Hart to Cornick, Wright & Frantz, November 2, 1909, both RGS 9-1-1, box 2, GDAH.
62 John C. Hart to Editor of the *Farmers Union News*, June 15, 1910; R. F. Duckworth to J. C. Hart, June 17, 1910; "Copper and Acid Deal—Tennessee Copper Company Makes a Contract with a Fertilizer Company," *New York Times*, November 24, 1908, 11; "Morgan to Finance Merger—Banking House Expected to Underwrite the New Fertilizer Company," *New York Times*, April 2, 1909, 11.
63 For IAC, see "Takes Entire Acid Output—Agricultural Corporation to Get 150,000 Tons a Year from Copper Company," *New York Times*, January 21, 1911, 20. Blakey, *The Florida Phosphate Industry*, 56.
64 R. F. Duckworth to J. C. Hart, June 17, 1910, RGS 9-1-1, box 3, GDAH.
65 John C. Hart to Jos. M. Brown, Governor, June 29, 1910, RGS 1-1-5, box 162, GDAH.
66 Timothy LeCain notes that the conversion of sulfur dioxide into sulfuric acid and ultimately fertilizer was not environmentally cost free, because overuse of phosphate fertil-

izer led to algae blooms and eutrophication of water bodies; see LeCain, "When Everybody Wins Does the Environment Lose?" This claim is true, but it must be measured against problems arising from millions of acres of abandoned, nutrient-poor southern cotton lands. The Soil Conservation Service later devoted much of its resources to the problem.

Chapter 8

1. "Fannin Stirred by the Victory of Republicans—Only the Senator Saved from the Political Revolution of October—Mass of Indictments for Buying of Votes," *Atlanta Constitution*, November 21, 1910, 1; "Use Apples to Buy Votes," *New York Times*, November 22, 1910, 1.
2. Hart, *Eighth Annual Report*, 29.
3. Walker, *Reports and Opinions of the Attorney-General of Georgia from June 15, 1917, to December 31, 1918*, 3; Matt. 26:11.
4. "To Ask Tennessee to Take Action," *Atlanta Constitution*, July 14, 1910, 3 (NFU initiative); "Committee Kills Ducktown Probe," *Atlanta Constitution*, July 20, 1910, 5; "No Fumes Now at Ducktown," *Atlanta Constitution*, July 21, 1910, 2 (report on the junket).
5. Report of William M. Bowron to Governor Joseph M. Brown, November 1910, RGS 9-1-1, box 3, GDAH.
6. Ibid., 4–5, 10–14.
7. Ibid., 7–8. TCC built another acid plant in 1916 when World War I increased demand for acid in the munitions industry; see Barclay, *Copper Basin*, 42. Montana companies were much less likely to adopt sulfur extraction out of concern that their distance from agricultural markets would increase transportation costs to the point that their product would be noncompetitive; see LeCain, "When Everybody Wins Does the Environment Lose?"
8. R. E. Stallings to Jos. M. Brown, November 9, 1910; R. E. Stallings to Hewlette Hall, December 10, 1910, both in H. A. Hall, *The Annual Report of H. A. Hall*, 10–13.
9. Hewlett Hall to Howard Cornick, December 12, 1910, RGS 9-1-1, box 3, GDAH; "Terms and Stipulations of Consent Agreement," filed February 27, 1911, *Georgia v. Tennessee Copper Co.*, U.S. Supreme Court, No. 1 Original, October Term, 1910.
10. Chittenden, "Report on a Preliminary Examination of the Effects of the Ducktown Sulphur Fumes on the Forests of Polk County, Tennessee," attached as Ex. F. to Original Bill of Complaint, 29–35, *Georgia v. Tennessee Copper Co.*, U.S. Supreme Court, No. 13 Original, October Term, 1905.
11. "$1,000,000 Spent by Tennessee Copper Company to Keep Sulphur Fumes From Injuring Vegetation," *Atlanta Constitution*, July 24, 1910, B7; "Conservation of Injurious Gases of Ducktown Copper Plants Results in Enormous Benefit to Farmers of the South," *Atlanta Constitution*, August 2, 1911, 5.
12. Report of William M. Bowron to Governor Joseph M. Brown, 30.
13. Ligon Johnson to T. S. Felder, June 21, 1912, GDAH, RGS 9-1-1, box 4. See the epilogue herein for a discussion of Johnson's role as counsel for the federal government in many cases against copper smelting firms in the West.
14. J. R. Kincaid and C. T. Owens (both state senators) to Joseph M. Brown, April 4, 1912; Will H. Shippen to Joseph M. Brown, April 9, 1912; Joseph M. Brown to T. S. Felder, April 13, 1912 (requesting meeting regarding Kincaid, Owens, and Shippen), all RGS

9-1-1, box 4, GDAH. Thomas S. Felder to DSC&I, October 11, 1912; C. W. Renwick to Thomas S. Felder, October 15, 1912 (the new DSC&I general manager); Thomas S. Felder to DSC&I, April 12, 1913, all RGS 9-1-1, box 4, GDAH. The 1913 TCC settlement appears in Motion for Leave to File and Agreement and Stipulation and the Cause Be Continued to October Term, 1916, October 20, 1913, *Georgia v. Tennessee Copper Co.*, No. 1 Original, October Term, 1913. For the resolutions, see Resolution of August 6, 1913, no. 2, 1913 *Ga. Laws*, vol. 2, 1295 (TCC agreement); Resolution of August 19, 1913, no. 16, 1913 *Ga. Laws*, vol. 2, 1293 (ultimatum). Victor Lamar Smith to John M. Slaton, September 15, 1913 (DSC&I response via counsel), RGS 1-1-5, box 204, GDAH; John M. Slaton to T. S. Felder, December 23, 1913, RGS 9-1-1, box 4, GDAH.

15 Victor Lamar Smith to John M. Slaton, December 20, 1913, RGS 9-1-1, box 4, GDAH; *Georgia v. Tennessee Copper Co.*, 206 U.S. 230, 237 (1907).

16 For the Weeks Act, see Act of March 1, 1911, Chap. 186, U.S. Stat. vol. 36, Part I, 961–63. The report, U.S. Department of Agriculture, *Message from the President of the United States*. The government's holdings are described in Transcript, deposition of E. B. Clarke, August 1914, 436, *Georgia v. Tennessee Copper Co.*, U.S. Supreme Court, No. 1 Original, October Term, 1914.

17 Glenn, *Denudation and Erosion in the Southern Appalachian Region*, 24–25, 78.

18 Ibid., 25, 77–79.

19 For Bain's findings, see "Brief and Argument on Behalf of the Defendant, Ducktown Sulphur, Copper & Iron Company, Ltd. on Motion for Final Decree," filed April 5, 1915, *Georgia v. Tennessee Copper Co.*, U.S. Supreme Court, No. 1 Original, October Term, 1914, 19–23.

20 Murrill, "A Serious Chestnut Disease"; Ashe, *Chestnut in Tennessee*. Ashe, a forester with the U.S. Forest Service, wrote before the blight became established in the Southern Appalachians. For assessments after the blight occurred, see Lutts, "Like Manna From God" (quotation at 497); Ronderos, "Where Giants Once Stood"; David M. Smith, "American Chestnut"; Youngs, "Right Smart Little Jolt"; Griffin, "Blight Control and Restoration of the American Chestnut."

21 Deposition of George W. Hedgecock, Transcript, vol. 1, 3–54, 44 ("I doubt it"), and continuing upon recall, 442–53; deposition of E. B. Clarke, Transcript, vol. 1, 442–53, both in *Georgia v. Tennessee Copper Co.*, U.S. Supreme Court, No. 1 Original, October Term, 1915. All subsequent citations to depositions in this chapter are to Transcript, vol. 1, filed in this case under this docket number.

22 Grice, *Report and Opinions of the Attorney General*, 16–18.

23 Deposition of B. H. Sebolt, Transcript, vol. 1, 265–72, quotations at 266–67.

24 Ibid.

25 Deposition of Will H. Shippen, Transcript, vol. 1, 514–15, 523, 533, and generally, 512–55. Two clever Georgia smoke suitors revived the jurisdiction by garnishment strategy when they attached TCC's shares in Shippen Brothers Lumber Co.; see "Two Suits Are Filed Against Ducktown Co.—Attachment Proceedings Are Filed on $224,000 of Stock for Alleged Damages," *Atlanta Constitution*, October 7, 1914, 11.

26 Brief and Argument on Behalf of the Defendant, Ducktown Sulphur, Copper & Iron Company, Limited, on Motion for Final Decree (quotation at 75), U.S. Supreme Court, No. 1 Original, October Term, 1914.

27 *Georgia v. Tennessee Copper Co.*, 237 U.S. 474, 478–79 (1915).

28 *Georgia v. Tennessee Copper Co.*, 237 U.S. 474 (1915) (opinion); 237 U.S. 678 (1915) (decree).

29 *Georgia v. Tennessee Copper Co.*, 240 U.S. 650 (1916) (final, post-McGill decree); Report of John T. McGill, January 1, 1916, 58, *Georgia v. Tennessee Copper Co.*, U.S. Supreme Court, No. 1 Original, October Term, 1914.

30 For early arbitration awards, see B. L. Smith to J. M. Slaton, Governor, December 15, 1913; Thomas H. Crawford to John M. Slaton, December 16, 1913; Allison S. Prince to John M. Slaton, December 18, 1913, all RGS 1-1-5, box 207, GDAH.

31 J. F Holden to Nat E. Harris, June 16, 1916, RGS 9-1-1, box 6, GDAH. The story is told in Woodward, *Tom Watson, Agrarian Rebel*, 320–42. For Brown's close political relationship with Watson, see Walter J. Brown, *J. J. Brown and Thomas E. Watson*.

32 See the *Atlanta Constitution*: "Governor to Visit Big Copper Plants—He Will Seek First-Hand Knowledge of the Alleged Damages to Georgia Crops—Shippen Answers Brown," June 9, 1916, 4; "Shippen Charges Branded as False—J. J. Brown Writes Letter to Governor on Copper Company Situation and Gives It Out for Publication," July 10, 1916, 9; "Every Foul Charge False, Says Shippen—Still Another Spicy Document Added to the Ducktown Controversy," June 17, 1916, 10.

33 For coverage in the *Atlanta Constitution*, see "Favor Continuing Smoke Contract—Only One Voter in Fannin Meeting Favors Cancelling Agreement with Copper Company," June 18, 1916, 12; Homer Legg, "Says It Was Copper Co. Employees Who Voted for Brown, Not Fannin Farmers," June 22, 1916, 8; "Governor Charged with Temporizing by Gilmer People—Such Conduct Makes for Mob Violence in Our Fair State, Declares Petition Passed by Farmers' Union," June 11, 1916, E1; "North Ga. Teachers Back Gilmer Farmers," July 9, 1916, 4.

34 *Atlanta Constitution*: "Copper Company Probe Is Opened," July 11, 1916, 12; "House Votes Today on the Usury Bill—Spirited Debate Ended by Adjournment on Tuesday, Renewal of Copper Company Contract Is Favored after Stormy Session," August 2, 1916, 9.

35 Annual Message of Nat E. Harris, June 28, 1916, *Ga. H. Journal*, 77; "1,040 Claims Are Made by Owners of Property against Copper Company," *Atlanta Constitution*, October 13, 1916, 9.

36 "Big War Orders for the South," Copperhill, Tenn., *Copper City Advance*, August 20, 1915, 1; the article is also found in the James Donaldson Clemmer Scrapbooks, vol. 6, 131, TSLA.

37 "Large War Orders Will Be Supplied by Atlanta Firms," *Atlanta Constitution*, August 15, 1915, 1; "Maker and Seller in War Acid Snarl—Tennessee Copper and International Agricultural See Shutdown Differently—Big Profits in Contract," *New York Times*, January 8, 1916, 13.

38 "Maker and Seller in War Acid Snarl"; "Copper Output All Sold," *New York Times*, April 28, 1916, 15; "Ducktown District Takes on Big Boom," *Atlanta Constitution*, May 14, 1916 (Hyatt quotation). TCC also entered into an agreement to manufacture and sell trinitrotoluol (TNT—another product made from sulfuric acid) to the Russian imperial government. The project collapsed along with Russia's military fortunes and the fall of the tsar to the Communists, but nonetheless remained in litigation for many years. See "Russia Sues U.S. Company," *Washington Post*, March 6, 1917, 14.

39 Petition of the State of Tennessee for Leave to Intervene, filed June 5, 1916, *Georgia v. Tennessee Copper Co.*, U.S. Supreme Court, No. 1 Original, October Term, 1915; Walker, *Reports and Opinions of the Attorney-General from June 15, 1917, to December 31, 1918*, 5–7; "Tennessee Enters the Ducktown Case—Intervention Seeks to Nullify Completely

Every Victory Won in Long Legal Battle," *Atlanta Constitution*, June 3, 1916, 1; J. A. Fowler to W. B. Miller, June 1, 1916, DBM; Order on Motion to Modify Former Orders and Decrees, filed April 16, 1917, *Georgia v. Tennessee Copper Co.*

40 War production requirements also impacted other aspects of conservation policy. The need for increased beef production led the Forest Service to permit grazing at levels far above previously determined capacity; see Langston, *Forest Dreams, Forest Nightmares*, 216–28.

41 Advisory Commission of the Council of National Defense, Committee on Chemicals to B. M. Baruch, Chairman, Committee on Raw Materials, August 11, 1917, 1, 7, RG 61, entry 1D1, box 89, and entry 21A-A4, NARA.

42 W. B. Miller to Jesse A. Drake, January 9, 1918; W. B. Miller to Commission on Car Service, January 9, 1918; W. B. Miller to Clifford Walker, February 7, 1918, all RGS 9-1-1, box 8, GDAH.

43 Josephus Daniels to Clifford Walker, February 14, 1918; W. B. Miller to Clifford Walker, May 6, 1918; Clifford Walker to Hugh M. Dorsey, May 7, 1918; Clifford Walker to Hugh M. Dorsey, May 11, 1918, all RGS 9-1-1, box 8, GDAH.

44 Brief of the Solicitor General as Amicus Curiae and in Behalf of the United States, *Georgia v. Tennessee Copper Co.*, No. 1 Original, October Term, 1917; "Ducktown Contract Charges Are Asked—Government Demands Abrogation of Clause Limiting Capacity of Plant," *Atlanta Constitution*, May 17, 1918, 5; "Copper Company Injunction Changes Are Fought," *Atlanta Constitution*, May 18, 1918, 9.

45 Will Shippen to Governor Hugh M. Dorsey and Attorney General Clifford Walker, "Memorandum for Shippen Brothers Lumber Company in the Matter of Petition of Ducktown Sulfur, Copper & Iron Company to Enter into a Contract with the State of Georgia" June 7, 1918, RGS 1-1-5, box 253, GDAH; Clifford Walker to Hugh M. Dorsey, June 28, 1918, RGS 9-1-1, box 7, GDAH.

46 Will H. Shippen, "Ducktown Companies Want an Edict of the U.S. Supreme Court Set Aside," *Atlanta Constitution*, June 18, 1918; W. F. Lamoreaux, "Copper Companies Declare Move to Increase Output Requested by Government," *Atlanta Constitution*, June 19, 1918. Shippen wrote another column to similar effect, Will H. Shippen, "Says Time to Call a Halt in Copper Company Litigation," *Atlanta Constitution*, July 16, 1918, 6.

47 Linda Pack to Hugh M. Dorsey, August 1918, RGS 1-1-5, box 254, GDAH.

48 Agreement between the State of Georgia and the Ducktown Sulphur, Copper & Iron Co. Ltd., signed August 20, 1918, by Hugh M. Dorsey, Governor for the state and by W. F. Lamoreaux, General Manager and William Butt, Attorney for the company, *Georgia v. Tennessee Copper Co.*, U.S. Supreme Court, No. 1 Original, October Term, 1918.

49 Ibid.

50 Walker, *Reports and Opinions of the Attorney-General of Georgia from June 15, 1917, to December 31, 1918*, 3.

51 For the arbitration claims and reports of the arbitrators, see RGS 9-1-1, box 8, and RGS 1-1-6, box 1, both GDAH.

52 See RGS 9-1-1, box 8, and RGS 1-1-6, box 1, both GDAH.

53 Eugene Talmadge to Birdie Suit, October 17, 1935, RGS 1-1-6, box 2, 1935, GDAH.

54 The terms are found in "Tennessee Copper Company Contract," Resolution No. 4, approved December 15, 1937, 1937 *Ga. Laws* 1426–34.

55 Scott and Guynn, "The Disappearance from Georgia of the Farm Union."
56 James G. Parks to Howard Cornick, July 11, 1908, DBM; Ward, *The Annals of Upper Georgia Centered in Gilmer County*, 431, 474.
57 "Mr. W. H. Shippen Will Maintain His Long Residence in Gilmer," *Ellijay Times-Courier*, January 6, 1933, 1.
58 Ibid.
59 The decline of Appalachian farming is considered in Eller, *Miners, Millhands, and Mountaineers*, 225–37; Salstrom, *Appalachia's Path to Dependency*, 60–82.
60 From casual conversations with lifetime residents during my visits to the Ducktown Basin Museum; see Quinn, "Tennessee Copper Basin."
61 The Supreme Court would later acknowledge its lack of scientific and technological competence in pollution cases. In 1971 Justice John Marshall Harlan expressed the "sense of futility that has accompanied this Court's attempts to treat with the complex technical and political matters that inhere in all disputes of this kind." *Ohio v. Wyandotte Chemicals Corp.*, 401 U.S. 493, 502 (1971).

Chapter 9

1 For the terms of the final settlement, see "Tennessee Copper Company Contract," Resolution No. 4, approved December 15, 1937, 1937 *Ga. Laws* 1426–34.
2 Estimates vary concerning the size of the desert, especially because the dimensions changed over time. The figures and descriptions presented here are from TVA field measurements and maps from the early 1940s and are repeated with minor variations in numerous TVA reports thereafter. (There are numerous TVA reports with similar titles, so to avoid confusion, they are referenced in short form by providing the named author, when present, or the entity, when not, then a short title, and then the year.) See Richard A. Wood, "Erosion Control and Reforestation"; Tennessee Valley Authority, Department of Forestry Relations, "A Proposal for Erosion Control" (1945), 1–2, 22; Muncy, "A Plan for Revegetation Completion" (1986), 1–2. The land area of Manhattan is 22.96 square miles; see United States Census, "New York—Place and County Subdivision GCT-PH1-R. Population, Housing Units, Area, and Density (geographies ranked by total population): 2000, Data Set: Census 2000 Summary File 1 (SF 1) 100-Percent Data," ⟨http://factfinder.census.gov/servlet/GCTTable?_bm=y&-geo_id=04000US36&-_box_head_nbr=GCT-PH1-R&-ds_name=DEC_2000_SF1_U&-_lang=en&-redoLog=false&-mt_name=PEP_2006_EST_GCTT1_ST2&-format=ST-7S&-_sse=on⟩ (accessed November 16, 2009).
3 Richard A. Wood, "Erosion Control and Reforestation," 5, pagination varies. The quotation is from a portion of the Wood report, "Recommendations for Future Action," 5; Seigworth, "Ducktown—a Postwar Challenge."
4 Tennessee Valley Authority, Department of Forestry Relations, "A Proposal for Erosion Control" (1942), 2–3; Richard A. Wood, "Erosion Control and Reforestation," 1, and the maps appended as figs. 1–4. For the size of the basin, see Emmons and Laney, *Geology and Ore Deposits of Ducktown Mining District*, 1–2, 13–14.
5 Smallshaw, "Denudation and Erosion in the Copper Basin," 5.
6 *Surface Mining Control and Reclamation Act of 1977*, 30 U.S.C. §§1234–1328; *Comprehensive Environmental Response, Compensation, and Liability Act of 1980*, 42 U.S.C. §9601 et. seq.
7 Tennessee Power Company and E. W. Clark & Co., *The Power of Water*, 1–10.

8 Lillard, "The Story of Ocoee No. 1 and Ocoee No. 2 Hydro Plants," 1–10.
9 Ibid., 11–13; Tennessee Power Company and E. W. Clark & Co., *The Power of Water*, 10; Switzer, "The Ocoee River Power Development," 42–47. For the link to the aluminum industry, see Pray, "The Ocoee No. 2 Power Development," 1.
10 Tennessee Valley Authority, Department of Forestry Relations, "A Proposal for Erosion Control" (1945), 3–4; Pray, "The Ocoee No. 2 Power Development," 2; Tennessee Valley Authority, Water Quality Department, "Status of the Ocoee Reservoirs" (1991); Nolt, *A Land Imperiled*, 86–87.
11 Case, *The Valley of East Tennessee*, 111; Dubé, *History of Tennessee Copper Company*, 34.
12 *Tennessee Valley Authority Act*, Public Law 73-17, 48 *Stat*. 58-72 (1933). The TVA Act, the Soil Conservation and Domestic Allotment Act (creating the Soil Conservation Service), and the Act for the Relief of Unemployment through Performance of Public Works and Other Purposes (creating the Civilian Conservation Corps) were the three central pillars of Roosevelt's New Deal conservation policy. See, e.g., Maher, *Nature's New Deal*.
13 *Tennessee Valley Authority Act*, preamble and section 23; Chase, *Rich Land, Poor Land*, 49–50.
14 Rothacher, "Soil Erosion in the Copper Basin"; Muncy, "A Plan for Revegetation Completion" (1986), 8; Tennessee Valley Authority, *The Hiwassee Valley Projects*, vol. 2: *The Appalachia, Ocoee No. 3, Nottely, and Chatuge Projects* (1948).
15 Lilienthal, *TVA: Democracy on the March*, 53–56. For a more recent approach to the South's colonial economy, see Wright, *Old South, New South*, 156–97.
16 Lilienthal, *TVA: Democracy on the March*, 56. The TVA eventually received severe criticism from environmentalists for its own activities. Its Tellico dam project threatened a tiny fish, the snail darter, provoking a spectacular national controversy over the Endangered Species Act. Foes of nuclear power and advocates of wild and scenic rivers also added their criticisms. See, e.g., Wheeler and McDonald, *TVA and the Tellico Dam*; Hargrove, *Prisoners of Myth: The Leadership of the Tennessee Valley Authority*, 155–241.
17 Brief and Argument of the Ducktown Sulphur, Copper & Iron Company, Limited on the Final Hearing, 21, *Georgia v. Tennessee Copper Co.*, U.S. Supreme Court, No. 5 Original (1905); Barclay, "The Great Copper Basin."
18 "The Ducktown Basin, Its Ore Deposits and Their Development"; Barclay, *Copper Basin*, 9–11, 42–45; Mouzon Peters, "Copper Basin Counts Its Blessings on 50th Anniversary," *Chattanooga Times*, January 25, 1958 (flotation mill).
19 "A Pictorial Outline of the Ducktown Basin in Which Is Located the Largest of the Major Industries of East Tennessee," *Knoxville Journal*, August 27, 1939, sec. 7; Dubé, *History of Tennessee Copper Company*, 59.
20 "Ducktown Basin's Desert Soon to Bloom Again," *Chattanooga Times*, July 3, 1932, magazine, 7; Claudia Beckler, quoted in Foehner, "The Historical Geography," 40; Mouzon Peters, "Healing the Scars in the Copper Basin," *Chattanooga Sunday Times*, September 2, 1956.
21 "Ducktown Basin's Desert Soon to Bloom Again"; Richard A. Wood, "Erosion Control and Reforestation," 2–3.
22 Tennessee Valley Authority, Department of Forestry Relations, "A Proposal for Erosion Control" (1945), 5.
23 Ibid., 2 (TCC motives); *Tennessee Electric Power Co. v. Tennessee Valley Authority*, 306 U.S. 118 (1939).

24 Fred Hixson, "Conservation Plan in Copper Basin Starts Tree Growth on Barren Land," *Chattanooga Times*, April 17, 1950 (Barclay quotation); Cities Service Company, "Engineering Report on Land Reclamation for the Copper Basin," 7. For the water table problems, see James W. Taylor, "Ducktown Desert," 97–107. On flood problems, see Tennessee Valley Authority, Division of Water Control Planning, "Floods on Toccoa-Ocoee River" (1958).

25 Marsh, *Man and Nature*.

26 Worster, *Dust Bowl*, 38–41.

27 F. E. Charles, Chief, Regional Division of Information, Department of Agriculture, to R. E. Barclay, December 5, 1940 (submitting the script to TCC); R. E. Barclay to F. E. Charles, December 21, 1940 (returning the script with edits). For the script, see U.S. Department of Agriculture, Soil Conservation Service, "The Badlands of Tennessee," program no. 141, *Fortunes Washed Away* (scheduled for broadcast on January 4, 1941), all in Barclay Papers, box 8, folder 24, TSLA.

28 U.S. Department of Agriculture, Soil Conservation Service, "The Badlands of Tennessee," 3–4.

29 Ibid., 5–7.

30 Ibid., 7–9.

31 Ibid., 9–12.

32 Richard A. Wood, "Erosion Control and Reforestation," 3.

33 The four maps are appended to ibid.

34 Ibid., 3. For representative overviews of the CCC, see, e.g., Salmond, *The Civilian Conservation Corps*; Minton, *The New Deal in Tennessee*, 57–62; "Tumbling Creek CCC Nestles in Foothills of Big Frog," *Cleveland Daily Banner*, May 1, 1936.

35 Richard A. Wood, "Erosion Control and Reforestation," 1–7.

36 Ibid., 1–7; Allen, "Pine Planting Tests in the Copper Basin."

37 Tennessee Valley Authority, Department of Forestry Relations, "A Proposal for Erosion Control" (1945), 4.

38 Ibid.; Muncy, "A Plan for Revegetation Completion" (1986), 5–8; Rothacher, "Soil Erosion in the Copper Basin," 41.

39 *Georgia v. Tennessee Copper Co.*, 240 U.S. 650 (1916) (permissible releases); Randall Higgins, "Copperhill Health Hazards Cited," *Chattanooga Times*, July 10, 1979, A5 (car anecdote).

40 Tennessee Valley Authority, Department of Forestry Relations, "A Proposal for Erosion Control" (1945), 8–9.

41 Muncy, "A Plan for Revegetation Completion" (1986), 5–8. A representative selection of technical papers includes Allen, "Pine Planting Tests in the Copper Basin"; Cummings, "Weeping Lovegrass"; Curlin, "Planted Pitch Pine Responds to Fertilization"; Berry, "Slit Application of Fertilizer Tablets and Sewage Sludge"; Tyre and Barton, "A Fresh Start at Old Problem."

42 Cities Services Company, "Engineering Report on Land Reclamation," 6–9.

43 Ibid., 9–18; Muncy, "A Plan for Revegetation Completion" (1986), 13–17; Tyre and Barton, "Fresh Start at Old Problem," 7–8.

44 Murray Wyche, "The Great Copper Desert Will Bloom Once More," *Chattanooga Times*, August 25, 1940; "TVA to Reclaim Copper Basin," *Paris (Tenn.) Post Intelligencer*, May 3, 1941; Richard A. Wallace, "Copper Basin Miracle," *Nashville Tennessean*, October 8, 1950; Worth Wilkerson, "Man and Nature Are Slowly Reclaiming the Tennessee Sahara at

Copperhill," *Chattanooga Times*, December 20, 1962; Ed Paxton, "Copperhill Is Again an Area of Green Land," *Paducah Sun-Democrat*, August 4, 1964; Jimmy Townsend, "Ducktown's Desert Beauty," *Atlanta Constitution*, December 27, 1978, 5; Dan George, "Barren Hills Are Beautiful to Copper Basin Natives," *Cleveland Daily Banner*, February 19, 1984; the same article appeared without naming George, but with the Beckler quotation in "Historic Copper Basin's Natives Find Beauty in Bare Red Hills," *Chattanooga News-Free Press*, February 19, 1984; "Basin People See Beauty in Barren Hills," *Maryville Alcoa Times*, February 20, 1984. For another laudatory article, see Aubrey Wilson, "The Greening of the Copper Basin," *Tennessee Conservationist* 60, no. 3 (May–June 1994): 21–26; Randall Higgins, "Copper Basin Residents Are Hoping to Bring Burra Burra to Life Again," *Chattanooga Times*, March 6, 1981. Another plea for preservation appears in M.-L. Quinn, "Tennessee's Copper Basin: A Case for Preserving an Abused Landscape," *Journal of Soil and Water Conservation* 43, no. 2 (March–April 1988): 140–44. For the TVA's response, see Muncy, "A Plan for Cooperatively Completing Revegetation" (1991), 19.

45 The postwar capital improvements are described in Mouzon Peters, "Tennessee Copper Company Shifts Easily from War to Peace Status," *Chattanooga Times*, April 4, 1946; Charles F. Crane, "Copper Basin Finds New Life, Prosperity: Tennessee Copper Firm in Midst of Gigantic Expansion Program," *Chattanooga News Free Press*, July 12, 1949; "New $5 Million Sulfuric Acid Plant Changes Skyline at Copperhill," *Chattanooga Times*, July 13, 1964; "$70 Million Expansion Program Revealed by Tennessee Copper," *Polk County News*, January 8, 1970; "$70 Million Cities Service Complex Highlight of Copperhill Development," *Chattanooga Free Press*, January 26, 1971. For a chronology of industrial improvements, see Dubé, *History of Tennessee Copper Company*, 59–89. For press coverage on industry decline, see Richard McCoy, "Cities Service Says It Is Losing Money on Copperhill Operations," *Polk County News*, February 3, 1977; Bobbie Rucker, "Copperhill Operations Planning to Lay Off 400 by Year's End," *Chattanooga Times*, March 14, 1979; Randall Higgins, "Copperhill Plant Cuts Production, Idles 170," *Chattanooga Times*, November 11, 1981. For the mine shaft, see Cities Service Company, "Calloway Mine Ventilation Proposal," circa 1975, DBM.

46 John Harmon, "Copper Slump Hurts 2 States," *Atlanta Constitution*, May 6, 1984.

47 Ibid.; "Polk to Halt School Buses Because of Money Crunch," *Chattanooga Times*, March 9, 1984.

48 Harmon, "Copper Slump Hurts 2 States" (Welch quotation); Bill Lohmann, "Copper Basin Drained: Company Pours in Aid to Keep Miners Afloat," *Memphis Commercial Appeal*, April 5, 1987 (Collis quotation). For the end of mining, see Karen Daniels, *Tennessee's Historic Copper Basin Area: An Overview*; Polk County Publishing Co., *Polk County Scrapbook: A Tribute to the Miners*; Polk County Publishing Co., *Polk County Scrapbook: 150 Years of Memories*; Polk County Publishing Co., *Polk County News: Reprints of Articles Relating to the History of Polk County*.

49 Harmon, "Copper Slump Hurts 2 States."

50 Muncy, "A Plan for Revegetation Completion" (1986), 19–30; Muncy, "A Plan for Cooperatively Completing Revegetation" (1991), 1–8.

51 Muncy, "A Plan for Cooperatively Completing Revegetation" (1991), 1–18.

52 For overviews regarding the creation of federal environmental law, see, e.g., Andrews, *Managing the Environment, Managing Ourselves*; Brooks, Jones, and Virginia, *Law and Ecology*; Flippen, *Nixon and the Environment*; Lazarus, *The Making of Environmental Law*.

53 Leonard Ray Teel and Mike Kautsch, "Sulfur Poisoning Town, McCaysville Residents

Complain," *Atlanta Constitution*, June 17, 1973, A1, 19; "Copperhill Plant Offers to Pay Gas Damages," *Atlanta Constitution*, June 19, 1973, A22; Sam Hopkins and Jeff Nesmith, "Copper Plant Freeing Excessive Sulfur Gas," *Atlanta Constitution*, June 21, 1973, 13; "Alarm to Sound for Pollution," *Chattanooga Times*, June 25, 1973, 2. For the enforcement action, see, e.g., John A. Allen Jr. to Rafael A. Ballagas (both in the Georgia Environmental Protection Division), June 6, 1974; Robert A. Collom Jr., Chief, Air Protection Branch, Georgia Environmental Protection Division to Paul Traina, Director, Enforcement Division, EPA, Atlanta office, October 3, 1975; Robert A. Collom Jr. to J. Leonard Ledbetter, Director Environmental Protection Division, May 17, 1976, all RGS 1-1-4.21, GDAH.

54 Tennessee Department of Public Health, Division of Water Quality Control, "Water Quality Management Plan for the Lower Tennessee River Basin" (1975), 168–69; Tennessee Department of Public Health, Division of Water Quality Control, "Water Quality Management Plan for the Lower Tennessee River Basin" (1978), V34–35.

55 Dick Kopper, "Cities Service at Copperhill Operations Sued by U.S. on Water Pollution Charges," *Chattanooga Times*, November 15, 1978, D12. Scouring is described in Cox, "Water Resources Review: Ocoee Reservoirs," 1, 25. The TVA continued to sluice Ocoee No. 3 up to January 2009. The release led to enforcement action by the Tennessee Department of Environment and Conservation, Division of Water Pollution Control. See Scott Barker, "TVA probed in Ocoee Release," *Knoxville News Sentinel*, January 14, 2009, ⟨www.knoxnews.com/2009/jan/14/tva-probed-in-ocoee-release/⟩ (accessed December 7, 2009).

56 Cox, "Water Resources Review: Ocoee Reservoirs," 1–3.

57 Ibid., 40–41; Libby Wann, "Revitalized Ocoee River Now Holds Raft of Wet Challenges," *Chattanooga Times*, August 3, 1977; Dick Kopper, "TVA Asks $5 Million to Leave Ocoee Open," *Chattanooga Times*, September 24, 1980; Eric Kaufmann, "Rafters File Again to Block TVA Plans for Ocoee Development," *Chattanooga Times*, October 9, 1981; "State, TVA Sign 35-Year Agreement Over White-Water Rafting on Ocoee," *Chattanooga Times*, March 20, 1984.

58 John Harmon, "Whitewater Mecca Is an Industrial Sewer," *Atlanta Constitution*, November 6, 1985.

59 U.S. Forest Service, Southern Region, *Ocoee Whitewater Center*.

60 Cox, "Water Resources Review: Ocoee Reservoirs," 33–38, 48–49; U.S. Forest Service, *Final Environmental Impact Statement, 1996 Olympic Whitewater Slalom Venue*, I-1-2, III-53–70 (quotations at III-61).

61 Muncy, "A Plan for Revegetation Completion," (1986), 17; Muncy, "A Plan for Cooperatively Completing Revegetation" (1991), 4.

62 Dubé, *History of Tennessee Copper Company*, 102–6.

63 42 *U.S.C.* §9607 (a). For the press release, see U.S. Department of Justice, "Oxy to Pay $129 Million in Love Canal Settlement," ⟨http://www.justice.gov/opa/pr/Pre_96/December95/638.txt.html⟩ (accessed December 13, 2009).

64 For a recent overview of Love Canal, see Blum, *Love Canal Revisited*.

65 The descriptions are drawn from U.S. Environmental Protection Agency, "Copper Basin Mining District"; Faulkner et al., "The Copper Basin Reclamation Project," *Basin Briefings, Copper Basin Mining District*, a newsletter published jointly on an irregular schedule by the Tennessee Department of Environment and Conservation, the U.S.

Environmental Protection Agency, and Glenn Springs Holdings, Inc. (various issues beginning April 2001).
66 Faulkner et al., "The Copper Basin Reclamation Project."
67 Ibid.
68 Memorandum of Understanding between Tennessee Department of Environment and Conservation, U.S. EPA and Oxy USA, Inc., January 11, 2001, In the Matter of Copper Basin Mining District, U.S. EPA Region 4, CERCLA Docket No. 01-10-C; U.S. Environmental Protection Agency, "Copper Basin Mining District," 7 (EPA opinion of Glenn Springs).
69 Barge, Waggoner, Sumner and Cannon, Inc., *A Future-Use Plan for Redevelopment of the Abandoned Mine Lands in the Copper Basin*, 2–13.
70 Ibid., 23–28.

Epilogue

1 Taliaferro, "Ducktown, by 'Skitt,' Who Has Been 'Thar,'" 31; Report of John T. McGill, January 1, 1916, 13, *Georgia v. Tennessee Copper Co.*, No. 1 Original, October Term, 1914; Southern Appalachian Man and the Biosphere Cooperative, *The Southern Appalachian Assessment*.
2 "Ligon Johnson Is Appointed—Atlanta Attorney Is Named Special United States Attorney," *Atlanta Constitution*, August 1, 1907, 4. A summary of Johnson's federal litigation appears in Ligon Johnson to W. J. Hughes, August 21, 1914, Dept. of Justice Central Files, RG 60, box 540, NARA; MacMillan, *Smoke Wars*, 145–56.
3 Ligon Johnson to W. J. Hughes, August 21, 1914. For scientific evidence in other smoke suits, see MacMillan, *Smoke Wars*, 101–24 (Deer Valley Farmers Assn); Ligon Johnson to Attorney General (George W. Wickersham), November 8, 1911, RG 60, NARA, box 540 (Shasta County Farmers Protective Assn.); Charry, "Defending the 'Great Barbecue.'"
4 Theodore Roosevelt to Charles Joseph Bonaparte, December 9, 1908, Dept. of Justice Central Files, RG 60, box 539, NARA.
5 Ligon Johnson to the Attorney General (Charles Joseph Bonaparte), January 25, 1909; Theodore Roosevelt to Charles Joseph Bonaparte, February 4, 1909, both from Department of Justice Central Files, RG 60, box 540, NARA.
6 Ligon Johnson to Attorney General (George W. Wickersham), November 8, 1911; W. J. Hughes to Attorney General (James C. McReynolds), August 25, 1914; Ligon Johnson to W. J. Hughes, August 21, 1914, all from Department of Justice Central Files, RG 60, box 540, NARA. Mammoth Copper installed a successful bag house. Balaklala shut down after control measures failed. The Bully Hill and Engels smelters closed when owners declined to incur the cost of corrective measures. See *Mountain Copper v. United States*, 142 F. 625 (9th Cir., 1906) (earlier ruling against the government). For the Anaconda settlement, see Memorandum of Caldwell to Attorney General (George W. Wickersham), May 1, 1911, Department of Justice Central Files, RG 60, box 540, NARA; MacMillan, *Smoke Wars*, 211–19, 228–56.
7 Andrews, *Managing the Environment, Managing Ourselves*, 373–93 (chronology).
8 Johnson summarized his career in the smelter cases in Ligon Johnson, "The History and Legal Phases of the Smoke Problem." For his career in entertainment law, see *National Cyclopaedia of American Biography*, s.v. Johnson, [Robert] Ligon.

9   For overviews of the federal common law of interstate nuisance, see Bryson and MacBeth, "Public Nuisance, the Restatement (Second) of Torts, and Environmental Law"; McCarthy, "The Federal Common Law of Nuisance"; Murchison, "Interstate Pollution: The Need for Federal Common Law"; Percival, "The Frictions of Federalism"; Walhbeck, "The Development of a Legal Rule."
10 *New York v. New Jersey*, 256 U.S. 296, 300–301 (1921); see Percival, "Frictions of Federalism," 39–44.
11 *New Jersey v. New York City*, 284 U.S. 237 (1933); Percival, "Frictions of Federalism," 44–50.
12 Wirth, *Smelter Smoke in North America*, 43, 119, 203 (influence of Tennessee Copper case upon Trail Smelter dispute); Dinwoodie, "The Politics of International Pollution Control"; Percival, "Frictions of Federalism," 50–52.
13 Percival, "Frictions of Federalism," 40–42 (delays).
14 Scholars debate the chronology of antipollution environmentalism, but judged in terms of national legislative policy, the movement gained traction after World War II and especially in the 1960s and 1970s. See, e.g., Andrews, *Managing the Environment, Managing Ourselves*; Brooks, Jones, and Virginia, *Law and Ecology*; Flippen, *Nixon and the Environment*; Gottlieb, *Forcing the Spring*; Samuel P. Hays and Barbara D. Hays, *Beauty, Health, and Permanence*. The cases are surveyed in Wahlbeck, "The Development of a New Rule"; and Percival, "Frictions of Federalism," 52–76.
15 *Ohio v. Wyandotte Chemicals Corp.*, 401 U.S. 493, 497–98 (1971).
16 *City of Milwaukee v. Illinois*, 451 U.S. 304 (1981). See discussion at Percival, "Frictions of Federalism," 52–69.
17 *Massachusetts v. Environmental Protection Agency*, 549 U.S. 497; Zdeb, "From *Georgia v. Tennessee Copper* to *Massachusetts v. EPA*." Zdeb explores the significance of the Georgia case to the standing issue in the Massachusetts decision.
18 *Massachusetts v. EPA*, 516–23.
19 Ibid., 518–21 (discussion and application of *Georgia v. Tennessee Copper Co.*). The vote was 5–4, with a dissent by Chief Justice John Roberts, joined by Justices Scalia, Thomas, and Alito. Roberts dismissed the Georgia case, writing, "The Court has to go back a full century in an attempt to justify its novel standing rule.... The Court's analysis hinges on *Georgia v. Tennessee Copper Co* ... a case that did indeed draw a distinction between a State and private litigants, but solely with respect to available remedies. The case has nothing to do with Article III standing." Ibid., dissent, 3. The participants in the Georgia case would have disagreed with Chief Justice Roberts because Georgia's claim of article 3 original jurisdiction was a key issue. See chapters 4 and 6 herein.
20 Zdeb, "From *Georgia v. Tennessee Copper* to *Massachusetts v. EPA*," 1082.

Bibliography

Manuscript Sources

Boulder, Colorado
    University of Colorado at Boulder Libraries, Archives
        National Farmers Union Collection
Cleveland, Tennessee
    Cleveland Public Library, History Branch
        Journal of Rosine Parmentier (1852), MSS 59
Ducktown, Tennessee
    Ducktown Basin Museum Archives
Morrow, Georgia
    Georgia Department of Archives and History
        Attorney General Annual Reports, RGS 9-1-8
        Attorney General Correspondence, Briefs, and Opinions, RGS 9-1-1
        Department of Natural Resources, Environmental Protection Division,
            Air Quality Control Section, RGS 88
        Executive Correspondence (general), RGS 1-1-5
        Governor's Unbound Correspondence, RGS 1-1-6
Nashville, Tennessee
    Tennessee State Library and Archives
        Robert Edward Barclay Papers, 1854–1977, Mf. 829
        James Donaldson Clemmer Scrapbooks on Polk County, Mf. 1525
        Governor James B. Frazier Papers, 1903–1905, GP 33
        Mayfield Papers, 1889–1929, Mf. 1529
        Supreme Court Trial Cases, 1796–1955, RG 170
Washington, D.C.
    National Archives and Records Administration
        Department of Justice Central Files, RG 60
        U.S. Supreme Court Cases—Original Jurisdiction Cases, RG 267.3.3

Published Primary Documents

Agency for Toxic Substances and Disease Registry. "Toxicological Profile for Sulfur Dioxide." Atlanta: U.S. Department of Health and Human Services, Public Health Service, 1998. ⟨http://www.atsdr.cdc.gov/toxprofiles/tp116.pdf⟩ (accessed June 30, 2007).

Allen, John C. "Pine Planting Tests in the Copper Basin." *Journal of the Tennessee Academy of Science* 25, no. 3 (July 1950): 199–216.

American Bureau of Mines and Union Consolidated Mining Company of Tennessee. *Report of the American Bureau of Mines.* New York: American Bureau of Mines, 1866.

American Institute of Mining Engineers. "A Brief Description of the Operations of the Tennessee Copper Company Prepared for the Ducktown Excursion." Paper prepared for a meeting of the American Institute of Mining Engineers, Chattanooga, Tenn., 1908.

Ashe, W. W. *Chestnut in Tennessee*. Tennessee Geological Survey Series, Bulletin 10. Nashville, Tenn.: Baird-Ward, 1911.

Ayres, H. B., W. W. Ashe, and George Otis Smith. *The Southern Appalachian Forests*. Washington, D.C.: Government Printing Office, 1905.

Baes, C. F., and S. B. McLaughlin. "Trace Elements in Tree Rings: Evidence of Recent and Historical Air Pollution." *Science* 224, no. 4648 (1984): 494–97.

Barge, Waggoner, Sumner and Cannon, Inc. *A Future-Use Plan for Redevelopment of the Abandoned Mine Lands in the Copper Basin, Polk County, Tennessee, Prepared for Glenn Springs Holdings, Inc., a Subsidiary of Occidental Petroleum Company*. Nashville, Tenn., July 2003.

Barrett, Charles Simon. *The Mission, History and Times of the Farmers' Union; a Narrative of the Greatest Industrial-Agricultural Organization in History and Its Makers*. Nashville, Tenn.: Marshall & Bruce, 1909.

Bartram, William. *Travels through North & South Carolina, Georgia, East & West Florida, the Cherokee Country, the Extensive Territories of the Muscogulges, or Creek Confederacy, and the Country of the Chactaws; Containing an Account of the Soil and Natural Productions of Those Regions, Together with Observations on the Manners of the Indians*. Philadelphia: James & Johnson, 1791.

Berry, C. R. "Slit Application of Fertilizer Tablets and Sewage Sludge Improve Initial Growth of Loblolly Pine Seedlings in Tennessee Copper Basin." *Reclamation Review* 2 (1979): 33–38.

Cities Service Company. "Engineering Report on Land Reclamation for the Copper Basin." Copperhill, Tenn.: Cities Service Company, Copperhill Operations, 1973.

Cox, Janice P. "Water Resources Review: Ocoee Reservoirs, 1990." Tennessee Valley Authority, River Basin Operations, Water Resources, Water Quality Department. Chattanooga, Tenn., 1990.

Cummings, W. H. "Weeping Lovegrass, *Eragrostis curvula*, Seeding Test Results in the Copper Basin." *Journal of the American Society of Agronomy* 39, no. 6 (June 1947): 522–29.

Curlin, J. W. "Planted Pitch Pine Responds to Fertilization." *Tree Planters' Notes* 55 (November 1962): 3–4.

Currey, Richard O. *A Sketch of the Geology of Tennessee: Embracing a Description of Its Minerals and Ores, Their Variety and Quality, Modes of Assaying and Value; with a Description of Its Soils and Productiveness, and Palaeontology*. Knoxville, Tenn.: Kinsloe & Rice, 1857.

Defense Mapping Agency Topographical Center. "Eastern United States 1:250,000 Chattanooga, Tenn. Series V501p, Ed. 4, Sheet Ni 16-3." Washington, D.C., 1972.

———. "Eastern United States 1:250,000 Rome, Ga. Series V501p, Ed. 4, Sheet Ni 16-6." Washington, D.C., 1972.

"The Ducktown Basin, Its Ore Deposits and Their Development." *Chemical Industries* 35 (May 1935): 81–83.

Emmons, S. F., and C. W. Hayes. *Contributions to Economic Geology*. U.S. Geological Survey, Bulletin no. 225. Washington, D.C.: Government Printing Office, 1904.

Emmons, W. H., and F. B. Laney. *Geology and Ore Deposits of the Ducktown Mining District, Tennessee*. U.S. Geological Survey, Professional Paper 139. Washington, D.C.: Government Printing Office, 1926.

"Farmer's Union Department." *Watson's Jeffersonian Magazine* 2, no. 2 (February 1907): 290–94.

Freeland, W. H., and C. W. Renwick. "Smeltery Smoke as a Source of Sulphuric Acid." *Engineering and Mining Journal* 89 (May 28, 1910): 1116–21.
———. "Smelting of Raw Sulphide Ores at Ducktown, Tenn." *Mineral Industry* 11 (1902): 191.
Galloway, Dora T. *Little Girl in Appalachia*. Newport, Ky.: n.p., 1964.
Gaussoin, Eugene. *The Ducktown Copper Mines of Tennessee, Their Value, Present Management, and Future*. New York: n.p., 1860.
Gilmer, George Rockingham. *Sketches of Some of the First Settlers of Upper Georgia, of the Cherokees, and the Author. Rev. and Corr. by the Author*. New York: D. Appleton, 1855. Reprint, Americus, Ga.: Americus Book Company, 1926.
Glenn, Leonidas Chalmers. *Denudation and Erosion in the Southern Appalachian Region and the Monongahela Basin*. U.S. Geological Survey, Professional Paper 72. Washington, D.C.: Government Printing Office, 1911.
Grice, Warren A. *Report and Opinions of the Attorney General for the Year 1914*. Atlanta: Byrd Printing, 1915.
Hall, H. A. *The Annual Report of H. A. Hall, Attorney-General of Georgia*. Atlanta: Chas. P. Byrd, 1911.
Hart, John C. *Second Annual Report of John C. Hart, Attorney-General of Georgia*. Atlanta: Geo. W. Harrison, 1904.
———. *Fifth Annual Report of John C. Hart, Attorney-General of Georgia*. Atlanta: Franklin-Turner, 1907.
———. *Sixth Annual Report of John C. Hart, Attorney-General of Georgia*. Atlanta: Franklin-Turner, 1908.
———. *Eighth Annual Report of John C. Hart, Attorney-General of Georgia*. Atlanta: Chas. P. Byrd, 1910.
Haywood, J. K. "Injury to Vegetation and Animal Life by Smelter Fumes." *Journal of American Chemical Society* 29, no. 7 (July 1907): 998–1009.
———. *Injury to Vegetation and Animal Life by Smelter Wastes*. U.S. Department of Agriculture, Bureau of Chemistry, Bulletin no. 110. Washington, D.C.: Government Printing Office, 1908.
———. *Injury to Vegetation and Animal Life by Smelter Wastes*. U.S. Department of Agriculture, Bureau of Chemistry, Bulletin no. 113. Washington, D.C.: Government Printing Office, 1910.
———. *Injury to Vegetation by Smelter Fumes*. U.S. Department of Agriculture, Bureau of Chemistry, Bulletin no. 89. Washington, D.C.: Government Printing Office, 1905.
———. "Smelter Smoke." *Science* 26, no. 667 (October 1907): 476–78.
Henrich, Carl. "The Ducktown Ore Deposits and the Treatment of the Ducktown Copper." *Transactions of the American Institute of Mining Engineers* 25 (1896): 173–245.
Holmes, Oliver Wendell. *The Common Law*. Boston: Little Brown, 1881.
Hursh, C. R. *Local Climate in the Copper Basin of Tennessee as Modified by the Removal of Vegetation*. U.S. Department of Agriculture, Circular 774. Washington, D.C.: Government Printing Office, 1948.
Johnson, Ligon. "The History and Legal Phases of the Smoke Problem." *Transactions of the American Institute of Mining Engineers* 58 (1918): 198–214.
LaForge, Laurence. *Physical Geography of Georgia*. Atlanta: Stein Printing Co., 1925.
Lilienthal, David Eli. *TVA: Democracy on the March*. New York: Harper & Brothers, 1944.
Marsh, George Perkins. *Man and Nature*. New York: Scribner, 1864. Reprint, Cambridge, Mass.: Belknap Press of Harvard University Press, 1965.

Maury, Matthew Fontaine, and Richard Owen Currey. *The Polk County Copper Company of Tennessee: Its Mineral Resources and Mining Prospects.* New Orleans: Bulletin Book and Job Office, 1859.

McCallie, S. W. "The Ducktown Copper Mining District." *Engineering and Mining Journal* 74 (October 4, 1902): 439–40.

McCandless, John M. "Report of Dr. Jno. M. McCandless to Hon. Jno. C. Hart, Attorney General of Georgia," May 1, 1908. In John C. Hart, *The Sixth Annual Report of John C. Hart, Attorney-General of Georgia*, 10–18. Atlanta: Franklin-Turner, 1908.

McKinney, Louise. "Bad Lands of Copperhill." Draft for the Work Progress Administration, Federal Writers' Project, Guidebook series, March 19, 1936. Hargrett Library, University of Georgia, Clipping Files, Fannin County.

Meierding, Thomas C. "Marble Tombstone Weathering and Air Pollution in North America." *Annals of the Association of American Geographers* 83, no. 4 (1993): 568–88.

"The Mining Interest at the South." *Russell's Magazine* 3 (1858): 442–47.

Muncy, Jack A. "A Plan for Cooperatively Completing Revegetation of Tennessee's Copper Basin by the Year 2000." Tennessee Valley Authority, Land Resources, Reclamation. Norris, Tenn., 1991.

———. "A Plan for Revegetation Completion of Tennessee's Copper Basin." Tennessee Valley Authority, Division of Land and Economic Resources. Norris, Tenn., 1986.

Murrill, William Alphonso. "A Serious Chestnut Disease." *Journal of the New York Botanical Garden* 7, no. 78 (June 1906): 143–53.

Olmsted, Frederick Law. *A Journey in the Back Country.* New York: Madison, 1860. Reprint, Williamstown, Mass.: Corner House, 1972.

Parks, James G. *A Manual of the Law of Pleading: Containing a Succinct Compilation of the Statutes and Decisions in Tennessee on That Subject.* Knoxville, Tenn.: Odgen Brothers, 1894.

Pray, Francis E. "The Ocoee No. 2 Power Development, A Concise History." Tennessee Valley Authority. Knoxville, Tenn., 1973.

Safford, James M. *A Geological Reconnoissance of the State of Tennessee: Being the Author's First Biennial Report. Presented to the Thirty-first General Assembly of Tennessee, December, 1855.* Nashville: G. C. Torbett, 1856.

Safford, James M., and Tennessee Geological Survey. *Geology of Tennessee.* Nashville, Tenn.: S. C. Mercer, 1869.

Schenck, Carl Alwin. *The Biltmore Story; Recollections of the Beginning of Forestry in the United States.* St. Paul: American Forest History Foundation, Minnesota Historical Society, 1955.

———. *Forest Protection; Guide to Lectures Delivered at the Biltmore Forest School.* Asheville, N.C.: Inland Press, 1909.

———. *Logging and Lumbering or Forest Utilization: A Textbook for Forest Schools.* Darmstadt, Germany: L. C. Wittich, 1912.

Shepard, Charles Upham. *Report of Charles Upham Shepard, on the Ducktown Copper Region and the Mines of the Union Consolidated Mining Company, of Tennessee.* Charleston: Walker, Evans, 1859.

Smallshaw, James. "Denudation and Erosion in the Copper Basin." Tennessee Valley Authority. Knoxville, Tenn., 1938.

Snow, John. *On the Mode of Communication of Cholera.* 2nd ed. London: J. Churchill, 1855.

Stephenson, Matthew F. *Geology and Mineralogy of Georgia, with a Particular Description of Her Rich Diamond District; the Process of Washing for Diamonds, Their Price and Mode of Cutting and Setting; Her Gold, Silver, Copper, Lead, Iron, Manganese, Graphite, Kaolin, Coal, Fire-Clay, Mica, Corundum, Slate, Marble, &C.* Atlanta: Globe Publishing Company, 1871.

Taliaferro, Hardin E. "Ducktown, by 'Skitt,' Who Has Been 'Thar.'" *Southern Literary Messenger* 31 (November 1860): 337–42.

Tennessee Department of Public Health, Division of Water Quality Control. "Water Quality Management Plan for the Lower Tennessee River Basin." Nashville, 1975.

———. "Water Quality Management Plan for the Lower Tennessee River Basin." Franklin, Tenn.: Enviro-Printers, 1978.

Tennessee Power Company and E. W. Clark & Co. Engineering and Management Department. *The Power of Water.* Columbus, Ohio: Lawrence Press, 1913.

Tennessee Valley Authority. *The Hiwassee Valley Projects.* Vol. 2: *The Appalachia, Ocoee No. 3, Nottely, and Chatuge Projects.* Technical Report No. 5. Washington, D.C.: Government Printing Office, 1948.

Tennessee Valley Authority, Department of Forestry Relations. "A Proposal for Erosion Control and Restoration of Vegetation in the Copper Basin." Norris, Tenn., 1945.

Tennessee Valley Authority, Division of Water Control Planning. "Floods on Toccoa-Ocoee River & Fightingtown Creek in the Vicinity of McCaysville, Ga.–Copperhill, Tenn." Knoxville, Tenn., 1958.

Tennessee Valley Authority, Water Quality Department. *Status of the Ocoee Reservoirs: An Overview of Reservoir Conditions and Uses.* OSTI ID: 5641599; DE91015961. Chattanooga, Tenn., 1991.

Tyre, Gary L., and Ronald G. Barton. "A Fresh Start at Old Problem." *Soil and Water Conservation News* 7, no. 4 (July 1986): 7–8.

United States. *American State Papers: Public Lands.* Washington, D.C.: Gales & Seaton, 1832.

———. *Articles and Agreement of Cession: Entered into on the 24th Day of April, 1802, between the Commissioners of the United States and those of Georgia.* Washington, D.C.: R. C. Weightman, 1807.

United States and Charles Joseph Kappler. *Indian Affairs: Laws and Treaties.* 2 vols. 2nd ed. Washington, D.C.: Government Printing Office, 1904.

U.S. Bureau of the Census. *Report on the Statistics of Agriculture in the United States at the Eleventh Census: 1890.* Washington, D.C.: Government Printing Office, 1895.

U.S. Census Office. *Twelfth Census of the United States, Taken in the Year 1900.* Vol. 5: *Agriculture,* part 1, *Farms, Livestock and Animal Products.* Washington, D.C.: Government Printing Office, 1902. Reprint, New York: Norman Ross Publishing, 1997.

U.S. Department of Agriculture, U.S. Forest Service, U.S. Geological Survey, and U.S. Weather Bureau. *Message from the President of the United States Transmitting a Report of the Secretary of Agriculture in Relation to the Forests, Rivers, and Mountains of the Southern Appalachian Region. December 19, 1901. Read, Referred to the Committee on Forest Reservations and the Protection of Game and Ordered to Be Printed.* Senate Document No. 84. 57th Congress, 2nd session, Washington, D.C.: Government Printing Office, 1902.

U.S. Department of the Interior, U.S. Geological Survey. "Epworth Quadrangle, Georgia-Tennessee, 7.5 Minute Series (Topographic) No. 35084-H4-TF-024." Washington, D.C., 1988.

———. "Mineral Bluff Quadrangle, Georgia-North Carolina-Tennessee, 7.5 Minute Series (Topographic) No. 134 NE." Washington, D.C., 1999.

U.S. Department of the Interior, U.S. Geological Survey, and Tennessee Valley Authority Mapping Services Branch. "Isabella Quadrangle, Tennessee-North Carolina, 7.5 Minute Series (Topographic), 133-SE, No. 35084-A4-TF-024." Washington, D.C., 1957; rev., 1978.

U.S. Department of the Interior, U.S. Geological Survey, and U.S. Department of Agriculture. "Ducktown Quadrangle, Tennessee-Polk County, 7.5 Minute Series (Topographic) No. 35084-A4-TF-024." Washington, D.C., 2003.

U.S. Environmental Protection Agency. "Copper Basin Mining District: Use of Cooperative Agreements Toward a Common Goal" (July 2005). ⟨www.epa.gov/aml/tech/copperbasin.pdf⟩ (accessed December 13, 2009).

U.S. Forest Service, Southern Region. *Final Environmental Impact Statement, 1996 Olympic Whitewater Slalom Venue, Ocoee River, Polk County, Tennessee, Ocoee Ranger District, Cherokee National Forest.* Cleveland, Tenn.: U.S. Department of Agriculture, U.S. Forest Service, Southern Region, 1994.

———. *Ocoee Whitewater Center: The Only Olympic Whitewater Site in the World Located on a Natural River.* Washington, D.C.: U.S. Department of Agriculture, U.S. Forest Service, Southern Region, 1991.

Walker, Clifford. *Reports and Opinions of the Attorney-General of Georgia from June 15, 1915, to December 31, 1916.* Atlanta: Index Printing, 1917.

———. *Reports and Opinions of the Attorney-General of Georgia from June 15, 1917, to December 31, 1918.* Atlanta: Byrd Printing Co., 1919.

Weed, Walter Harvey. "Copper Deposits in Georgia." In *Contributions to Economic Geology*, ed. S. F. Emmons and C. W. Hayes. U.S. Geological Survey, Bulletin no. 225. Washington, D.C.: Government Printing Office, 1904.

Wood, Richard A. "Erosion Control and Reforestation of the Copper Basin, a Progress Report." Tennessee Valley Authority, Watershed Protection Division, Department of Forestry Relations. Knoxville, Tenn., 1942.

Worcester, S. A., and Elias Boudinot. *The Acts of the Apostles.* New Echota, Ga.: John F. Wheeler and John Candy, 1833.

———. *Cherokee Hymns.* 2nd ed. New Echota, Ga.: J. F. Wheeler, 1830.

Secondary Books and Journals

Adams, Jane. *Calendar of Incoming Correspondence, Governor Hugh M. Dorsey, 1917–1921.* Atlanta: Georgia Department of Archives and History, 1972.

Anderson, William L. *Cherokee Removal: Before and After.* Athens: University of Georgia Press, 1991.

Andrews, Richard N. L. *Managing the Environment, Managing Ourselves: A History of American Environmental Policy.* New Haven, Conn.: Yale University Press, 1999.

Arnett, Alex Mathews. *The Populist Movement in Georgia: A View of the "Agrarian Crusade" in the Light of Solid-South Politics.* New York: Columbia University, 1922.

Ayers, Edward L. *The Promise of the New South: Life after Reconstruction.* New York: Oxford University Press, 1992.

Baker, Andrew J. "Charcoal." In *Encyclopedia of American Forest and Conservation History*, ed. Richard C. Davis, 1:73–77. New York: Macmillan, 1983.

Barclay, R. E. *The Copper Basin, 1890 to 1963*. Knoxville, Tenn.: Cole Printing and Thesis Service, 1975.
———. *Ducktown Back in Raht's Time*. Chapel Hill: University of North Carolina Press, 1946.
———. "The Great Copper Basin of Tennessee." *L & N Employee's Magazine* 10, no. 9 (1934): 4–7.
———. *Introduction to the Copper Basin*. Ducktown, Tenn.: The Copper Station, n.d. Also found in Barclay Papers, box 8, folder 1, Tennessee State Library and Archives, Nashville.
———. *The Railroad Comes to Ducktown*. Knoxville, Tenn.: Cole Printing & Thesis Service, 1973.
Barnhardt, Wilton. "The Death of Ducktown." *Discover* 8 (October 1997): 35–43.
Bartley, Numan V. *The Creation of Modern Georgia*. Athens: University of Georgia Press, 1983.
Beth, Loren P. *John Marshall Harlan: The Last Whig Justice*. Lexington: University Press of Kentucky, 1922.
Bierce, Ambrose. *The Unabridged Devil's Dictionary*. Edited by David E. Schultz and S. T. Joshi. Athens: University of Georgia Press, 2000.
Bird, Sharon Iowa. "Exchange Networks in the Prehistoric Southeastern United States." M.A. thesis, University of Georgia, 1978.
Blair, Walter. *Native American Humor*. San Francisco: Chandler Publishing, 1960.
Blakey, Arch Fredric. *The Florida Phosphate Industry: A History of the Development and Use of a Vital Mineral*. Cambridge, Mass.: Harvard University Press, 1973.
Blum, Elizabeth D. *Love Canal Revisited*. Lawrence: University Press of Kansas, 2008.
Boles, John B. *A Companion to the American South*. Malden, Mass.: Blackwell, 2002.
Bolgiano, Chris. *The Appalachian Forest*. Mechanicsburg, Pa.: Stackpole Books, 1995.
Bone, Robert G. "Normative Theory and Legal Doctrine in American Nuisance Law: 1850–1920." *Southern California Law Review* 59 (September 1986): 1104–1226.
Brecher, Jeremy, Jerry Lombardi, Jan Stackhouse, and Brass Workers History Project. *Brass Valley: The Story of Working People's Lives and Struggles in an American Industrial Region*. Philadelphia: Temple University Press, 1982.
Brenner, Joel Franklin. "Nuisance Law and the Industrial Revolution." *Journal of Legal Studies* 3 (1974): 403–33.
Brooks, Richard Oliver, Ross Jones, and Ross A. Virginia. *Law and Ecology: The Rise of the Ecosystem Regime*. Ecology and Law in Modern Society. Aldershot: Ashgate, 2002.
Brown, R. Harold. *The Greening of Georgia: The Improvement of the Environment in the Twentieth Century*. Macon, Ga.: Mercer University Press, 2002.
Brown, Walter J. *J. J. Brown and Thomas E. Watson, Georgia Politics: 1912–1928*. Macon, Ga.: Mercer University Press, 1989.
Brundage, W. Fitzhugh. "Racial Violence, Lynchings, and Modernization in the Mountain South." In *Appalachians and Race: The Mountain South from Slavery to Segregation*, ed. John C. Inscoe. Lexington: University Press of Kentucky, 2001.
Bryson, John E., and Angus MacBeth. "Public Nuisance, the Restatement (Second) of Torts, and Environmental Law." *Ecology Law Quarterly* (1972): 241–81.
Burt, Jesse C. "Desert in the Appalachians." *Nature Magazine* 48 (November 1956): 486–88, 499.
Buxton, Barry M., and Malinda L. Crutchfield, eds. *The Great Appalachian Forest: An Appalachian Story*. Boone, N.C.: Appalachian Consortium Press, 1985.

Cadle, Farris W. *Georgia Land Surveying History and Law*. Athens: University of Georgia Press, 1991.

Case, Earl C. *The Valley of East Tennessee, the Adjustment of Industry to Natural Environment*. Tennessee Department of Education, Division of Geology, Bulletin 36. Nashville, Tenn., 1925.

Charry, Stephen W. "Defending the 'Great Barbecue': W. Lon Johnson and the 1921 Northport Smelter Pollution Suits." *Pacific Northwest Quarterly* 91, no. 2 (Spring 2000): 59–69.

Chase, Stuart. *Rich Land, Poor Land: A Study of Waste in the Natural Resources of America*. New York: Whittlesey House McGraw-Hill, 1936.

Church, Michael A. "Smoke Farming: Smelting and Agricultural Reform in Utah, 1900–1945." *Utah Historical Quarterly* 72, no. 3 (Summer 2004): 196–218.

Clay, Grady. "Copper-Basin Cover-Up." *Landscape Architecture* 73, no. 4 (1983): 49–55, 94.

Clepper, Henry. *Professional Forestry in the United States*. Baltimore: Johns Hopkins University Press, 1971.

Cobb, James C. *Away Down South: A History of Southern Identity*. New York: Oxford University Press, 2005.

———. *Industrialization and Southern Society, 1877–1984*. New Perspectives on the South. Lexington: University Press of Kentucky, 1984.

Cobb, James C., and Thomas G. Dyer. "Economic Prosperity or Environmental Protection: Georgia, Tennessee, and the Tennessee Copper Companies, 1903–1975." Paper presented at Organization of American Historians, New Orleans, April 1979.

Cohen, Norm. "Robert W. Gordon and the Second Wreck of the Old '97." *Journal of American Folklore* 87, no. 343 (January–March 1974): 12–38.

Coleman, Kenneth, ed. *A History of Georgia*. Athens: University of Georgia Press, 1991.

Coleman, Kenneth, and Charles Stephen Gurr, eds. *Dictionary of Georgia Biography*. Athens: University of Georgia Press, 1983.

Cook, James F. "Terrell, Joseph Meriwether." In *Dictionary of Georgia Biography*, ed. Kenneth Coleman and Charles Stephen Gurr. Athens: University of Georgia Press, 1983.

"Copper Ore and Cotton: Dangerous Freight." *Hunt's Merchant Magazine and Commercial Review* 33 (July–December 1855): 394.

Coulter, E. Merton. "The Georgia-Tennessee Boundary Line." *Georgia Historical Quarterly* 35 (1951): 269–306.

Cox, Thomas, Robert S. Maxwell, Phillip Brennon Thomas, and Joseph J. Malone. *This Well-Wooded Land: Americans and Their Forests from Colonial Times*. Lincoln: University of Nebraska Press, 1985.

Craig, Robin K. *The Clean Water Act and the Commerce Clause*. Washington, D.C.: Environmental Law Institute, 2004.

Crampton, John A. *The National Farmers Union: Ideology of a Pressure Group*. Lincoln: University of Nebraska Press, 1965.

Cronon, William. *Nature's Metropolis: Chicago and the Great West*. New York: W. W. Norton, 1991.

———. *Uncommon Ground: Toward Reinventing Nature*. New York: W. W. Norton, 1995.

Daniels, Karen. *Tennessee's Historic Copper Basin Area: An Overview*. Benton, Tenn.: Polk County Publishing, 1985.

Daniels, Pete. *Breaking the Land: The Transformation of Cotton, Tobacco, and Rice Cultures since 1980*. Urbana: University of Illinois Press, 1985.

Davis, Donald E. "Living on the Land: Blue Ridge Life and Culture." *Georgia Wildlife* 6, no. 2 (1997): 38–53.

———. *Where There Are Mountains: An Environmental History of the Southern Appalachians.* Athens: University of Georgia Press, 2000.

Day, Joan. *Bristol Brass: A History of the Industry.* Newton Abbot, England: David & Charles, 1973.

Dinwoodie, D. H. "The Politics of International Pollution Control: The Trail Smelter Case." *International Journal* 27, no. 2 (1972): 219–35.

Dodds, Gordon B. "The Stream-Flow Controversy." *Journal of American History* 56, no. 1 (June 1969): 59–69.

Donnelly, Ralph W. "Confederate Copper." *Civil War History* 1, no. 4 (1955): 355–70.

Dubé, Tom. *History of Tennessee Copper Company and Successor Firms at the Copperhill Plant and the Ducktown Mining District, Copper Basin, Tennessee.* Bothell, Wash.: SAIC, 2008.

Duggan, Betty J. "Being Cherokee in a White World: The Ethnic Persistence of a Post-Removal American Indian Enclave." Ph.D. diss., University of Tennessee, Knoxville, 1998.

Dunaway, Wilma A. *The First American Frontier: Transition to Capitalism in Southern Appalachia, 1700–1860.* Chapel Hill: University of North Carolina Press, 1996.

Ehle, John. *Trail of Tears: The Rise and Fall of the Cherokee Nation.* New York: Doubleday, 1988.

Eller, Ronald D. "Land as Commodity: Industrialization of the Appalachian Forests." In *The Great Appalachian Forest: An Appalachian Story*, ed. Barry M. Buxton and Malinda L. Crutchfield, 15–23. Boone, N.C.: Appalachian Consortium Press, 1985.

———. *Miners, Millhands, and Mountaineers: Industrialization of the Appalachian South, 1880–1930.* Knoxville: University of Tennessee Press, 1982.

Ellison, George. "Talking about Mountain Balds and 'Laurel Hells.'" *Smoky Mountain News*, March 21, 2001. ⟨http://www.smokymountainnews.com/issues/3-01-/3-21-01/back_then.shtml⟩ (accessed August 26, 2009).

Evans, E. Raymond. "Fort Marr Blockhouse: The Last Evidence of America's First Concentration Camps." *Journal of Cherokee Studies* 2, no. 2 (1977): 256–62.

Faulkner, Ben B., Franklin Miller, Frank Russell, and Ken Faulk. "The Copper Basin Reclamation Project." *Reclamation Matters* 2, no. 1 (Spring–Summer 2005): 24–27.

Finger, John R. *The Eastern Band of the Cherokees, 1819–1900.* Knoxville: University of Tennessee Press, 1984.

Fisher, Commodore B. *The Farmers' Union.* Lexington: University Press of Kentucky, 1920.

Flagg, Samuel B. *City Smoke Ordinances and Smoke Abatement.* U.S. Bureau of Mines, Bulletin no. 49. Washington, D.C.: Government Printing Office, 1912.

Flippen, J. Brooks. *Nixon and the Environment.* Albuquerque: University of New Mexico Press, 2000.

Floyd, Robert J. *Tennessee Rock and Mineral Resources.* Tennessee Department of Conservation, Division of Geology, Bulletin no. 66. Nashville, Tenn., 1966.

Foehner, Nora Lynn. "The Historical Geography of Environmental Change in the Copper Basin." M.S. thesis, University of Tennessee, Knoxville, 1980.

Friedman, Lawrence M. *A History of American Law.* 3rd ed. New York: Simon & Schuster, 2005.

Gaston, Paul M. *The New South Creed: A Study in Southern Mythmaking*. New York: Knopf, 1970.

Gates, William B. *Michigan Copper and Boston Dollars: An Economic History of the Michigan Copper Mining Industry*. Studies in Economic History. Cambridge, Mass.: Harvard University Press, 1951.

Geisen, James Conrad. "The South's Greatest Enemy? The Cotton Boll Weevil and Its Lost Revolution, 1892–1930." Ph.D. diss., University of Georgia, 2004.

Gennett, Andrew. *Sound Wormy: Memoir of Andrew Gennett, Lumberman*. Edited by Nicole Hayler. Athens: University of Georgia Press, 2002.

Gigantino, Jim. "Land Lottery System." In *New Georgia Encyclopedia*. ⟨http://www.georgiaencyclopedia.org⟩ (accessed July 31, 2007).

Gilbert, Grove Karl. *Hydraulic-Mining Debris in the Sierra Nevada*. U.S. Geological Survey, Professional Paper no. 105. Washington, D.C.: Government Printing Office, 1917.

Goad, Sharon Iowa. "Chemical Analysis of Native Copper Artifacts from the Southeastern United States." *Current Anthropology* 21, no. 2 (April 1980): 270–71.

———. "Exchange Networks in the Prehistoric Southeastern United States." M.A. thesis, University of Georgia, 1978.

Goodman, Claire Garber, and Anne-Marie E. Cantwell. *Copper Artifacts in Late Eastern Woodlands Prehistory*. Evanston, Ill.: Center for American Archeology at Northwestern University, 1984.

Goodwyn, Lawrence. *Democratic Promise: The Populist Moment in America*. New York: Oxford University Press, 1976.

Gottlieb, Robert. *Forcing the Spring: The Transformation of the American Environmental Movement*. Washington, D.C.: Island Press, 1993.

Graves, Henry Solon. *Forest Mensuration*. New York: John Wiley & Sons, 1906.

Griffin, Gary J. "Blight Control and Restoration of the American Chestnut." *Journal of Forestry* 98, no. 2 (February 2000): 22–27.

Grimsley, Mark. *The Hard Hand of War: Union Military Policy toward Southern Civilians, 1861–1865*. Cambridge: Cambridge University Press, 1995.

Grinder, R. Dale. "The Battle for Clean Air: The Smoke Problem in Post-Civil War America." In *Pollution and Reform in American Cities, 1870–1930*, ed. Martin V. Melosi, 83–103. Austin: University of Texas Press, 1980.

Gugliotta, Angela. "Class, Gender, and Coal Smoke: Gender Ideology and Environmental Injustice in Pittsburgh, 1868–1914." *Environmental History* 5, no. 2 (2000): 165–93.

Hahn, Steven. *The Roots of Southern Populism: Yeoman Farmers and the Transformation of the Georgia Upcountry, 1850–1890*. New York: Oxford University Press, 1983.

Hall, Kermit. *The Magic Mirror: Law in American History*. New York: Oxford University Press, 1989.

———. *The Oxford Companion to the Supreme Court of the United States*. New York: Oxford University Press, 1992.

Hall, Marcus. "The Provincial Nature of George Perkins Marsh." *Environmental History* 10 (2004): 191–204.

Halliday, Stephen. *The Great Stink of London: Sir Joseph Bazalgette and the Cleansing of the Victorian Capital*. London: Alan Sutton, 1998.

Hamilton, Daniel. "The Confederate Sequestration Act." *Civil War History* 52, no. 4 (December 2006): 373–408.

Hamilton, Henry. *The English Brass and Copper Industries to 1800.* 2nd ed. New York: A. M. Kelley, 1967.
Hargrove, Erwin C. *Prisoners of Myth: The Leadership of the Tennessee Valley Authority, 1933–1990.* Princeton Studies in American Politics. Princeton, N.J.: Princeton University Press, 1994.
Hays, Samuel P. *The American People and the National Forests: The First Century of the U.S. Forest Service.* Pittsburgh, Pa.: University of Pittsburgh Press, 2009.
———. *Conservation and the Gospel of Efficiency: The Progressive Conservation Movement, 1890–1920.* Harvard Historical Monographs, 40. Cambridge, Mass.: Harvard University Press, 1959.
Hays, Samuel P., and Barbara D. Hays. *Beauty, Health, and Permanence: Environmental Politics in the United States, 1955–1985.* Studies in Environment and History. Cambridge: Cambridge University Press, 1987.
Heffington, Douglas. "Tennessee's Copper Basin: An Ethnic Overview." *Tennessee Genealogical Magazine* 44, no. 2 (1997): 17–23.
Hempel, Sandra. *The Strange Case of the Broad Street Pump: John Snow and the Mystery of Cholera.* Berkeley: University of California Press, 2007.
Hicks, John Donald. *The Populist Revolt: A History of the Farmers' Alliance and the People's Party.* Minneapolis: University of Minnesota Press, 1931.
Highsaw, Robert B. *Edward Douglass White: Defender of the Conservative Faith.* Baton Rouge: Louisiana State University Press, 1981.
Hirt, Paul W. *A Conspiracy of Optimism: Management of the National Forests since World War II.* Lincoln: University of Nebraska Press, 1994.
Hoffer, Peter Charles. *The Law's Conscience: Equitable Constitutionalism in America.* Chapel Hill: University of North Carolina Press, 1990.
Hofstadter, Richard. *The Age of Reform: From Bryan to F.D.R.* New York: Knopf, 1955.
Hoig, Stan. *The Cherokees and Their Chiefs: In the Wake of Empire.* Fayetteville, Ark.: University of Arkansas Press, 1998.
Horwitz, Morton J. *The Transformation of American Law, 1780–1860.* Studies in Legal History. Cambridge, Mass.: Harvard University Press, 1977.
Howe, Mark DeWolfe. *Oliver Wendell Holmes: The Shaping Years, 1841–1870.* Cambridge, Mass.: Harvard University Press, 1957.
Hurley, Andrew. *Environmental Inequalities: Class, Race, and Industrial Pollution in Gary, Indiana, 1945–1980.* Chapel Hill: University of North Carolina Press, 1995.
Hurst, Vernon J., and Lewis H. Larson Jr. "On the Source of Copper at the Etowah Site, Georgia." *American Antiquity* 24, no. 2 (October 1958): 177–81.
Hyde, Charles K. *Copper for America: The United States Copper Industry from Colonial Times to the 1990s.* Tucson: University of Arizona Press, 1998.
Inscoe, John C. *Appalachians and Race: The Mountain South from Slavery to Segregation.* Lexington: University Press of Kentucky, 2001.
———. "The Discovery of Appalachia: Regional Revisionism as Scholarly Renaissance." In *A Companion to the American South,* ed. John B. Boles, 369–86. Malden, Mass.: Blackwell, 2002.
———. *Mountain Masters, Slavery, and the Sectional Crisis in Western North Carolina.* Knoxville: University of Tennessee Press, 1989.
Isenberg, Andrew C. *The Destruction of the Bison: An Environmental History, 1750–1920.* Cambridge: Cambridge University Press, 2000.

Jacoby, Karl. *Crimes against Nature: Squatters, Poachers, Thieves, and the Hidden History of American Conservation.* Berkeley: University of California Press, 2001.

Johnson, Mary Elizabeth. "Box and Container Industry." In *Encyclopedia of American Forest and Conservation History*, vol. 1, ed. Richard C. Davis, 45–47. New York: Macmillan, 1983.

Jolly, Harley E. "The Cradle of Forestry, Where Tree Power Started." *American Forests* 76 (October, 1970): 16–21.

Jones, Alton DuMar. "The Administration of Governor Joseph M. Terrell in Light of the Progressive Movement." *Georgia Historical Quarterly* 48, no. 3 (September 1964): 271–90.

Joyce, Joseph A. *Treatise on the Law Governing Nuisances: With Particular Reference to Its Application to Modern Conditions and Covering the Entire Law Relating to Public and Private Nuisances, Including Statutory and Municipal Powers and Remedies, Legal and Equitable.* Albany, N.Y.: Matthew Bender, 1906.

Judd, Richard William. *Common Lands, Common People: The Origins of Conservation in Northern New England.* Cambridge, Mass.: Harvard University Press, 1997.

———. "George Perkins Marsh: The Man and Their Man." *Environmental History* 10 (2004): 169–90.

Kiefer, David M. "Sulfuric Acid: Pumping Up the Volume: An 18th-Century English Physician's Lead Cathedrals Helped Launch a Chemical Industry." *Today's Chemist at Work* 10, no. 9 (2001): 57–58.

Keiter, Robert B. *Keeping Faith with Nature: Ecosystems, Democracy, and America's Public Lands.* New Haven, Conn.: Yale University Press, 2003.

Kephart, Horace. *Camping and Woodcraft: A Handbook for Vacation Campers and for Travelers in the Wilderness.* New York: Outdoor Publishing Company, 1910. Reprint, Knoxville: University of Tennessee Press, 1988.

———. *Our Southern Highlanders.* New York: Outing Publishing, 1913.

King, Duane H. *The Cherokee Indian Nation: A Troubled History.* Knoxville: University of Tennessee Press, 1979.

Klein, Randolph Shipley. *Portrait of an Early American Family: The Shippens of Pennsylvania across Five Generations.* Philadelphia: University of Pennsylvania Press, 1975.

Koch, Tom. *Cartographies of Disease: Maps, Mapping, and Medicine.* Redland, Calif.: ESRI Press, 2005.

Kurtz, Paul M. "Nineteenth-Century Anti-entrepreneurial Nuisance Injunctions—Avoiding the Chancellor." *William & Mary Law Review* 17 (Summer 1976): 621–70.

Lamborn, John E., and Peterson Charles S. "The Substance of the Land: Agriculture v. Industry in the Smelter Cases of 1904 and 1906." *Utah Historical Quarterly* 53, no. 4 (Fall 1985): 308–25.

Lamplugh, George R. "Yazoo Land Fraud." In *New Georgia Encyclopedia.* ⟨http://www.georgiaencyclopedia.org⟩ (accessed September 4, 2005).

Landis, Susan Hill. *You Have Stept Out of Your Place: A History of Women and Religion in America.* Louisville, Ky.: Westminster John Knox Press, 1996.

Langston, Nancy. *Forest Dreams, Forest Nightmares: The Paradox of Old Growth in the Inland West.* Seattle: University of Washington Press, 1995.

Lankton, Larry D. *Beyond the Boundaries: Life and Landscape at the Lake Superior Copper Mines, 1840–1875.* New York: Oxford University Press, 1997.

Larson, Sylvia B. "Marsh, George Perkins." *American National Biography*. Edited by Mark C. Carnes and John A. Garraty. ⟨http//www.anb.org/⟩ (accessed February 25, 2007).

Lathrop, William Gilbert. *The Brass Industry in the United States: A Study of the Origin and the Development of the Brass Industry in the Naugatuck Valley and Its Subsequent Extension over the Nation*. Rev. ed. Mount Carmel, Conn.: W. G. Lathrop, 1926.

Lazarus, Richard J. *The Making of Environmental Law*. Chicago: University of Chicago Press, 2004.

LeCain, Timothy J. "When Everybody Wins Does the Environment Lose? The Environmental Techno-Fix in Twentieth-Century American Mining." In *The Technological Fix: How People Use Technology to Create and Solve Problems*, ed. Linda Rosner, 137–54. New York: Routledge, 2004.

Lewis, Ronald L. *Transforming the Appalachian Countryside: Railroads, Deforestation, and Social Change in West Virginia, 1880–1920*. Chapel Hill: University of North Carolina Press, 1998.

Lewis, Sinclair. *Babbitt*. New York: Harcourt, Brace, 1922. Reprint, New York: Penguin, 1996.

Lichtenstein, Andrew. *Twice the Work of Free Labor: The Political Economy of Convict Labor in the New South*. New York: Verso, 1996.

Lillard, Roy G. *The History of Polk County, Tennessee*. Benton, Tenn.: Polk County Historical & Genealogical Society, 1999.

———. "The Story of Ocoee No. 1 and Ocoee No. 2 Hydro Plants." Tennessee Valley Authority. Knoxville, Tenn., 1978.

Lowenthal, David. *George Perkins Marsh, Prophet of Conservation*. Seattle: University of Washington Press, 2000.

Luconi, Stefano. "The Enforcement of the 1941 Smoke-Control Ordinance and Italian Americans in Pittsburgh." *Pennsylvania History* 66, no. 4 (1999): 580–94.

Lutts, Ralph H. "Like Manna from God: The American Chestnut Trade in Southwestern Virginia." *Environmental History* 9, no. 3 (July 2004): 497–525.

MacMillan, Donald. *Smoke Wars: Anaconda Copper, Montana Air Pollution, and the Courts, 1890–1924*. Helena: Montana Historical Society Press, 2000.

Maher, Neil M. *Nature's New Deal: The Civilian Conservation Corps and the Roots of the American Environmental Movement*. New York: Oxford University Press, 2007.

Marszalek, John F. *Sherman's March to the Sea*. Abilene, Tex.: McWhiney Foundation Press, McMurray University, 2005.

Mathew, William M. *Edmund Ruffin and the Crisis of Slavery in the Old South: The Failure of Agricultural Reform*. Athens: University of Georgia Press, 1988.

McCarthy, J. P. "The Federal Common Law of Nuisance." *Tennessee Law Review* 49 (1982): 919–54.

McCrary, Ben H., and LeRoy P. Graf, eds. "Vineland in Tennessee, 1852: The Journal of Rosine Parmentier." *East Tennessee Historical Society Publications* 31 (1959): 95–111.

McEvoy, Arthur F. *The Fisherman's Problem: Ecology and the Law in the California Fisheries*. Cambridge: Cambridge University Press, 1986.

McLaren, John P. S. "Nuisance Law and the Industrial Revolution—Some Lessons from Social History." *Oxford Journal of Legal Studies* 3 (1983): 155–221.

McLeod, Kari S. "Our Sense of Snow: The Myth of John Snow in Medical Geography." *Social Science & Medicine* 50, nos. 7–8 (2000): 923–35.

McLoughlin, William Gerald. *Cherokee Renascence in the New Republic*. Princeton, N.J.: Princeton University Press, 1986.

McMath, Robert C. *American Populism: A Social History, 1877–1898*. New York: Hill & Wang, Noonday Press, 1993.

Melosi, Martin V. *The Sanitary City: Urban Infrastructure in America from Colonial Times to the Present*. Creating the North American Landscape. Baltimore: Johns Hopkins University Press, 2000.

Merrill, Karen R. *Public Lands and Political Meaning: Ranchers, the Government, and the Property between Them*. Berkeley: University of California Press, 2002.

Miller, Char. *American Forests: Nature, Culture, and Politics*. Development of Western Resources. Lawrence: University Press of Kansas, 1997.

———. *Gifford Pinchot and the Making of Modern Environmentalism*. Washington, D.C.: Island Press, Shearwater Books, 2001.

Minton, John Dean. *The New Deal in Tennessee, 1932–1938*. New York: Garland, 1979.

Montgomery, Michael B., and Joseph S. Hall. *Dictionary of Smoky Mountain English*. Knoxville: University of Tennessee Press, 2004.

Moore, James William. *Moore's Federal Practice, Third Edition*. New York: LexisNexis Matthew Bender, 2007.

Morris, Christopher. "A More Southern Environmental History." *Journal of Southern History* 75, no. 3 (August 2009): 581–98.

Murchison, Kenneth M. "Interstate Pollution: The Need for Federal Common Law." *Virginia Journal of Natural Resources* 6 (1986): 1–51.

Nash, Roderick. *Wilderness and the American Mind*. 4th ed. New Haven, Conn.: Yale University Press, 2001.

The National Cyclopedia of American Biography. Vol. F. Clifton, N.J.: J. T. White, 1934–42.

"National Wilderness Preservation System—Tennessee." ⟨http://www.wilderness.net/index.cfm?fuse=NWPS&sec=stateView&state=tn&map=tn⟩ (map); ⟨http://www.wilderness.net/index.cfm?fuse=NWPS⟩ (data for each wilderness area) (both sites accessed August 5, 2005).

Navin, Thomas R. *Copper Mining and Management*. Tucson: University of Arizona Press, 1978.

Nelson, Lewis B. *History of the U.S. Fertilizer Industry*. Muscle Shoals, Ala.: Tennessee Valley Authority, 1990.

Nelson, Scott Reynolds. *Iron Confederacies: Southern Railways, Klan Violence, and Reconstruction*. Chapel Hill: University of North Carolina Press, 1999.

Nelson, Wilbur. "The Manufacture of Sulphuric Acid in Tennessee for 1911." *Resources of Tennessee* 2, no. 1 (January 1912): 23–37.

Newell, Edmund. "Atmospheric Pollution and the British Copper Industry, 1690–1920." *Technology and Culture* 38, no. 3 (July 1997): 655–89.

Noe, Kenneth W., and Shannon H. Wilson, eds. *The Civil War in Appalachia: Collected Essays*. Knoxville: University of Tennessee Press, 1997.

Nolt, John. *A Land Imperiled: The Declining Health of the Southern Appalachian Bioregion*. Knoxville: University of Tennessee Press, 2005.

Norgren, Jill. *The Cherokee Cases: Two Landmark Federal Decisions in the Fight for Sovereignty*. Norman: University of Oklahoma Press, 2004.

Note. *Virginia Law Review* 10, no. 2 (December 1923): 147–50.

Orth, John V. *The Judicial Power of the United States: The Eleventh Amendment in American History*. New York: Oxford University Press, 1987.
Ottewell, Guy. "There Are No Ducks in Ducktown." *Defenders Magazine* 50, no. 4 (August 1975): 338–39.
Parish, Thurman. *The Old Home Place: Pioneer Mountain Life in Polk County, Tennessee*. Benton, Tenn.: Polk County Publishing Co., 1994.
Parker, Ernest. *Days Gone By: Early Gilmer County, Georgia*. Ellijay, Ga.: Gilmer County Genealogical Society, 1999.
Percival, Robert V. "The Frictions of Federalism: The Rise and Fall of the Federal Common Law of Interstate Nuisance." University of Maryland, Pub-Law Research Paper No. 2003-02. 2003. ⟨http://ssrn.com/abstract=452992⟩ (accessed September 6, 2007).
Perdue, Theda. *Slavery and the Evolution of Cherokee Society, 1540–1866*. Knoxville: University of Tennessee Press, 1979.
Perdue, Theda, and Michael D. Green. *The Columbia Guide to American Indians of the Southeast*. Columbia Guides to American Indian History and Culture. New York: Columbia University Press, 2001.
Persico, V. Richard, Jr. "Early Nineteenth-Century Cherokee Political Organization." In *The Cherokee Indian Nation: A Troubled History*, ed. Duane H. King, 92–109. Knoxville: University of Tennessee Press, 1979.
Phillips, Ulrich Bonnell. *Georgia and State Rights: A Study of the Political History of Georgia from the Revolution to the Civil War, with Particular Regard to Federal Relations*. Washington, D.C.: Government Printing Office, 1902.
Pinchot, Gifford. *Breaking New Ground*. New York: Harcourt Brace, 1947.
Pisani, Donald J. "Forests and Conservation, 1865–1890." *Journal of American History* 72, no. 2 (September, 1985): 340–59.
Platt, Harold L. "Invisible Gases: Smoke, Gender, and the Redefinition of Environmental Policy in Chicago, 1900–1920." *Planning Perspectives* 10, no. 1 (1995): 67–97.
Polk County Publishing Co. *Polk County News: Reprints of Articles Relating to the History of Polk County*. Benton, Tenn.: Polk County Publishing Co., 1990.
———. *Polk County Scrapbook: 150 Years of Memories*. 2 vols. Benton, Tenn.: Polk County Publishing Co., 1989.
———. *Polk County Scrapbook: A Tribute to the Miners; Copper Basin Ore Mining, 1843–1987*. Benton, Tenn.: Polk County Publishing Co., 1989.
Powers, Ron. *Mark Twain: A Life*. New York: Free Press, 2005.
Provine, D. M. "Balancing Pollution and Property Rights: A Comparison of the Development of English and American Nuisance Law." *Anglo-American Law Review* 7 (January–March 1978): 761–821.
Quinn, M.-L. "The Appalachian Mountains' Copper Basin and the Concept of Environmental Susceptibility." *Environmental Management* 15, no. 2 (March–April 1991): 179–94.
———. "Early Smelter Sites: A Neglected Chapter in the History and Geography of Acid Rain in the United States." *Atmospheric Environment* 23, no. 6 (1989): 1281–92.
———. "Industry and Environment in the Appalachian Copper Basin, 1890–1930." *Technology and Culture* 34, no. 3 (1993): 575–612.
———. "Tennessee's Copper Basin: A Case for Preserving an Abused Landscape." *Journal of Soil and Water Conservation* 43, no. 2 (March–April 1988): 140–44.

Rees, Ronald. "The South Wales Copper-Smoke Dispute, 1833–95." *Welsh History Review* 10, no. 4 (1981): 480–96.

Righter, Robert W. *The Battle over Hetch-Hetchy: America's Most Controversial Dam and the Birth of Modern Environmentalism*. New York: Oxford University Press, 2005.

Robards, M. J. "Tennessee's Wealthy Wasteland." *L & N Magazine* 26, no. 11 (November 1950): 11–14, 44.

Rome, Adam W. "Coming to Terms with Pollution: The Language of Environmental Reform, 1865–1915." *Environmental History* 1, no. 3 (1996): 6–28.

Ronderos, Ana. "Where Giants Once Stood: The Demise of the American Chestnut and Efforts to Bring It Back." *Journal of Forestry* 98, no. 2 (February 2000): 10–11.

Roper, Daniel M. "Logging the Cohutta Wilderness." *Georgia Backroads* 1, no. 1 (Spring 2002): 6–18.

Rosen, Christine. "Differing Perceptions of the Value of Pollution Abatement across Time and Place: Balancing Doctrine in Pollution Nuisance Law, 1840–1906." *Law and History Review* 11, no. 2 (1993): 303–81.

———. "Knowing Industrial Pollution: Nuisance Law and the Power of Tradition in a Time of Rapid Economic Change, 1840–1864." *Environmental History* 8, no. 4 (2003): 565–97.

Rosner, Lisa, ed. *The Technological Fix: How People Use Technology to Create and Solve Problems*. New York: Routledge, 2004.

Rothacher, Jack S. "Soil Erosion in Copper Basin." *Journal of Forestry* 52, no. 1 (1954): 41.

Royster, Charles. *The Destructive War: William Tecumseh Sherman, Stonewall Jackson, and the Americans*. New York: Knopf, 1991.

Salmond, John A. *The Civilian Conservation Corps, 1933–1942: A New Deal Case Study*. Durham, N.C.: Duke University Press, 1967.

Salstrom, Paul. *Appalachia's Path to Dependency: Rethinking a Region's Economic History, 1730–1940*. Lexington: University Press of Kentucky, 1994.

Sanders, M. Elizabeth. *Roots of Reform: Farmers, Workers, and the American State, 1877–1917*. American Politics and Political Economy. Chicago: University of Chicago Press, 1999.

Scott, Carole E., and Richard D. Guynn. "The Disappearance from Georgia of the Farm Union." ⟨http://www.westga.edu/~bquest/1997/farmer html⟩ (accessed May 19, 2004).

Seigworth, Kenneth J. "Ducktown—a Postwar Challenge." *American Forests* 49, no. 11 (1943): 521–23, 558.

Sellers, Christopher C. *Hazards of the Job: From Industrial Disease to Environmental Health Science*. Chapel Hill: University of North Carolina Press, 1997.

Shabecoff, Philip. *A Fierce Green Fire: An American Environmental Movement*. Rev. ed. Washington, D.C.: Island Press, 2003.

Shands, William E., Robert G. Healy, and Conservation Foundation. *The Lands Nobody Wanted: Policy for National Forests in the Eastern United States: A Conservation Foundation Report*. Washington, D.C.: The Foundation, 1977.

Shaw, Barton C. *The Wool-Hat Boys: Georgia's Populist Party*. Baton Rouge: Louisiana State University Press, 1984.

Silber, Nina. *The Romance of Reunion: Northerners and the South, 1865–1900*. Chapel Hill: University of North Carolina Press, 1993.

Simpson, John W. *Dam! Water, Power, Politics, and Preservation in Hetch Hetchy and Yosemite National Park*. New York: Pantheon Books, 2005.

Simson, William. "Parades amid the Standoff in the Old Red Scar: Interpreting

Film Images of Striking Industrial Operatives in the East Tennessee Copper Basin, 1939–1940." *Journal of Appalachian Studies* 7, no. 3 (Fall 2000): 227–55.
Smith, Charles Dennis. "The Appalachian National Park Movement, 1885–1901." *North Carolina Historical Review* 37 (January 1960): 38–65.
Smith, David M. "American Chestnut: Ill-Fated Monarch of the Eastern Hardwood Forest." *Journal of Forestry* 98, no. 2 (February 2000): 12–15.
Southern Appalachian Man and the Biosphere Cooperative. *The Southern Appalachian Assessment: Prepared by Federal and State Agencies Coordinated through the Southern Man and the Biosphere Cooperative.* Vol. 5: *Atmospheric Technical Report.* Atlanta: U.S. Forest Service, Southern Region, 1996.
Steen, Harold K. *The U.S. Forest Service: A History.* Seattle: University of Washington Press, 1976.
Steen, Harold K., and Forest History Society, eds. *The Origins of the National Forests: A Centennial Symposium.* Durham, N.C.: Forest History Society, 1992.
Steinberg, Theodore. *Nature Incorporated: Industrialization and the Waters of New England.* Studies in Environment and History. Cambridge: Cambridge University Press, 1991.
———. *Slide Mountain, or The Folly of Owning Nature.* Berkeley: University of California Press, 1995.
Stevens, Robert Bocking. *Law School: Legal Education in America from the 1850s to the 1980s.* Studies in Legal History. Chapel Hill: University of North Carolina Press, 1983.
Stewart, Mart A. "Southern Environmental History." In *A Companion to the American South*, ed. John B. Boles, 409–23. Malden, Mass.: Blackwell, 2002.
Stoll, Steven. *Larding the Lean Earth: Soil and Society in Nineteenth-Century America.* New York: Hill and Wang, 2002.
Stradling, David. *Smokestacks and Progressives: Environmentalists, Engineers and Air Quality in America, 1881–1951.* Baltimore: Johns Hopkins University Press, 1999.
Summers, Harry G. "Desolation and War: Necessity and Choice." Paper presented at the First International Conference Addressing Environmental Consequences of War: Legal, Economic, and Scientific Perspectives, Washington, D.C., June 1998.
Surrency, Erwin C. *History of the Federal Courts.* Dobbs Ferry, N.Y.: Oceana Publications, 2002.
Sutter, Paul. *Driven Wild: How the Fight against Automobiles Launched the Modern Wilderness Movement.* Seattle: University of Washington Press, 2002.
———. "What Gullies Mean: Georgia's 'Little Grand Canyon' and Southern Environmental History." *Journal of Southern History* 76, no. 3 (August 2010): 579–616.
Switzer, J. A. "Economic Aspects of the Smoke Nuisance." *Resources of Tennessee* 1, no. 5 (November 1911): 170–81.
———. "The Ocoee River Power Development." *Resources of Tennessee* 2, no. 2 (February 1913): 42–47.
Taliaferro, Hardin E., and Raymond C. Craig. *The Humor of H. E. Taliaferro.* Knoxville: University of Tennessee Press, 1987.
Tarr, Joel A. *The Search for the Ultimate Sink: Urban Pollution in Historical Perspective.* Akron, Ohio: University of Akron Press, 1996.
Tarr, Joel A. E., and Carl Zimring. "The Struggle for Smoke Control in St. Louis: Achievement and Emulation." In *Common Fields: An Environmental History of St. Louis*, ed. Andrew Hurley, 199–220. St. Louis: Missouri Historical Society Press, 1997.
Tate, Lucius Eugene. *History of Pickens County.* Spartanburg, S.C.: Reprint Co., 1978.

Tate, William. *Documents and Memoirs, Genealogical Tables, the Tates of Pickens County*. Marietta, Ga.: Continental Book Co., 1953.

Taylor, A. Elizabeth. "The Origin and Development of the Convict Leasing System in Georgia." *Georgia Historical Quarterly* 26 (1942): 113-28.

Taylor, James W. "Ducktown Desert: A Study of the Impact of Industrial Culture upon the Physical Development of a Secluded Area of the Southern Appalachians." M.A. thesis, Syracuse University, 1950.

Teale, Edwin Way. "The Murder of a Landscape: Shortsighted Exploitation Poisoned and Denuded the Hills, and 100 Square Miles of Southern Countryside Have Become an Almost Hopeless Desert." *Natural History* 60 (October 1951): 352-56.

Thornton, Russell. "Demography of the Trail of Tears." In *Cherokee Removal: Before and After*, ed. William L. Anderson, 75-95. Athens: University of Georgia Press, 1991.

Twain, Mark. *Life on the Mississippi*. Boston: J. R. Osgood and Company, 1883.

Tyner, James W. *Those Who Cried: The 16,000; A Record of the Individual Cherokees Listed in the United States Official Census of the Cherokee Nation Conducted in 1835*. N.p.: Chi-ga-u Inc., 1974.

Tyrrell, Ian. "Acclimatisation and Environmental Renovation: Australian Perspectives on George Perkins Marsh." *Environmental History* 10 (2004): 153-67.

Uekoetter, Frank. "Solving Air Pollution Problems Once and for All: The Potential and Limits of Technological Fixes." In *The Technological Fix: How People Use Technology to Create and Solve Problems*, ed. Linda Rosner, 155-74. New York: Routledge, 2004.

Urbinato, David. "London's Historic 'Pea-Soupers.'" *EPA Journal* 20, nos. 1-2 (1994): 44.

Vinten-Johansen, Peter, Howard Brody, Nigel Paneth, Stephen Rachman, Michael Rip, and David Zuck. *Cholera, Chloroform, and the Science of Medicine: A Life of John Snow*. New York: Oxford University Press, 2003.

Walhbeck, Paul J. "The Development of a Legal Rule: The Federal Common Law of Public Nuisance." *Law & Society Review* 32, no. 3 (1998): 613-38.

Ward, George Gordon. *The Annals of Upper Georgia Centered in Gilmer County*. Carrollton, Ga.: Thomasson Printing & Office Equipment Co., 1965.

Waselkov, Gregory A., and Kathryn E. Holland Braund. *William Bartram on the Southeastern Indians*. Indians of the Southeast. Lincoln: University of Nebraska Press, 1995.

Watson, Harry L. "The Common Rights of Mankind: Subsistence, Shad, and Commerce in the Early Republican South." *Journal of American History* 83, no. 1 (1996): 13-43.

Weed, Walter Harvey. *The Copper Mines of the World*. New York: Hill Publishing Company, 1907.

Weiner, Douglas R. "A Death-Defying Attempt to Articulate a Coherent Definition of Environmental History." *Environmental History* 10, no. 3 (July 2005): 404-20.

Weller, Francis. "Economic Aspects of the Smoke Nuisance." *Resources of Tennessee* 1, no. 5 (November 1911): 170-81.

Wheeler, William Bruce, and Michael J. McDonald. *TVA and the Tellico Dam, 1936-1979: A Bureaucratic Crisis in Post-Industrial America*. Knoxville: University of Tennessee Press, 1986.

White, Richard. "Are You an Environmentalist or Do You Work for a Living?" In *Uncommon Ground: Rethinking the Human Place in Nature*, ed. William Cronon, 171-85. New York: Norton, 1996.

Wilkins, Thurman. *Cherokee Tragedy: The Story of the Ridge Family and the Decimation of a People*. New York: Macmillan, 1970.

Williams, A. J. *A Confederate History of Polk County, Tenn., 1860–1866*. Nashville, Tenn.: McQuiddy, 1923. Reprint, Signal Mountain, Tenn.: Mountain Press, 2002.

Williams, David. *The Georgia Gold Rush: Twenty-Niners, Cherokees, and Gold Fever*. Columbia: University of South Carolina Press, 1993.

Williams, Michael A. *Americans and Their Forests: A Historical Geography*. Cambridge: Cambridge University Press, 1989.

Wilms, Douglas C. "Cherokee Land Use in Georgia before Removal." In *Cherokee Removal Before and After*, ed. William L. Anderson, 1–28. Athens: University of Georgia Press, 1991.

Wilson, Aubrey. "The Greening of the Copper Basin." *Tennessee Conservationist* 60, no. 3 (May–June 1994): 21–26.

Wirth, John D. *Smelter Smoke in North America: The Politics of Transborder Pollution*. Lawrence: University Press of Kansas, 2000.

Wood, H. G. *A Practical Treatise on the Law of Nuisances in Their Various Forms: Including Remedies Therefore at Law and in Equity*. 3rd ed. San Francisco: Bancroft-Whitney, 1893.

Woodward, C. Vann. *Origins of the New South, 1877–1913*. Baton Rouge: Louisiana State University Press, 1971.

———. *Tom Watson, Agrarian Rebel*. New York: Macmillan, 1938.

Workers of the Writers' Program of the Works Progress Administration in Georgia. *Georgia: The WPA Guide to Its Towns and Countryside*. With an introduction by Phinizy Spalding. Columbia: University of South Carolina Press, 1940.

Worster, Donald. *Dust Bowl: The Southern Plains in the 1930s*. New York: Oxford University Press, 1979.

Wright, Gavin. *Old South, New South: Revolutions in the Southern Economy since the Civil War*. New York: Basic Books, 1986.

Wynn, Graeme. "'On Heroes, Hero-Worship, and the Heroic' in Environmental History." *Environmental History* 10 (2004): 133–51.

Young, Timothy R. "Criminal Liability under the Refuse Act of 1899 and the Refuse Act Permit Program." *Journal of Criminal Law, Criminology, and Police Science* 63, no. 3 (September 1972): 366–76.

Youngs, Robert L. "Right Smart Little Jolt: Loss of the Chestnut and a Way of Life." *Journal of Forestry* 98, no. 2 (February 2000): 17–21.

Zdeb, Sara. "From *Georgia v. Tennessee Copper* to *Massachusetts v. EPA*: Parens Patriae Standing for State Global Warming Plaintiffs." *Georgetown Law Journal* 96, no. 3 (March 2008): 1060–82.

Index

Abernathy, L. H., 73
Absentee ownership, 45, 51, 145
Acid condensation plants, 125; acid production at, 196, 197; of DSC&I, 177–78, 179, 196; facilities closed in 2001, 245; Georgia smoke suit spurred, 7, 177, 194, 218, 221; hopes for, 7, 173–74, 178, 180, 187, 195–96; profitability of, 7, 157–58, 211–12; science of, 173–74; and sulfur dioxide emissions, 194, 197, 198, 199, 221, 235; of TCC, 158, 177, 179, 187, 192, 196, 201, 290 (n. 7)
Acid rain, 5, 43, 127
Adams, Randolph, 122, 232
Agriculture, 51; of Cherokees, 18, 19; in Ducktown Basin, 46–47, 219–20; impact of sulfur dioxide emissions on, 13, 43, 49, 84, 158–59, 171, 198. *See also* Farmers
Air Quality Act (1967), 43
Allen, John, 62
Allen, N. Q., 75
American Association for the Advancement of Science (AAAS), 150, 151
American Federation of Labor (AFL), 50
American Forestry Association, 150–51
American Smelting (ASARCO), 171, 172
Amount in controversy requirement, 73, 272 (n. 37)
Anaconda Copper, 44, 251, 252–53
Anderson, J. W., 85, 183
Andrews, Richard N. L., 8
Angelico Peak, 15, 16
Antipauper law, 62–63, 65, 78, 81–82
Appalachian Mountain Club of New England, 150–51
Appalachian Mountains. *See* Southern Appalachian Mountains
Appalachian National Forest Association, 154
Appalachian National Park Association of the South Atlantic States, 150–51

Arbitration, 208, 216, 217, 218, 221
Arsenic, 44, 171, 243
Akin, James, 159
Atlanta Constitution, 39–40, 101, 147, 154, 176; on copper industry, 66, 95–96, 239; and Georgia smoke suit, 89–90, 94, 97, 107–8, 110, 155; on pollution, 240–41, 244; Shippen columns for, 154, 208, 283 (n. 36)
Attorney fees, 207–8; contingent-fee contracts, 52, 69–70, 271 (n. 28)

Bain, S. M., 203
Balancing-of-interests tests: chancery courts and, 62, 63, 75, 76, 81; and interstate pollution, 94; in *Mountain Copper* case, 162, 166; and nuisance law, 61, 79; Supreme Court and, 166
Barclay, Robert Edward, 119, 122, 177, 230, 264 (n. 36); calls Ducktown a desert, 4; *Ducktown Back in Raht's Time*, 32, 261 (n. 23); on Ducktown's ethnic composition, 45; on *Fortunes Washed Away*, 231; *The Railroad Comes to Ducktown*, 35, 261 (n. 23)
Barnes, J. H., 71
Barnes lawsuits. *See Ducktown Sulphur, Copper & Iron Co. v. Barnes*
Barnhardt, Wilton, 4
Barrett, Charles S., 188, 189, 218, 288 (nn. 49, 52)
Barton, R. M., Jr., 75–76
Bartram, William, 20–21
Baruch, Bernard, 9, 212, 214
Beckler, David, 237–38
Bell, A. J., 43, 47
Bell Creek, 247
Bierce, Ambrose, 81
Big Frog Mountain, 15, 16
Blacks, 39–40, 45, 96
Blue Ridge Dam, 243
Blue Ridge Mountains, 15, 36, 100, 152

Blue Ridge Post, 191
Boliden Intertrade AG (BIT), 240, 245
Bonaparte, Charles J., 9
Bowron, William M., 196–98, 199
Brandis, Dietrich, 156
Brandt, Hennig, 175
Brown, J. J., 208–9
Brown, Joseph M., 193, 196, 198, 199, 200
Buchanan, W. T., 192
Burke, George L., 67, 68, 78
Burra-Burra Copper Company of Tennessee, 31–32, 33
Burra Burra Hill, 238
Burra Burra mine, 37, 122
Burt, Jesse C., 4
Burtz, A. H., 209
Butt, William, 191
Butte, Mont., 12, 44

Caldwell, John, 23, 25–26
Canada, 10, 254, 263 (n. 9)
Candler, W. E. "Buck," 84
Carbon dioxide, 10, 257
Cartecay River, 152
Carter, Jimmy, 241–42
Cash, Johnny, 148
Champerty laws, 52, 69
Chancery courts, 56, 62, 64, 74–75; and circuit courts, 52, 55, 269 (n. 40)
Channing, J. Parke, 155, 157–58
Charcoal: and coke, 54; copper industry's use of, 31, 144; and deforestation, 5, 87; making of, 30
Chase, Stuart, 3, 226, 228, 231
Chastine, O. F., 159
Chattahoochee National Forest, 202
Chattanooga News, 108
Chattanooga Times, 228–29
Cherokee National Forest, 8, 152, 202
Cherokee Nation v. Georgia, 112
Cherokees, 17–23; agriculture of, 18, 19; "civilization" program toward, 18–19, 21, 22; knowledge of copper deposits, 24; population, 19, 20; removal of, 9, 22–23; sale of lands of, 24; and tribal sovereignty, 20, 21–22, 112; villages of, 18, 19; and *Worcester v. Georgia*, 9, 22

Chestnut blight, 185, 203
Chicago: sewage disposal by, 113–14, 117; stockyards in, 61
Chisholm, Alexander, 109
Chisholm v. Georgia, 11, 109, 112
Chittenden, Alfred K., 148–49, 151, 154, 198
Chittenden, H. M., 88
Cities Service Company, 236, 237, 238; and Clean Air Act standards, 241, 242; EPA suit against, 242–43
Civilian Conservation Corps (CCC), 234
Civil War, 112; copper as resource in, 32–33; Sherman's March through Georgia, 97–98, 275 (n. 39)
Clarke, E. B., 202
Class actions, 64
Clay, Grady, 3
Clean Air Act (1963), 10, 43, 241, 253, 257
Clean Water Act (Federal Water Pollution Control Act, 1972), 241, 255, 256
Clemmer, James Donaldson, 261 (n. 23)
Climate, 18, 27, 46, 125, 127
Coal, 98–99, 100
Cohutta Mountains, 15, 16, 26
Coke, 54, 87
Collis, Frank, 239
Commercial viability, 96, 166, 252
Common Law, The (Holmes), 167–68
Constitution, U.S.: article 3 of, 72, 108, 109, 110, 165, 278 (n. 12); article 4 of, 111–12; Eleventh Amendment, 110, 112, 278 (n. 13); Fifth Amendment, 150; and original jurisdiction, 108, 109, 114
Contingent-fee contracts, 52, 69–70, 271 (n. 28)
Convict labor, 39, 95
Coosawatee River Valley, 152
Copeland, R. Meigs, 63
Copper, 5, 65; Cherokee knowledge of deposits, 24; commercial uses of, 25; demand for, 25, 36; extracting marketable sulfur from, 7, 118, 162, 173, 194, 223, 228; military uses of, 32–33, 210; ore grades, 27–28, 42, 54–55, 117; prices, 33, 34, 37, 238; rediscovery of Ducktown deposits, 24–25, 37
Copper City Advance, 210

Copper mining and smelting: beginning of in Ducktown, 11, 28; of black ore, 28, 29, 33, 37; blast furnaces used in, 41, 123, 155; boom in, 24, 27, 264 (n. 36); early methods of, 28; end of in Ducktown, 8, 238–40; in Georgia, 101; initial stage of, 5, 27–35, 37; investment in, 27, 31, 36–37; long-term viability of, 221; in Michigan, 127; number of employees in, 28, 32, 96–97, 179; and ore shipments, 29, 117; production costs, 34, 119, 238; production output, 28, 55, 90–91, 97, 123, 124, 197; restoration of after Civil War, 33; as source of employment, 7, 94, 108, 159, 195, 220, 227–28. *See also* Open heap roasting; Pyritic smelting

Copper Road, 27, 29, 35, 40, 65, 74; building of, 8, 14, 23, 37

Cornick, Howard: and Georgia smoke suit, 90, 122, 155, 160, 161, 162, 198; political calculations of, 76; in Tennessee court litigation, 67, 81, 105, 182, 185, 214

Cornick, Tully, 63, 74

Corporate law, 93

Cotton, 51, 103, 176, 178; Cherokees and, 18; NFU and, 189, 191

Crofts. See *Ducktown Sulphur, Copper & Iron Co. Ltd. v. Crofts*

Currey, R. O., 24, 27

Curtis, W. A., 49–50

Damage awards: under arbitration system, 217; from *Barnes* suits, 57; by chancery and circuit courts, 56, 181, 182; did little to stop environmental damage, 186; from *Fain* case, 59, 64, 77; for Hot House Creek farmers, 183, 186; from jury verdicts, 78, 181, 182, 183; and law of equity, 161–62; under nuisance law, 56; paid by TCC, 147, 186, 207–8, 210, 218; proving entitlement to, 69; seen as cost of doing business, 186, 201; in timber lawsuits, 183, 186

Daniels, Josephus, 9, 213, 214

Daves, W. A., 172

Davis, John W., 213–14

Davis Mill Creek, 5, 225, 247

Demand letters, 92

DeWeese, J. T., 192

Dickens, Charles, 170

Dickey, A. B., 85, 119, 141, 183

Dispersion, 7, 42, 117–18, 124

Diversity jurisdiction, 72–73, 185

Donora, Penn., 117

Dorsey, Hugh M., 213

Douglass, Edward, 206

Drake, Jesse A., 189, 192, 196, 203–4, 209

Dual federalism, 113, 165–66

Ducktown Basin: agriculture in, 46–47, 182, 219–20; during Civil War, 32–33; climate, 18, 27, 46, 125, 127; as Copper Basin, 4, 15, 219, 226; discovery of copper in, 24–25, 37; "Ducktown" name, 15, 19; end of mining and acid industries in, 8, 238–40; ethnic composition of, 45, 268 (n. 22); following collapse of mining's first stage, 37, 38; geographic location, 5, 15–16; industrialization of, 15, 37; pattern of ownership in, 223; political geography of, 16–17; prior to copper mining, 5, 14, 24; railroad arrival in, 36, 40–41, 65; roads and footpaths, 17–18, 23, 26–27, 38, 47, 65, 161; soil erosion in, 5, 38, 126–27, 151, 202, 226, 235; sources of employment in, 7, 94, 108, 159, 195, 220, 227–28; topography of, 46, 101–2, 126–27; as vacation destination, 4, 8, 16, 249–50; white settlers in, 18, 23, 24, 144; wildlife, 5, 24

Ducktown Basin Museum, 16, 237–38

Ducktown Desert: concentric rings of, 85–86, 222–23; "desert" label given to, 3–4, 87; and desert silt, 202–3, 223, 225, 240, 242, 243, 245; forest fires given as cause of, 126; geographic size of, 125, 152–53, 154, 222, 223, 294 (n. 1); as iconic emblem, 221, 231; livestock grazing given as cause of, 126; logging as cause of, 5, 126; as "man-made desert of folly," 228–29; microclimate effects in, 125; subsoil in, 235; sulfur dioxide emissions as cause of, 31, 86, 125–26; water table under, 230; wildlife flees from, 5. *See also* Reforestation and reclamation

Ducktown Gazette, 118
Ducktown smoke. *See* Sulfur dioxide emissions
Ducktown smoke litigation, 6, 44–45, 180–81, 260 (n. 9); challenge facing farmers in, 43; co-defendant question in, 68, 69, 181, 186; companies' avoidance of jury trials in, 49, 55, 57, 67; companies' multiplicity-of-suits argument in, 55, 56, 64, 67, 181; companies' strategy of delay in, 49, 51–52, 57, 66–67, 68, 72, 74, 77–78, 79, 81, 181; contingent-fee suits, 69–70; court location as issue in, 73–74, 181; damage awards in, 57, 58, 59, 69, 77, 78, 181, 182, 183, 186; filed before justices of the peace, 67–68; *in forma pauperis* claims in, 62–63, 65, 67, 82; and injunctions, 59, 64–65, 66; liability question in, 68, 69, 79, 161, 180, 183–84; motivations in, 12–13, 22, 59; and nuisance law, 45, 47, 48–49, 69, 161, 180; Tennessee Supreme Court decisions on, 57, 61, 67, 68–69, 76–77, 86, 99, 161, 180, 181, 186. See also *Ducktown Sulphur, Copper & Iron Co. v. Barnes*; *Ducktown Sulphur, Copper & Iron Co. v. Fain*; Hot House Creek: lawsuits by farmers of; *Madison v. Ducktown, Sulphur, Copper & Iron Co.*; Timber industry: lawsuits by
Ducktown Sulphur, Copper & Iron Company (DSC&I): acid conversion plant of, 177–78, 179, 196; acquires Union Consolidated Mining Co., 36; *Barnes* suits against, 48, 49, 51, 52–53, 55–56, 57, 58; creates demonstration farm, 229; establishes itself in Ducktown, 36, 96; flotation mill installed by, 228, 253; Isabella works, 63, 122, 124, 178, 179, 196, 222; labor conflicts of, 49–50; legislative strategy of, 63; London-based, 51, 96, 109, 178, 181, 201, 211; open heap roasting by, 41, 42, 120, 123; property taxes of, 51, 103; public relations work of, 215; pyritic smelting adopted, 42, 119, 120; rail access and, 6, 36; refuses to settle Georgia smoke suit, 199, 200, 201, 207; refuses to settle with Shippen, 147; signs Georgia smoke suit settlement, 216; Supreme Court injunction against, 206–7, 212, 213–15; swallowed up by TCC, 7, 218, 223; during World War I, 211–12
Ducktown Sulphur, Copper & Iron Co. Ltd. v. Crofts, 67–68, 73
Ducktown Sulphur, Copper & Iron Co. v. Barnes, 44–45; composition of suitors, 45, 46; damage awards from, 57, 58, 69; filing of, 47–48; legal strategy of DSC&I in, 49, 52–53, 55–56, 57, 58; and nuisance law, 45, 47, 48–49, 69, 161, 180; Tennessee Supreme Court decision on, 57, 61, 69, 86, 99, 161, 180
Ducktown Sulphur, Copper & Iron Co. v. Fain, 64–66, 181; copper companies' lawyers in, 64, 65–66; damage awards from, 59, 77; and injunctions, 64–65, 66; Tennessee Supreme Court on, 67
Duckworth, R. F., 188, 193
DuPont Company, 210
Dust Bowl, 227, 230–31

Early, W. R., 159
Eberth, Karl, 115
Edward I (king of England), 113
Ellijay (town), 146
Ellijay Valley, 15, 36, 152
Ellison, George, 59
Eminent domain, 75, 78
"Empty chair" defense, 205
Endangered Species Act (1973), 255
Engineering and Mining Journal, 121, 157
Environmentalism, 8–9, 10, 255, 300 (n. 14). See also Forest conservation movement
Environmental laws: and balancing test, 61; Butte smoke ordinance of 1890, 44; in Great Britain, 113; of 1960s and '70s, 43, 220, 241, 249, 253, 255–56; Refuse Act of 1899, 113; against strip mining, 223–24; water pollution act of 1948, 113
Environmental Protection Agency (EPA): creation of, 121, 220, 241, 249, 255; and Glenn Springs Holdings, Inc., 245, 246; and greenhouse gas emissions, 10, 256–57; suit against Cities Service by,

242–43; and superfund, 224, 246, 248; and TVA, 241, 249
Equity, 179; ancient maxims of, 92, 161–62; and law, 52, 166, 269 (n. 40)

Fain cases. See *Ducktown Sulphur, Copper & Iron Co. v. Fain*
Falding, P. J., 177, 179
Fannin County, 101, 191, 195
Farmers: affidavits from in Georgia smoke suit, 158–59; cotton, 51, 189, 191; fence raising by, 38–39; lawsuits by, 13, 43, 52, 58, 64–65, 70, 74–75, 182–83, 186; market provided by copper industry, 27, 37–38, 47, 56, 62; mine employment of, 71–72; movements by, 79, 188, 189; and populism, 12; squatters, 45–46; and tenancy, 45, 51; as witnesses, 69, 78, 80. See also Agriculture; National Farmers Union
Farmers' Alliance, 188
Farmers Union News, 188, 193
Farner, Isaac, 74, 90, 122, 162
Farner, P. J., 74, 75
Farner v. Ducktown Sulphur, Copper & Iron Co. and Tennessee Copper Co., 75, 90, 122–23
Federal Water Pollution Control Act (1948), 113
Felder, T. S., 199, 200, 209
Felton, Rebecca Latimer, 188–89, 288 (n. 50)
Fernow, Bernhard Eduard, 156
Fertilizer, 175–76, 193; overuse of, 289–90 (n. 66); sulfuric acid for, 158, 174, 176, 178
Fiat justitia, ruat coelum principle, 59, 76
Firewood, 5, 30, 126, 128, 144, 222
Fish, 5, 245
Fishing, 34, 61
Flatt and Scruggs, 148
Flotation milling, 228, 253
Forest conservation movement, 12, 88, 151, 153–54, 168, 253; Georgia smoke suit aligned with, 153, 168, 214, 221, 230; Marsh theories and, 149–50, 230; T. Roosevelt and, 11, 150, 152
Forest fires, 126

Forest Mensuration (Graves), 184
Forest Protection (Schenck), 157
Forest reserves, 150–51, 152, 202, 282 (n. 25)
Forests: chestnut blight destruction of, 185, 203; logging impact on, 31, 33, 34, 38, 87, 144, 149–50; national, 11, 150, 234, 249, 251, 253; prior to copper mining, 5, 14, 24; sulfur dioxide emissions and, 86, 124, 128, 148–49, 152–53, 157, 198; tree counting in, 184; and watersheds, 88. See also Reforestation and reclamation; Timber industry
Forest Service, U.S. (Bureau of Forestry), 142, 168, 244–45; administers national forests, 150; and Ducktown reclamation program, 233; and Georgia smoke suit, 148, 152, 153; name changes of, 281 (n. 20)
Fortner, J. A., 57, 58
Fortunes Washed Away, 231–33
Foster, Harold Day, 142, 148, 151, 168
Fowler, James A., 203, 212
Frantz, John H., 167
Frazier, James B., 93–94, 212
Freeland, W. H., 147, 178, 191, 211, 232; and pyritic smelting, 118, 121; on smoke litigation, 50, 66, 77
Fry, J. T., 183

Gair, Robert, 143
Galloway, Dora, 102, 127–28
Gaussoin, Eugene, 31
Gennett, Andrew, 146
George, Dan, 237
Georgia: Articles of Agreement and Cession, 112; border with Tennessee, 16, 262 (n. 8); during Civil War, 97–98, 112, 275 (n. 39); convict labor used by, 39, 95; copper mining in, 101; cozy relationship to industry by, 34, 94–95, 98–99; and Ducktown reclamation, 223; Ducktown smoke as threat to, 11, 84–85, 96–98, 102–3, 214; economy of, 103, 239; quasi-sovereignty recognized by Supreme Court, 6, 9, 165, 180, 200; resists "invasion" by Tennessee, 97–98, 111–13, 120; state sovereignty as doctrine of, 11, 20,

21, 112, 214; taxes and tax revenue in, 103, 160–61, 276 (n. 53). *See also* Interstate air pollution

Georgia General Assembly, 91, 103, 154–55, 172, 173; attacks on Cherokee sovereignty by, 21–22; authorizes railroads, 94–95, 275 (n. 31); on Georgia smoke suit settlement, 199, 200, 209–10, 215–16; on injunction question, 191, 196; on mining, 100–101; political calculations of, 11, 95, 120; study commission of, 83–84, 85, 89; subordination of citizen interests by, 34, 94–95

Georgia Supreme Court, 99–100, 276 (n. 43)

Georgia v. Tennessee Copper Co.: affidavits in, 158–60, 203–5; as antipollution legal precedent, 6, 10, 220–21, 251, 253–54, 255, 256, 257, 300 (n. 19); arbitration system set up under, 208, 216, 217, 218; consideration of injunction in, 7, 168–69, 170–71, 179–80, 195, 201–6, 212; copper companies' legal case in, 156–57, 160–64, 168, 203, 205; demand letter prior to filing, 91–93; docket numbers of, 259 (n. 1), 283 (n. 30); DSC&I refusal to settle, 199, 200, 201, 207; DSC&I signs settlement of, 216; final arguments of (1907), 164; initial settlement of (1904), 120, 141; injunction issued against DSC&I, 206–7; length of litigation in, 120, 216–18, 254–55, 284 (n. 38); original filing of, 94, 97, 103, 105; and original jurisdiction, 108–9, 110, 112, 117, 154; plaintiffs' alignment with forest conservation movement, 153, 168, 214, 221, 230; plaintiffs' reliance on facts, 154, 155–56, 159, 185, 202; political context of, 120, 201; property damage issue in, 111, 116; public support for, 107–8, 111, 120; refiling of as second lawsuit (1905), 154–55; as regional conflict, 159; ruling on interstate air pollution, 6–7, 164–65, 166–67; ruling on states as quasi-sovereign, 6, 9, 165, 180, 200, 254, 255; Supreme Court ruling on (1907), 6–7, 9, 164–67, 168–69, 179–80, 200, 253–54, 255, 257; TCC settlement in, 198–99, 200, 201, 207; World War I impact on, 214

Gilbert, William, 276 (n. 47)
Gilliam, W. Y., 209
Gilmer, George, 21
Glenn, Leonidas Chalmers, 202–3
Glenn Springs Holdings, Inc., 245, 247, 248
Global warming, 10, 251, 256–57
Godfrey v. American Smelting & Refining Co., 171, 285 (n. 4)
Gold mining, 20, 21, 100, 101
Gordon, J. G., 172
Grady, Henry, 96
Graham, N. B., 70
Graves, Henry Solon, 184
Great Britain, 113
Great Depression, 228, 234
Great Smoky Mountains, 15, 17, 124
Greenhouse gases, 10, 256
Gribble, J. S., 66
Grice, Warren, 202, 203
Guthrie, Woody, 148

Hahn, Steven, 269 (n. 39)
Hale, W. D., 184
Hall, Hewlett, 196, 198
Hamilton, Alice, 115
Hammond, James Henry, 174
Hankivell, Octavus, 43
Harbison, E. M., 74
Harlan, John Marshall, 112, 164, 294 (n. 61)
Harper, F. M., 183
Harper, W. L., 183
Harris, Nat, 208, 209–10, 212
Hart, John C., 11, 91, 105–6, 120, 141, 179; accomplishments in Georgia smoke suit, 116, 194, 214; aligns Georgia smoke suit with conservation movement, 153, 168, 214, 221, 230; argues facts about smoke damage, 154, 155–56, 158, 185; *Atlanta Constitution* on, 89, 147–48; background, 98, 106–7; counsels patient delay on injunction, 7, 173, 178, 179, 186–87, 195–96, 201; fees received by, 170; files suit in U.S. Supreme Court,

94, 97, 103, 105; and NFU, 189, 192–93, 194; Parks on, 120–21, 164; political objectives of, 120–21, 172–73; political pressures on, 141, 173, 187–88, 191–92; receives help from Shippen, 142, 148; resignation of, 193–94
Haywood, J. K., 71, 152–53, 154, 162, 185
Hedgecock, George W., 202, 203
Henrich, Carl, 37, 118
Henry V (king of England), 113
Henson, L. Y., 50
Hetch-Hetchy dam, 190, 289 (n. 54)
Hill, Lamar, 209
Hiwassee River, 15, 40
Holden, J. F., 208
Holmes, J. S., 152, 154
Holmes, Oliver Wendell: background, 112, 168; *The Common Law*, 167–68; and injunction decision, 206; on *Missouri II*, 163; opinion in Georgia smoke suit (1907), 6, 7, 9, 164–67, 168–69, 179–80, 200, 253–54, 255, 257
Holston Treaty (1791), 20
Hot House Creek, 46, 59; lawsuits by farmers of, 70, 182–83, 186
Hughes, Charles Evans, 206
Humphrey, Billy, 70
Hunt's Merchant's Magazine, 29
Hursh, C. R., 125, 229
Huxley, Aldous, 101
Hyatt, G. G., 159
Hyatt, Virgil, 211
Hydraulic mining, 100–101, 276 (n. 47)
Hydroelectric industry, 224–25, 226, 233, 240–41

Incidental-benefits rule, 62, 63, 74–75, 76, 78, 79, 99
Indian Removal Act (1830), 22
Indian Trade and Intercourse Acts (1790), 18
Injunction: against DSC&I, 206–7, 212, 213–15; *Fain* cases and, 64–65; flexibility in remedies of, 179, 287 (n. 27); as goal of Ducktown farmers, 59; Hart's delay in seeking, 7, 173, 178, 179, 186–87, 195–96, 201; *Madison v. Ducktown* and, 74–75, 76–77, 90, 122–23; Supreme Court consideration of in Georgia smoke suit, 7, 168–69, 170–71, 179–80, 195, 201–6, 212
International Agricultural Corporation (IAC), 193, 198, 210–11
International Shoe Co. v. Washington, 53
Interstate air pollution, 17, 97, 102, 251, 254; and balancing-of-interests tests, 94; as invasion, 97–98, 112–13; and nuisance law, 80; and original jurisdiction, 108–9, 163; Supreme Court decision on, 6–7, 164–65, 166–67
Intertrade Holdings, 245
Invasion, 97–98, 111–13, 120

Jackson, Andrew, 9, 22
Jarvis, Thomas, 95
Jeffersonian, 208
Jobs vs. environment dichotomy, 7, 108, 159, 171–72, 195
Johnes, F. M., 159
Johnson, H. L., 202
Johnson, Ligon, 102, 105, 111, 120, 164; accomplishments of, 116, 194; argues facts about smoke damage, 154, 155–56, 159; background, 106–7; and forest conservation movement, 153–54, 160, 168, 214, 221, 230; as special counsel for T. Roosevelt administration, 9, 199–200, 251–52; subsequent history of, 253
J.P. Vestal Lumber Co., 183
Jurisdiction, 53, 165–66, 181, 182, 278 (n. 13); diversity, 72–73, 185; in interstate disputes, 82; *in personam* and *in rem*, 53, 181–82, 269–70 (n. 42); and nuisance law, 56–57. *See also* Original jurisdiction
Jury-of-view process, 78–79
Jury trials, 74; copper companies' avoidance of, 49, 55, 57, 67; damages from, 78, 181, 182, 183; in nuisance actions, 52, 79–80
Justices of the peace, 67–68, 69, 73, 160

Keffer, Charles, 152, 282–83 (n. 30)
Kephart, Horace, 58–59

Kimsey, Fred M., 115, 116
Kimsey, L. E., 115, 116
Kisselburg, J. V., 71
Knoxville Journal and Tribune, 66, 74
Knoxville Tribune, 108
Koch, Robert, 115

Labor conflict, 50–51
Laches (inexcusable delay), 161, 167
Ladew, J. Harvey, 185
Ladew v. Tennessee, 72–73
Ladson, C. T., 189, 192, 196
LaForge, Laurence, 38, 42–43
Lamoreaux, W. F., 215
Lawes, John Bennett, 176
Lazarus, Richard, 7
LeCain, Timothy J., 279 (n. 30), 289–90 (n. 66)
Ledford, Mercer, 83
Legg, Homer, 209
Lewisohn, Samuel, 51
Liability: *Barnes* decision on, 69, 161, 180; collective, 68; incidental benefits bill on, 62, 63; and jury-of-view process, 79; in timber lawsuits, 183–84
Liebig, Justus von, 175
Lilienthal, David, 227, 228, 231
Lillard, Roy, 224, 261 (n. 23)
Little, John D., 209
Little Frog Mountain, 15, 16, 26
Livestock, 24, 38–39, 126
Lockett, Virginia, 224
Logging, 95, 144; amount cleared, 33, 149–50; as cause of Ducktown Desert, 5, 126; commercial, 143, 145; deforestation caused by, 31, 33, 34, 38, 87, 149–50; and soil erosion, 151; and transportation, 33–34. *See also* Forests; Timber industry
Logging and Lumbering or Forest Utilization (Schenck), 143, 144
London: DSC&I based in, 51, 96, 109, 178, 181, 201, 211; Great Stink (1858), 113; killer smog (1952), 117
Louisiana v. Texas, 110, 160, 165
Louisville & Nashville Railroad, 76
Love Canal, 246, 248

Lumpkin, Joseph H., 99–100
Lutts, Ralph H., 203

Madison, Margaret, 45, 48–49, 51, 54, 62; damages paid to, 57, 58
Madison, William, 54, 64, 90, 122, 162, 280 (n. 44); damages paid to, 57, 58
Madison v. Ducktown, Sulphur, Copper & Iron Co., 74, 90, 122–23, 162, 214; companies' victory in, 79, 180; decision, 75, 76–77, 180; filing of, 64–65; newspapers on, 66
Man and Nature (Marsh), 11, 88, 149–50, 230
Manual of the Law of Pleading, A (Parks), 66
Marcy, Henry O., 150
Marietta & North Georgia Railroad (M&NGRR), 39, 104
Marsh, George Perkins, 88, 111, 149–50, 202, 230
Marshall, John, 22
Massachusetts v. Environmental Protection Agency, 10, 256–57
Mayfield, J. E., 67, 71, 73
Mayfield, P. B., 50, 58, 62, 63, 214
Mayfield & Sons, 49, 64, 66
McCandless, John M., 84, 187
McConnell, T. M., 66
McCormick, Jack, 244
McGhee, Avery, 71, 72, 74, 90, 122, 162
McGhee v. Tennessee Copper Co., 75, 90, 122–23
McGill, John T., 123–24, 125–26, 206, 251
McKinney, Louise, 4
McMath, Robert C., 12, 189
McReynolds, James Clark, 205–6
Merrimack River, 61
Miller, W. B, 212, 213, 215
Missouri v. Illinois and the Sanitary District of Chicago (*Missouri I*), 113–14, 163, 165, 167, 278 (n. 25)
Missouri v. Illinois and the Sanitary District of Chicago (*Missouri II*), 163, 167
Mountain Copper Co. v. United States, 162–63, 166, 168, 251, 253, 299 (n. 6)
Mountain laurel, 58–59
Multiplicity of suits argument, 55, 56, 64, 67, 181

Muncy, Jack, 240
Murray, James, 176
Murrill, William Alphonso, 185

National Environmental Policy Act (1970), 255
National Farmers Union (NFU), 190, 198, 204, 219; and J. J. Brown, 208, 209; cotton agenda of, 189, 191; Fannin and Gilmer chapters of, 12, 188, 189, 191, 208, 209, 215, 218; financial interests of, 12, 193; founding of, 188; and Georgia smoke suit, 9, 12, 189–91, 196, 199, 201, 217; Hart and, 189, 192–93, 194; officers of, 188, 218, 288 (nn. 49, 50); political clout of, 12, 171, 188–89, 196
National Field, 190
Newell, Wilmon, 85, 86, 87–89, 97, 111
New Hampshire v. Louisiana, 110
New Jersey v. New York City, 254
New York Times, 89, 211
New York v. New Jersey, 254
Nixon, Richard M., 241, 255
Noise pollution, 60, 99–100
North Potato Creek, 247
Nuisance law, 48, 60, 82; and air pollution, 7, 61, 80, 100, 220, 254; and balancing-of-interests tests, 61, 79; *Barnes* suits and, 45, 47–49, 69, 161, 180; claims against railroads under, 99, 276 (n. 43); damages under, 56; and environmental policy, 47; jurisdiction and, 56–57; and jury trials, 52, 79–80; and modern industry, 60–61, 68–69; remedies under, 52, 166; rural populists and, 79–80; and substantive law, 83; Tennessee General Assembly and, 60, 62–63, 81; Tennessee Supreme Court on, 57, 61; TEPCO lawsuit under, 249; transformation of, 6, 10, 80

Occidental Petroleum Corporation, 245
Ocoee dams: No. 1 dam, 224; No. 2 dam, 224–25, 243; No. 3 dam, 226, 233, 243, 244; TVA acquires, 241
Ocoee Gorge, 8, 37, 40, 224

Ocoee River, 15, 16; as clogged with silt, 5, 202–3, 223, 240; flooding from, 230; water quality in, 241, 242, 244. *See also* Toccoa-Ocoee River
Ocoee Timber Company, 183
Ocoee Tract, 24
Ohio v. Wyandotte Chemicals Corp., 255
Olmsted, Frederick Law, 28
On the Mode of Communication of Cholera (Snow), 114–15
Open heap roasting: delay in abandoning, 122; ending of, 120, 123; environmental damage from, 16, 149; Georgia smoke suit aim of stopping, 6, 93, 105–6; long history of using, 41, 53; process described, 29–30, 41; and pyritic smelting, 119; sulfur dioxide emissions produced by, 5, 14–15
Original jurisdiction: applied to interstate pollution, 108–9, 163; Georgia smoke suit and, 108–9, 110, 112, 117, 154, 254; Supreme Court and, 6, 108, 110, 114, 254, 255; U.S. Constitution on, 108, 109, 114
Osborne, D. B., 158
Oxy USA, Inc., 245–47, 248

Pack, Linda, 215–16
Padgett v. Ducktown Sulphur, Copper & Iron Co., 180
Parker, James, 181
Parks, James G., 70, 121–22, 123, 182, 185; on acid conversion, 158, 187; as expert on circuit court procedure, 66–67, 79; and Georgia smoke suit, 90, 160, 164, 191; on Hart, 120–21, 164; legal stratagems of, 49, 68, 77, 81, 105, 181, 214; political efforts of, 76, 78; on Will Shippen, 146–47
Parmentier, Rosine, 145, 183
Pasteur, Louis, 115
Patterson, A. B., 152, 154, 282 (n. 30)
Percival, Robert, 163
Philips, Peregrine, 173
Pinchot, Gifford, 11, 150, 190
Polk County: Board of Education, 239; Chancery Court, 55, 57, 66, 122–23, 260

(n. 9); Circuit Court, 59, 70, 74, 81, 119, 162, 181, 182, 260 (n. 9); tax revenue in, 50
Polk County Copper Company, 31–32, 33
Populism, 12, 66, 72, 79–80; and NFU, 188; republicanism and, 269 (n. 39); in Tennessee, 63, 79
Potato Creek, 38, 202
Poverty. *See* Antipauper law
Pratt, N. P., 176
Preemption doctrine, 256
Prince, G. W., 43
Profits: from pyritic smelting, 123, 157; from sulfur and sulfuric acid, 7, 42, 157–58, 211–12; of TCC, 90–91, 182, 197–98, 211
Property damage, 111, 116
Property taxes, 50–51, 103, 239
Public domain, 92–93, 150, 173
Public good, 61, 190
Pumpkin Creek, 5, 16, 24
Pyritic smelting: copper companies' adoption of, 42, 119, 120, 121, 176–77; discovery of successful method for, 118–19; Georgia smoke suit spurred, 218, 221; hopes for, 7, 141, 173; and open heap roasting, 119; profitability of, 123, 157; sulfur dioxide emissions under, 123–25, 141

Quinn, M.-L., 4, 126, 259 (n. 6)
Quintell, John, 55

Racism: against blacks, 39–40, 45, 96; against Cherokee Indians, 20–21
Raht, August, 118
Raht, Julius Eckhardt, 32, 33, 34, 35
Railroad Comes to Ducktown, The (Barclay), 35
Railroads, 11, 26, 33, 34, 37, 95, 104; arrival in Ducktown, 36, 40–41, 65; construction of, 40; dangers of passenger travel, 147–48; Georgia legislature and, 39, 94–95; impact on copper mining, 6, 36, 37, 41, 65; initial failures in building track for, 35; nuisance claims against, 99, 276 (n. 43); route of through Ducktown, 15; and timber industry, 145

Rainfall, 4, 16, 27, 88, 127, 151, 202; acid, 5, 43, 127
Reese, S. M., 50
Reforestation and reclamation, 229, 230, 248, 249; CCC and, 234; choice of trees in, 234–35, 236; and common law nuisance, 240–41; Newell on, 88–89; planning for, 233–35; statistics of by 1986, 240; sulfur gases as hindrance to, 235; survival rate of trees in, 236–37, 249; technological advances required for, 7–8, 237; TEPCO lawsuit as harbinger of, 225–26; TVA and, 227, 233–34, 235, 236, 249; and water, 242–43, 247–48. *See also* Forests
Refuse Act, 113
Revere, Paul, 25
Richard III (king of England), 113
Rich Land, Poor Land (Chase), 3
Ridge, Major, 22–23
Rivers, E. D., 218
Roberts, John, 300 (n. 19)
Roebuck, John, 173–74
Rogers, H. A., 115, 116, 158
Rogers, L. D., 156
Roosevelt, Franklin D., 226, 229–30, 231, 234
Roosevelt, Theodore, 151, 154, 168, 252; and Ducktown, 9, 251; and forest conservation, 11, 150, 152
Ross, John, 23
Ruffin Edmund, 286 (n. 13)
Russell, Frank, 239, 247
Russell's Magazine, 27, 30–31
Rye, Tom C., 212

Sales, law of, 182
Salt Lake Tribune, 172
Salus populi est suprema lex principle, 76
Sanders, Carl, 262 (n. 8)
Sargent, Charles S., 150
Schenck, Carl Alwin, 143, 144, 156–57, 184–85
Schurz, Carl, 149
Scott, Winfield, 23
Sebolt, B. H., 3, 204
Seigworth, Kenneth, 126, 223

Seymour, Charles, 46, 282–83 (n. 30)
Sherman, William Tecumseh, 97–98, 275 (n. 39)
Shippen, Frank, 142–43, 144, 146, 156, 205, 284 (n. 41)
Shippen, W. H. (Will), 155, 168, 208–9, 284 (n. 41); affidavits by, 156, 159, 204–5; gives help to Hart, 142, 148; organizes opposition to copper companies, 146–47; presses for injunction, 171, 199, 201; subsequent history of, 219; as timber magnate, 142–43, 144, 146; writes newspaper columns, 154, 208, 215, 283 (n. 36)
Shiras, George, 114
Silva of North America (Sargent), 150
Slaton, John M., 200
Smallshaw, James, 43, 126
Smith, Hoke, 188–89, 208
Smith, James M., 95
Smokestacks, 7, 124, 141, 164, 173
Snow, John, 114–15
Soil conservation, 230–31
Soil Conservation Service (SCS), 231, 233, 237, 249
Soil erosion, 88, 152, 161, 202, 226, 235; causes of, 5, 38, 126–27, 151
Soil replenishment, 174, 286 (n. 13). *See also* Fertilizer
Sound Wormy (Gennett), 146
South: as "colonial region," 227; Confederacy in, 32–33; cotton production in, 51, 176; Ducktown rooted in, 10–11; and industrialization, 37, 96, 227; Jim Crow segregation in, 96; and New South ideology, 12, 95, 96, 264–65 (n. 36)
Southern Appalachian Mountains, 15; environmental history of, 11; railroads' impact on, 35, 37; timber industry in, 142, 143; transportation inaccessibility in, 17–18, 73
Southern Literary Messenger, 14, 31
Sovereign immunity, 109–10, 112
Sovereignty: definition of, 20; divided, 83; national, 82; as part of Georgia ideology, 11, 21, 109; and Supreme Court on states as quasi-sovereign, 6, 9, 165, 180, 200, 254, 255, 257

Stallings, R. E., 198
Stansbury Mountain, 15, 102, 152
Statute of limitations, 161, 184, 186
Steam power, 41, 99–100, 144–45
Steinberg, Ted, 61
Stevens, John Paul, 257
Stevens, O. B., 84
Stevenson, Paul E., 183, 186
Stradling, David, 44
Stream-flow theories, 230
Subsoiling technique, 237
Substantive law, 83
Sulfur, 5, 54–55, 118, 197–98; profitability of, 7, 42
Sulfur dioxide emissions, 5; and acid condensation plants, 194, 197, 198, 199, 221, 235; and acid rain, 5, 43, 127; as cause of Ducktown desert, 31, 86, 125–26; coal burning and, 98–99, 100; descriptions of Ducktown fogs, 14–15, 42–43, 251; extraction of, 7, 118, 162, 173, 194, 223, 228; formation of, 153; health effect of, 115–16; from high-sulfur ore, 54–55; as hindrance to reforestation, 235; Oliver Wendell Holmes on, 167; along Hot House Creek, 46, 70; impact on farmers and agriculture, 13, 43, 49, 84, 158–59, 171, 198; injunctions against, 171, 206–7; as invasion from Tennessee, 97–98, 112–13; killing of vegetation by, 3, 31, 44, 48–49, 81, 86, 89, 97, 119, 148–49, 153, 154, 157; odor of, 3, 64, 98, 127; from open heap roasting, 5, 14–15; as political issue, 104; as prime toxic agent in smelter smoke, 178; under pyritic smelting, 123–25, 141; as threat to timber industry, 84, 145; wind dispersion of, 42, 117–18
Sulfuric acid: in acid rain, 43; beginning production of, 228; environmental effects in producing, 153, 289–90 (n. 66); for fertilizer, 158, 174, 176, 178; marketing of, 125; military uses of, 210, 212–13, 292 (n. 38); profits from, 211–12; qualities of, 154; uses of, 174. *See also* Acid condensation plants
Sulfur trioxide, 153

Superfund law (Comprehensive Environmental Response, Compensation, and Liability Act, 1980), 224, 245–46, 248

Supreme Court, U.S.: modus operandi of, 109

—decisions: *Cherokee Nation v. Georgia* (1831), 112; *Chisholm v. Georgia* (1793), 11, 109, 112; *International Shoe Co. v. Washington* (1945), 53; *Ladew v. Tennessee* (1910), 72–73; *Louisiana v. Texas* (1899), 110, 160, 165; *Massachusetts v. Environmental Protection Agency* (2007), 10, 256–57; *Missouri v. Illinois and the Sanitary District of Chicago (Missouri I*, 1901), 113–14, 163, 165, 167, 278 (n. 25); *Missouri v. Illinois and the Sanitary District of Chicago (Missouri II*, 1906), 163, 167; *New Hampshire v. Louisiana* (1882), 110; *New Jersey v. New York City* (1933), 254; *New York v. New Jersey* (1908), 254; *Ohio v. Wyandotte Chemicals Corp.* (1971), 255; *Wetmore v. Tennessee Copper Co.* (1910), 72–73, 185; *Worcester v. Georgia* (1832), 9, 22, 112. See also *Georgia v. Tennessee Copper Co.*

Surface Mining Control and Reclamation Act (1978), 223–24

Sussman, John H., 158

Swain v. Tennessee Copper Co., 68–69, 181, 186

Taliaferro, Hardin, 14–15, 31, 251

Talmadge, Eugene, 218

Tarr, Joel, 118

Tate, Farish Carter, 103–4

Tennessee: border with Georgia, 16, 262 (n. 8); jurisprudence in, 52, 269 (n. 40); populism in, 63; state court location, 73–74

Tennessee Chemical Company, 236

Tennessee Chemical Corporation, 238, 239, 240

Tennessee Copper Company (TCC): acid conversion plant of, 158, 177, 179, 187, 192, 196, 201, 290 (n. 7); becomes chemical manufacturer, 228; begins Ducktown smelting, 37, 161; blast furnaces of, 123, 155; chartered in New Jersey, 72–73, 181, 185; Copperhill works, 63, 122, 124, 141, 158, 177, 179, 196, 222, 225; flotation mill installed by, 228, 253; giant smokestack of, 124, 141, 164, 173; history of, 36–37, 51; and IAC fertilizer trust, 193; legislative strategy of, 63; merges with Cities Service Company, 236; number of employees, 179, 211; open heap roasting by, 41, 42, 120, 123; opposes jury-of-view process, 78–79; payment of damages and settlements by, 147, 186, 207–8, 210, 218; Pittsburgh and Tennessee Copper Co. purchased by, 105; production output, 90–91, 197–98, 211; profits of, 90–91, 182, 197–98, 211; property holdings of, 223; pyritic smelting adopted by, 119, 120; rail access and, 6, 37; and reclamation efforts, 229, 230, 233, 236; settlement of Georgia smoke suit, 198–99, 200, 201, 207; surpasses DSC&I in size, 68, 200–201; swallows DSC&I, 7, 218, 223; TEPCO suit against, 225; and TVA, 223, 229, 233, 235, 248; *Wall Street Journal* on, 96; during World War I, 211, 292 (n. 38)

Tennessee Department of Environment and Conservation (TDEC), 247, 248

Tennessee Electric Power Company (TEPCO), 224–25, 249

Tennessee General Assembly, 60, 62–64, 81–82, 214

Tennessee River Valley, 15, 40

Tennessee Supreme Court, 77–78, 123; *Ducktown Sulphur, Copper & Iron Co. v. Barnes* (1900), 57, 61, 69, 86, 99, 161, 180; *Ducktown Sulphur, Copper & Iron Co. v. Fain* (1902), 67; *Madison v. Ducktown, Sulphur, Copper & Iron Co.* (1904), 76–77, 180; *Swain v. Tennessee Copper Co.* (1903), 68–69, 181, 186

Tennessee Valley Authority (TVA), 125, 231, 238, 239–40; criticized by environmentalists, 243, 295 (n. 16); Ducktown reclamation efforts of, 226–27, 229, 233–34, 235, 236, 245, 249; and EPA, 241, 249; establishment of, 226, 295 (n. 12); and

Ocoee dams, 226, 241, 243; F. Roosevelt's support for, 229–30; and TCC, 223, 229, 233, 235, 248
Terrell, Joseph M., 89, 106, 120, 153, 155, 214; demand letter by, 91–93; political aims of, 84, 121; as skilled lawyer, 84, 86–87
Thompson, Frank M., 212
Timber industry: absentee owners in, 145; high costs of, 33–34, 145; interest of in Georgia smoke suit, 282–83 (n. 30); lawsuits by, 183–86; and rail transportation, 33–34, 37, 145; Shippen Brothers' interests, 142–43; steam power in, 144–45; sulfur dioxide emissions as threat to, 84, 145; use of trees by, 143. *See also* Forests; Logging
Toccoa-Ocoee River, 102, 224
Toccoa River, 16, 34, 36, 95
Tourism: Ducktown as vacation destination, 4, 8, 16, 249–50; whitewater rafting, 8, 243–44, 245
Townsend, Jimmy, 237
Treaty of New Echota (1835), 23
Twain, Mark, 108, 193

Uekoetter, Frank, 279 (n. 30)
Unaka Mountains, 15, 16, 202
"Unclean hands" doctrine, 163, 167
Union Consolidated Mining Company of Tennessee, 31–32, 33, 34, 35, 36, 54
Unions, 50, 238
U.S. Forest Service. *See* Forest Service, U.S.
U.S. Geological Survey (USGS), 151, 202

Vanderbilt, George, 156
Verner, J. H., 98, 115
Vestal, J. P., 200
Vogel, Martin H., 160

Walker, Clifford, 196, 211–12, 213, 215, 217
Wallace, E. R., 43
Wall Street Journal, 90, 96
War Industries Board, 212, 213
Washington Post, 74

Water: desert silt in, 202–3, 223, 225, 240, 242, 243, 245; lack of, 230; quality, 225, 241, 242, 243–44, 245
Water pollution, 243–44, 245; and disease, 114–15, 163; Supreme Court on, 254
Water Quality Improvement Act (1970), 255
Watson, Tom, 12, 189, 208, 210, 288 (n. 50)
Watson's Jeffersonian Magazine, 189
Weaver, A. J., 25
Wedge, Utley, 177
Weed, Walter Harvey, 101
Weeks Act (1911), 31, 152, 202
Welch, Larry, 239
Wetmore, George Peabody, 45–46, 145, 183, 185, 283 (n. 30)
Wetmore v. Tennessee Copper Co., 72–73, 185
White, Edward Douglas, 112
Whitewater rafting, 8, 243–44, 245
Whitney, J. D., 29
Williams, John, 184
Willson, T. J., 183
Wilson, J., 85
Wilson, James, 11, 151, 152
Wilson, S. F., 75
Wilson, Woodrow, 212
Winds, 71, 82, 149, 247; microclimate effect on, 125; and smoke dispersal, 42, 117–18
Withrow, John J., 183
Witnesses, 69, 78, 80
Witt, J. B., 159
Witzel, J. H., 84, 85
Wood, Richard A., 126, 222–23
Worcester, Samuel, 22
Worcester v. Georgia, 9, 22, 112
World War I, 210–11, 215, 292 (n. 38); production requirements during, 212, 293 (n. 40)
World War II, 233
Worster, Donald, 230–31
Wright, Gavin, 96

Yeats, W. S., 84